Microbial Processes for Bioremediation

BIOREMEDIATION

The *Bioremediation* series contains collections of articles derived from many of the presentations made at the First, Second, and Third International In Situ and On-Site Bioreclamation Symposia, which were held in 1991, 1993, and 1995 in San Diego, California.

First International In Situ and On-Site Bioreclamation Symposium
1(1) *On-Site Bioreclamation: Processes for Xenobiotic and Hydrocarbon Treatment*
1(2) *In Situ Bioreclamation: Applications and Investigations for Hydrocarbon and Contaminated Site Remediation*

Second International In Situ and On-Site Bioreclamation Symposium
2(1) *Bioremediation of Chlorinated and Polycyclic Aromatic Hydrocarbon Compounds*
2(2) *Hydrocarbon Bioremediation*
2(3) *Applied Biotechnology for Site Remediation*
2(4) *Emerging Technology for Bioremediation of Metals*
2(5) *Air Sparging for Site Bioremediation*

Third International In Situ and On-Site Bioreclamation Symposium
3(1) *Intrinsic Bioremediation*
3(2) *In Situ Aeration: Air Sparging, Bioventing, and Related Remediation Processes*
3(3) *Bioaugmentation for Site Remediation*
3(4) *Bioremediation of Chlorinated Solvents*
3(5) *Monitoring and Verification of Bioremediation*
3(6) *Applied Bioremediation of Petroleum Hydrocarbons*
3(7) *Bioremediation of Recalcitrant Organics*
3(8) *Microbial Processes for Bioremediation*
3(9) *Biological Unit Processes for Hazardous Waste Treatment*
3(10) *Bioremediation of Inorganics*

Bioremediation Series Cumulative Indices: 1991-1995

For information about ordering books in the Bioremediation series, contact Battelle Press. Telephone: 800-451-3543 or 614-424-6393. Fax: 614-424-3819. Internet: sheldric@battelle.org.

Microbial Processes for Bioremediation

Edited by

Robert E. Hinchee and Fred J. Brockman
Battelle Memorial Institute

Catherine M. Vogel
U.S. Air Force Armstrong Laboratory

BATTELLE PRESS
Columbus • Richland

Library of Congress Cataloging-in-Publication Data

Hinchee, Robert E.
    Microbial processes for bioremediation / edited by Robert E. Hinchee,
Fred J. Brockman, Catherine M. Vogel.
      p.   cm.
    Includes bibliographical references and index.
    ISBN 1-57477-009-8 (hc : acid-free paper)
    1. Bioremediation—Congresses. 2. Microbial biotechnology—
Congresses. I. Hinchee, Robert E. II. Brockman, Fred J. III. Vogel,
Catherine M.
    TD192.5.M325 1995
    628.5′2—dc20   95-32253
                                                              CIP

Printed in the United States of America

Additional copies may be ordered through:
Battelle Press
505 King Avenue
Columbus, Ohio 43201, USA
1-614-424-6393 or 1-800-451-3543
Fax: 1-614-424-3819
Internet: sheldric@battelle.org

CONTENTS

FOREWORD

This book and its companion volumes (see overleaf) comprise a collection of papers derived from the Third International In Situ and On-Site Bioreclamation Symposium, held in San Diego, California, in April 1995. The 375 papers that appear in these volumes are those that were accepted after peer review. The editors believe that this collection is the most comprehensive and up-to-date work available in the field of bioremediation.

Significant advances have been made in bioremediation since the First and Second Symposia were held in 1991 and 1993. Bioremediation as a whole remains a rapidly advancing field, and new technologies continue to emerge. As the industry matures, the emphasis for some technologies shifts to application and refinement of proven methods, whereas the emphasis for emerging technologies moves from the laboratory to the field. For example, many technologies that can be applied to sites contaminated with petroleum hydrocarbons are now commercially available and have been applied to thousands of sites. In contrast, there are as yet no commercial technologies commonly used to remediate most recalcitrant compounds.

The articles in these volumes report on field and laboratory research conducted both to develop promising new technologies and to improve existing technologies for remediation of a wide spectrum of compounds.

The editors would like to recognize the substantial contribution of the peer reviewers who read and provided written comments to the authors of the draft articles that were considered for this volume. Thoughtful, insightful review is crucial for the production of a high-quality technical publication. The peer reviewers for this volume were:

Nelly Abboud, *University of Connecticut*
Jacqueline A.M. Abou-Rizk, *Union Carbide Corp.*
Peter Adriaens, *University of Michigan*
C. Marjorie Aelion, *University of South Carolina*
Bobby Allen, *Memphis Environmental Center*
Lisa Alvarez-Cohen, *University of California–Berkeley*
Rolf Arneberg, *Biogénie, inc.* (Canada)
Mick Arthur, *Battelle Columbus*
Erica S.K. Becvar, *U.S. Air Force*
Christian F. Bocard, *Institut Français du Pétrole*
Robert W. Botto, *Argonne National Laboratory*
Joan Braddock, *University of Alaska Fairbanks*
Alec Breen, *U.S. Environmental Protection Agency*
Barry Christian, *Earth Technology*
Steven D. Comfort, *University of Nebraska-Lincoln*
Mike Davidson, *Orange County, California*
C. Di Leo, *Eniricerche S.p.A.* (Italy)

Ken Doe, *Environment Canada*
J. R. Doyle, *Woodward-Clyde Consultants*
Murali Dronamraju (USA)
Jean Ducreux, *Institut Français du Pétrole*
Michael J. Dybas, *Michigan State University*
David D. Emery, *Bioremediation Service, Inc.*
Paul E. Flathman, *OHM Remediation Services Corp.*
Cheryl D. Gullett, *Battelle Pacific Northwest*
Kevin R. Hosler, *Wastewater Technology Centre* (Canada)
Don Hudgins, *Memphis Environmental Center, Inc.*
Michael H. Huesemann, *Battelle Pacific Northwest*
Stephen R. Hutchins, *U.S. Environmental Protection Agency*
Aloys Hüttermann, *Universität Göttingen*
Ursula Jenal-Wanner, *Stanford University*
Douglas E. Jerger, *OHM Remediation Services Corp.*
Colin Johnston, *CSIRO* (Australia)
Jeffrey Kavanaugh, *University of Western Florida/CEDB*
Soo-Youn Lee, *Washington State University*
Joe Lepo, *University of Western Florida*
Leonard Lion, *Cornell University*
Raina Miller, *University of Arizona*
Scott Miller, *AGRA Earth & Environmental, Inc.*
David K. Nicholson, *Woodward-Clyde Consultants*
Andrew Ogram, *Washington State University*
Fusako Okada, *EBARA Research Co Ltd* (Japan)
Anders Persson, *ANOX AB* (Sweden)
Tommy J. Phelps, *Oak Ridge National Laboratory*
A. Robertiello, *Eniricerche S.p.A* (Italy)
Joseph Rouse, *University of Oklahoma*
David Sabatini, *University of Oklahoma*
Diane L. Saber, *Saber Environmental Consultants, Inc.*
Pasquale Saccedau, *Eniricerche S.p.A* (Italy)
Bruce Sass, *Battelle Columbus*
Eric Schmitt, *ESE Biosciences, Inc.*
Alan Seech, *Grace Dearborn, Inc.* (Canada)
Douglas Selby, *Las Vegas Valley Water District*
Tatsuo Shimomura, *EBARA Research Co., Ltd.* (Japan)
Greg Smith, *Great Lakes Environmental Center*
Jim Spain, *U.S. Air Force*
Robert J. Steffan, *Envirogen, Inc.*
H. David Stensel, *University of Washington*
David Stevens, *Utah State University*
Kevin Strohmeier, *Environmental Technologies & Solutions, Inc.*
Alison Thomas, *U.S. Air Force*
Sarah Tremaine, *Hydrosystems, Inc.*
Jean-Paul Vandecasteele, *Institut Français du Pétrole*

Peter Vanneck, *University of Ghent*
David White, *University of Tennessee*
Peter Wilderer, *Technische Universität München*
Brian Wrenn, *University of Cincinnati*
Robert Wyza, *Battelle Columbus*

The figure that appears on the cover of this volume was adapted from the article by Di Palma et al. (see page 336).

Finally, I want to recognize the key members of the production staff, who put forth significant effort in assembling this book and its companion volumes. Carol Young, the Symposium Administrator, was responsible for the administrative effort necessary to produce the ten volumes. She was assisted by Gina Melaragno, who tracked draft manuscripts through the review process and generated much of the correspondence with the authors, co-editors, and peer reviewers. Lynn Copley-Graves oversaw text editing and directed the layout of the book, compilation of the keyword indices, and production of the camera-ready copy. She was assisted by technical editors Bea Weaver and Ann Elliot. Loretta Bahn was responsible for text processing and worked many long hours incorporating editors' revisions, laying out the camera-ready pages and figures, and maintaining the keyword list. She was assisted by Sherry Galford and Cleta Richey; additional support was provided by Susan Vianna and her staff at Fishergate, Inc. Darlene Whyte and Mike Steve proofread the final copy. Judy Ward, Gina Melaragno, Bonnie Snodgrass, and Carol Young carried out final production tasks. Karl Nehring, who served as Symposium Administrator in 1991 and 1993, provided valuable insight and advice.

The symposium was sponsored by Battelle Memorial Institute with support from many organizations. The following organizations cosponsored or otherwise supported the Third Symposium.

Ajou University–College of Engineering (Korea)
American Petroleum Institute
Asian Institute of Technology (Thailand)
Biotreatment News
Castalia
ENEA (Italy)
Environment Canada
Environmental Protection
Gas Research Institute
Groundwater Technology, Inc.
Institut Français du Pétrole
Mitsubishi Corporation
OHM Remediation Services Corporation
Parsons Engineering Science, Inc.
RIVM–National Institute of Public Health and the Environment
(The Netherlands)
The Japan Research Institute, Limited

Umweltbundesamt (Germany)
U.S. Air Force Armstrong Laboratory–Environics Directorate
U.S. Air Force Center for Environmental Excellence
U.S. Department of Energy Office of Technology Development
    (OTD)
U.S. Environmental Protection Agency
U.S. Naval Facilities Engineering Services Center
Western Region Hazardous Substance Research Center–
    Stanford and Oregon State Universities

Neither Battelle nor the cosponsoring or supporting organizations reviewed this book, and their support for the Symposium should not be construed as an endorsement of the book's content. I conducted the final review and selection of all papers published in this volume, making use of the essential input provided by the peer reviewers and other editors. I take responsibility for any errors or omissions in the final publication.

Rob Hinchee
June 1995

Dissolution and Cosubstrate Effects in the Transfer of Biodegradation Kinetics to the Field

Bruce E. Rittmann

ABSTRACT

Translation of biodegradation kinetics to the field requires consideration of dissolution kinetics from nonaqueous-phase sources and that the biodegradation kinetics can be controlled by the concentrations of biomass, primary electron donor, and primary electron acceptor. This paper presents quantitative tools for linking these phenomena. An example application for a ganglia nonaqueous-phase liquid (NAPL) demonstrates that the dissolution rate can be increased roughly 100-fold when biomass, primary electron donor, and primary electron acceptor are at high levels. Increases in the dissolution rate are much less significant when a primary electron donor is not present to enrich the cells in intracellular-reduced electron carriers.

INTRODUCTION

Most of the common organic pollutants in groundwater and soils share two important characteristics: (1) they have limited water solubility, and (2) the key initial steps of their biodegradation require cosubstrates (Rittmann et al. 1994). The low solubility means that their in situ removal depends on the rate of interphase transfer from a trapped nonaqueous source, such as dissolution of a NAPL or desorption from the solid phase. The rate of interphase mass transport is increased when rapid biodegradation of dissolved solute occurs near the nonaqueous-phase source, increasing the mass transport driving force. However, the cosubstrate requirement means that concentrations of materials other than the contaminant itself control the rate of biodegradation.

INTERPHASE MASS TRANSPORT AND BIODEGRADATION

Seagren et al. (1993, 1994) developed simple models to quantify the interaction between the rate of NAPL dissolution and biodegradation rates. Biodegradation rates, represented by a lumped, first-order rate coefficient (K_{B1}), increase the dissolution rate of NAPL ganglia or an NAPL pool when the biodegradation rate near the NAPL interface is sufficiently large compared to advective and dispersive rates. For one-dimensional flow of groundwater through a porous medium contaminated with NAPL ganglia, Seagren et al. (1993) derived the following equation for how biodegradation, advection, and longitudinal dispersion control the dissolution rate:

$$r_d = AvC_s Da_1 [1 - \left(\frac{Da_1}{w}\right) - \left(\frac{\left(\frac{Da_1}{w}\right)\left(\frac{2}{1 + (1 + 4w/Pe)^{1/2}}\right)}{\frac{1}{2} Pe[1 - (1 + 4w/Pe)^{1/2}]}\right) \tag{1}$$

$$\times (1 - \exp[(\tfrac{1}{2} Pe)\{1 - (1 + 4w/Pe)^{1/2}\}])]$$

in which r_d = average NAPL dissolution rate [MT^{-1}], A = domain cross-sectional area [L^2], v = average pore-water velocity [LT^{-1}], C_s = equilibrium concentration of the NAPL solute [ML^{-3}], $Da_1 = K_1L/v$, $w = Da_1 + Da_2$, $Da_2 = K_{B1}L/v$, $Pe = vL/D_x$, K_1 = lumped interfacial mass-transport coefficient [T^{-1}], L = length of the contaminated domain [L], K_{B1} = lumped first-order biodegradation coefficient [T^{-1}], and D_x = longitudinal dispersion coefficient [L^2T^{-1}]. For equation 1, r_d is increased by in situ biodegradation only when Da_2 is greater than 0.1 (Seagren et al. 1993).

COSUBSTRATE EFFECTS ON K_{B1}

The cosubstrate effects come about because initial biochemical reactions for degradation of almost all of the common organic contaminants require direct participation by a reduced electron carrier (ECH_2), such as nicotinamide adenine dinucleotide (NADH) or a reduced metalloenzyme (Rittmann et al. 1994; Bae and Rittmann 1994). The need for a reduced electron carrier is obvious for reductive dehalogenation, in which the organic carbon is reduced with the release of a halogen anion. A good example is the first reductive dechlorination of trichloroethene:

$$Cl_2CCClH + ECH_2 \rightarrow HClCCClH + EC + H^+ + Cl^-$$

Net oxidations, monooxygenations and dioxygenations, which initiate aerobic biodegradation of aromatic and aliphatic hydrocarbons, require a reduced cosubstrate. A good example is the monooxygenation of toluene:

$$C_6H_5CH_3 + ECH_2 + O_2 \rightarrow C_6H_5CH_2OH + EC + H_2O$$

Even though toluene is net-oxidized, a reduced electron carrier is needed, because reduction of O_2 requires four electrons, only two of which can come from the oxidation of toluene to its alcohol; the other two electrons must be supplied by the cosubstrate, ECH_2.

The toluene monooxygenation also illustrates that molecular O_2 is a cosubstrate in oxygenation reactions. In general, oxygenation reactions appear to be more sensitive to oxygen limitation than are respiratory reactions using O_2 as a terminal electron acceptor (Malmstead et al. 1994).

Bae (1992), using experiments with carefully controlled concentrations of acetate and dissolved oxygen, showed that the intracellular concentration of NADH responded dramatically and systematically to changes in the concentrations of the cell's primary electron-donor and electron-acceptor substrates. The cells became enriched in NADH when the acetate concentration was high or the oxygen concentration was low (Bae 1992; Bae and Rittmann 1994).

Experimental work by Bae and Rittmann (1990), Wrenn (1992), and Saéz and Rittmann (1993) documented that manipulations to the intracellular concentrations of reduced electron carriers could greatly accelerate or decelerate detoxification reactions. Bae and Rittmann (1990) saw major increases in the rate of reductive dechlorination of carbon tetrachloride by denitrifying biofilms when the electron donor (acetate) was augmented or the electron acceptor (nitrate) was removed. Wrenn (1992) documented dramatic increases in the rate of reductive dechlorination of 1, 1, 1-trichloroethane by methanogenic and sulfate-reducing biofilms when the critical electron donor (formate) was augmented. Saéz and Rittmann (1993) showed that the initial monooxygenation of 4-chlorophenol was rapid only when the cosubstrate phenol was being oxidized to create a high intracellular level of nicotinamide adenine dinucleotide phosphate (NADPH).

Bae and Rittmann (1994) employed the measured response of NADH to the external concentrations of acetate and oxygen to develop biodegradation rate equations that explicitly consider the effects of cosubstrates. For the case of oxygenation reactions, the rate express is:

$$-r_s = kSX \{\log(NADH/NAD) + M\} \frac{S_a}{K_a + S_a} \tag{2}$$

in which r_s = rate of contaminant detoxification ($ML^{-3}T^{-1}$), k = an overall first-order (in S) rate coefficient ($L^3M^{-1}T^{-1}$), S = concentration of the contaminant being detoxified (ML^{-3}), X = biomass concentration (ML^{-3}), $NADH/NAD$ = ratio of intracellular concentrations of reduced and oxidized electron carriers, $M = -1.827$, S_a = dissolved-oxygen concentration (ML^{-3}), and K_a = half-maximum-rate concentration for dissolved oxygen (ML^{-3}). Furthermore,

$$\log(NADH/NAD) = 0.273(-q_d) - \frac{1.11 S_a}{0.019 + S_a} - 0.717 \tag{3}$$

where

$$-q_d = 2.13 \left(\frac{S_d}{0.184 + S_d} \right) \left(\frac{S_a}{0.019 + S_a} \right) \tag{4}$$

and $-q_d$ = specific rate of electron-donor utilization, S_d = concentration of the primary electron donor, S_a and S_d are in units of mg/L and $-q_d$ is in units of mg COD/mg cells-hour, where COD stands for chemical oxygen demand.

The effects of cosubstrate concentrations on the NAPL-dissolution rate can be quantified by computing K_{B1} and Da_2 (needed in equation 1) from equations 2 through 4. In particular,

$$K_{B1} = kX\{log(NADH/NAD) + M\} \frac{S_a}{K_a + S_a} \tag{5}$$

APPLICATION

The effects of k, X, S_d, and S_a on the NAPL dissolution rate are evaluated by a simple example. A field site is contaminated with NAPL ganglia. The NAPL has an equilibrium concentration of $C_s = 1$ g/m^3 = 1 mg/L. The extent of contamination is L = 10 m, and the cross-sectional area is A = 1 m^2. The flow velocity of v = 1 m/day is held constant. The lumped interfacial mass-transfer coefficient is estimated as $K_1 = 200$ day^{-1}, whereas the longitudinal dispersion coefficient is estimated as $D_x = 1 \times 10^{-3}$ m^2/day. These parameters make Pe = 10^4, $Da_1 = 2 \times 10^3$, and $Da_2 = 10$ K_{B1}.

Laboratory biodegradation studies with a high dissolved-oxygen concentration (8.0 mg/L) showed an apparent k value (denoted k') of 5 L/g-day when the bacteria were not stimulated with an electron donor (i.e., $-q_d = 0$. Adapting equation 5 gives

$$k' = 5 \text{ L/g-day} = k\{log(NADH/NAD) + M\} \frac{S_a}{K_a + S_a} \tag{6}$$

Using equation 3 to compute log(NADH/NAD) and letting $K_a = 2$ mg O_2/L (Malmstead et al. 1994) give k = 2,380 L/g-day.

The rate of dissolution (r_d, in g/day) is computed for a range of X, S_d, and S_a concentrations; each parameter has units of mg/L of pore water. The computation procedure is as follows:

1. X, S_d, and S_a are specified.
2. $-q_d$ is computed from S_d, S_a, and equation 4.
3. log(NADH/NAD) is computed from S_a, $-q_d$, and equation 3.

4. k' is computed from log(NADH/NAD), S_a, and equation 6.
5. K_{B1} is computed as $K_{B1} = k'X$.
6. Finally, r_d is computed from equation 1 and the definitions of Da_2 and w given below equation 1.

Table 1 summarizes how the dissolution rates respond to changes in X, S_a, and S_d. When no microbial activity exists (case 1), the background dissolution rate is 1 g/d. The presence of a modest microbial community (X = 1 mg/L) does not accelerate dissolution when oxygen is absent (case 2), because oxygen is required as a cosubstrate for the initial monooxygenase reactions. Providing a low oxygen concentration (case 3) increases the rate slightly to 1.2 g/d, but stimulating microbial activity to have an increased bacterial concentration (10 mg/L case 4) accelerates the dissolution rate to 2.6 g/d.

Comparing cases 4, 5, and 6 shows that too much oxygen can reduce the dissolution rate when the primary electron donor is not present, because the high oxygen depletes the cells' intracellular concentration of reduced electron carrier more than can be compensated for by increased availability of O_2 as a cosubstrate itself.

Cases 7 through 10 demonstrate the major impact that can be obtained by augmenting with a primary electron donor that can enrich the cells in reduced electron carrier. Even when the oxygen concentration is low (case 7), maintaining 10 mg/L of primary electron donor accelerates the dissolution rate 46-fold over the background. When sufficient donor can be supplied, the fastest dissolution rates occur when the primary donor and acceptor concentrations are simultaneously high (case 10).

TABLE 1. Summary of impacts on dissolution rate when monooxygenation controls the biodegradation.

Case	X	S_a	S_d	r_d g/day
1.	0	0	0	1.0
2.	1	0	0	1.0
3.	1	1	0	1.2
4.	10	1	0	2.6
5.	10	8	0	1.5
6.	10	3	0	2.0
7.	10	1	10	46.0
8.	10	8	1	60
9.	10	3	10	80
10.	10	8	10	104

ACKNOWLEDGMENTS

This work was supported by the Subsurface Sciences Program of the U.S. Department of Energy.

REFERENCES

Bae, W. 1992. "Modeling Dual-Limitation Kinetics Incorporating Cofactor Responses." Ph.D. dissertation, Dept. Civil Engineering, Univ. of Illinois, Urbana, IL.

Bae, W., and B. E. Rittmann. 1990. "Effects of Electron Acceptor and Electron Donor on Bio-degradation of CCl$_4$ by Biofilms." *Proc. 1990 Spec. Conf. on Environ. Engr.*, pp. 390-397, Amer. Soc. Civil Engrs., New York, NY.

Bae, W., and B. E. Rittmann. 1994. "Accelerating the Rate of Cometabolic Degradations Requir-ing an Intracellular Electron Source — Model and Biofilm Application." *Water Sci. Technol.*, in press.

Malmstead, M. J., F. Brockman, A. J. Valocchi, and B. E. Rittmann. 1994. "Modeling Biofilms Biodegradation Requiring Cosubstrate: The Quinoline Example." *Water Sci. Technol.*, in press.

Rittmann, B. E., E. Seagren, B. Wrenn, A. J. Valocchi, C. Ray, and L. Raskin. 1994. *In Situ Bioremediation*, 2nd ed., Noyes Publishers, Inc., Park Ridge, NY.

Saéz, P. B., and B. E. Rittmann. 1993. "Biodegradation kinetics of a mixture containing a pri-mary substrate (phenol) and an inhibitory co-metabolite (4-chlorophenol)." *Biodegradation* 4:3-23.

Seagren, E. A. 1994. "Quantitative Evaluation of Flushing and Biodegradation for Enhancing In Situ Dissolution of Nonaqueous-Phase Liquids." Ph.D. dissertation, Dept. of Civil Engr., Univ. of Illinois, Urbana, IL.

Seagren, E. A., B. E. Rittmann, and A. J. Valocchi. 1993. "Quantitative Evaluation of Flushing and Biodegradation for Enhancing In Situ Dissolution of Nonaqueous Phase Liquids." *J. Contam. Hydrol.* 12:103-112.

Seagren, E. A., B. E. Rittmann, and A. J. Valocchi. 1994. "Quantitative Evaluation of the Enhancement of NAPL-Pool Dissolution by Flushing and Biodegradation." *Environ. Sci. Technol.* 28:833-839.

Wrenn, B. A. 1992. "Substrate Interactions During the Anaerobic Biodegradation of 1,1,1-Tri-chloroethane." Ph.D. dissertation, Dept. of Civil Engr., Univ. of Illinois, Urbana, IL.

Effects of Surfactants on Fluoranthene Mineralization by *Sphingomonas paucimobilis* Strain EPA 505

*Suzanne Lantz, Jian-Er Lin,
James G. Mueller, and Parmely H. Pritchard*

ABSTRACT

Past results from surfactant-enhanced biodegradation studies have been equivocal because of inhibitory effects of the surfactants and a poor understanding of the characteristics of PAH-degrading microorganisms that make them responsive to surfactants. We have studied the mineralization of ^{14}C-radiolabeled fluoranthene by high cell masses of *Sphingomonas paucimobilis*, strain EPA 505, and have shown that initial rates of mineralization can be enhanced by concentrations of the surfactant Triton X-100 as high as 2%. Mass balances are reported that show complete degradation of fluoranthene. The presence of soil stimulated biodegradation of fluoranthene in the same manner as surfactants, presumably because of increased dissolution rates from soil particulates. The usefulness of this bacterium in the bioremediation of PAH-contaminated soil is discussed.

INTRODUCTION

PAHs exist naturally in the environment from petrogenic and phytogenic sources (Mueller et al. 1989; NAS 1983). Environments contaminated with large amounts of these chemicals are considered hazardous owing to potential carcinogenic, mutagenic, and teratogenic effects of specific PAHs (Moore et al. 1989). Generally, higher-molecular-weight PAHs, containing four or more fused rings, pose the greatest hazard to both the environment and human health (U.S. EPA 1982). There is, consequently, much interest in developing remedial methods, such as bioremediation, to selectively remove these chemicals from contaminated environmental materials.

When suitable environmental conditions are present, biodegradation of low-molecular-weight PAHs by indigenous microorganisms readily occurs (Mueller et al. 1991; Mueller et al. 1993; Cerniglia 1993). However, under the same conditions, biotransformation of high-molecular-weight (HMW) PAHs is less likely.

Although bacteria have been isolated in pure cultures that grow on HMW PAHs such as fluoranthene and pyrene (Mueller et al. 1990; Weissenfels et al. 1990; Walter et al. 1991; Grosser et al. 1991; Heitkamp et al. 1988), strategies for stimulating this activity and the degradation of other HMW PAHs in contaminated soils are not readily available.

HMW PAHs usually occur as contaminants in natural ecosystems and waste treatment systems at mass levels that exceed their water solubility and are sorbed to particulate surfaces. These characteristics largely account for their slow biodegradation. Because pure cultures of bacteria grow on PAH compounds primarily in the dissolved state (Stucki & Alexander 1987; Thomas et al. 1986; Volkerling et al. 1992; Wodzinsky & Coyle 1974), the dissolution of PAHs into the aqueous phase will be a prerequisite for microbial degradation. Surfactants can enhance PAH solubilization and dissolution and increase the equilibrium concentration of the compound in the aqueous phase (Edwards et al. 1991). Thus the use of surfactants as a possible strategy for enhancing bioremediation has been discussed (West & Harwell 1992). Several laboratory studies have verified that surfactants can, in some cases, enhance degradation and mineralization of PAHs, while in other cases, inhibition has been observed (Laha & Luthy 1992; Tiehm 1994). Accordingly, successes in the field have been limited (Mueller et al. 1993; West & Harwell 1992).

Before surfactants can be used routinely to promote bioavailability and bioremediation, considerable new research is needed to more fully understand the mechanisms by which surfactants stimulate PAH bioremediation. Our approach has been to emphasize the characteristics of the microorganisms that are necessary to optimize surfactant-enhanced bioavailability of PAHs from coal tar-contaminated soils. We have, consequently, studied the mineralization of ^{14}C-radiolabeled fluoranthene by *Sphingomonas paucimobilis*, strain EPA 505, and we show that mineralization can be enhanced by the presence of high concentrations of the surfactant Triton X-100.

MATERIALS AND METHODS

Sphingomonas paucimobilis EPA 505 was originally isolated from creosote-contaminated soil as an organism that would grow on fluoranthene as a source of carbon and energy (Mueller et al. 1990). For studies with surfactants, it was grown for 48 hr at 30°C with shaking in complex medium (Luria-Bertani broth). Stationary phase cells were harvested by centrifugation, washed in 0.025M phosphate buffer, and then used as inocula in the mineralization experiments. Cell concentrations were determined by plate counts. Fluoranthene mineralization experiments were initiated by placing labeled (approximately 60,000 DPM of ^{14}C-3-fluoranthene; Sigma Chemical) and unlabeled fluoranthene dissolved in acetone into the bottom of 300-mL Biometer flasks (Bellco) and allowing the acetone to evaporate. Then 50 mL of basal salts medium were added to each flask and the liquid sonicated to break up the fluoranthene crystals. A desired volume of sterile surfactant and of washed cell suspension were then added to the flasks. The surfactant, Triton X-100, is a liquid and the neat chemical was designated as 100% concentration. Next, 1.0 mL of 2N NaOH solution was placed in the side arm of the flasks to trap any radioactive CO_2 produced. Flasks were incubated at 30°C with shaking and periodically sampled for concentration of fluoranthene and radioactivity associated with cells and CO_2. Fluoranthene was determined by filtering aliquots of the culture fluid onto 7.0 µm filters to remove fluoranthene crystals and allow cells to pass through. The filters were then extracted with methylene chloride. The filtrate was centrifuged to remove cells and the supernatant extracted with methylene chloride. Extracts were combined and fluoranthene concentrations measured by gas chromatography (Mueller et al. 1990). Radioactivity was determined by scintillation counting of aqueous filtrates and filters (cells). The effect of different surfactant concentrations on the solubility of fluoranthene was determined using GC analysis.

RESULTS AND DISCUSSION

Figure 1 shows the mineralization of fluoranthene by high cell densities (8×10^{10} cells per mL) of EPA 505 in the absence of surfactant. Approximately 50% of the added radioactivity appeared as CO_2. Carbon dioxide production was linear, indicating that the degradation rate was limited by fluoranthene dissolution from the crystals (Tiehm 1994; Volkerling et al. 1992). When smaller size crystals were used (longer period of sonication), greater crystal surface area was available, and degradation rates, although still linear, were greater.

The addition of different concentrations of Triton X-100 increased the apparent solubility of fluoranthene by more than 100 fold, giving a concentration of 226 mg/L at 2% Triton X-100 concentration (aqueous solubility of fluoranthene is approximately 0.26 mg/L at 25°C). The critical micelle concentration (CMC) of Triton X-100 is <0.01%(v/v) (Laha & Luthy 1992), and thus concentrations of fluoranthene in all of our tests were pseudo-solubilized in surfactant micelles. EPA 505, when added in high concentrations to media containing different concentrations of surfactant, was able to substantially mineralize fluoranthene at all concentrations (Figure 2). Note that mineralization rates are less linear, suggesting that fluoranthene dissolution from crystals was no longer limiting growth.

It has been previously observed that use of surfactants reduced or inhibited biodegradation because of surfactant toxicity to the bacteria or the mixed culture used in the study (Lin et al. 1995). Tiehm (1994) has shown, for example, that a variety of nonionic surfactants, at relatively low concentrations (around 4 to 10 mM), inhibited the mineralization of fluoranthene and pyrene by a pure culture, but not a mixed microbial community that had been enriched for its ability to mineralize these PAHs. Laha and Luthy (1992) showed that low concentrations of Triton X-100 (0.8 mM) inhibited phenanthrene mineralization by an enriched microbial community. Our results show that fluoranthene mineralization by *Sphingomonas paucimobilis*, strain EPA 505 was enhanced by the presence of the surfactant Triton X-100 up to 34 mM concentration. This organism, and others like it, could be excellent candidates for optimizing the bioremediation of PAH-contaminated soils through the use of bioaugmentation procedures, because at

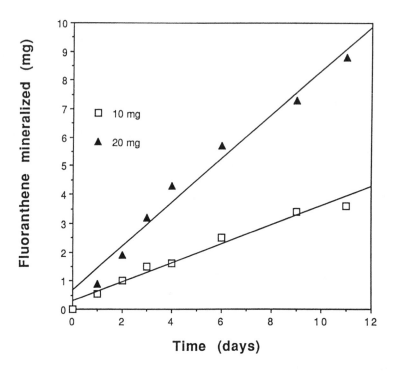

FIGURE 1. Mineralization profiles of fluoranthene, expressed as mg equivalents, at two different masses by strain EPA 505 (inoculum size = 8 × 10^{10} cells/mL) in minimal salts medium without surfactant.

high surfactant concentrations, toxicity to the indigenous microorganisms will reduce competition from within the microbial community.

The mineralization of fluoranthene at high surfactant concentrations was, however, not without some effect on the mineralization. From Figure 2, it can be seen that, at the two highest concentrations of surfactant, less than 50% of the original radioactivity (expressed as equivalent mass in the figure) appeared as CO_2. This was accompanied by the production of a red-colored metabolite in the medium. To more fully understand these results, the experiment was repeated and after 48 hours of incubation, the cell culture was sacrificed and a mass balance of fluoranthene concentration and radioactivity determined. These results are shown in Table 1. At all surfactant concentrations, fluoranthene was almost completely consumed. The effect of the high surfactant concentration appears to be on the production of radioactive products ("soluble" fraction in Table 1) with less incorporation into biomass and carbon dioxide. Although we are performing further experiments to determine the exact details of this response, it is tempting to suggest that the surfactants caused the leaching of partially degraded fluoranthene products from the cells before they could be completely mineralized to CO_2 and cell biomass. The appearance of the red color may be indicative of this phenomenon.

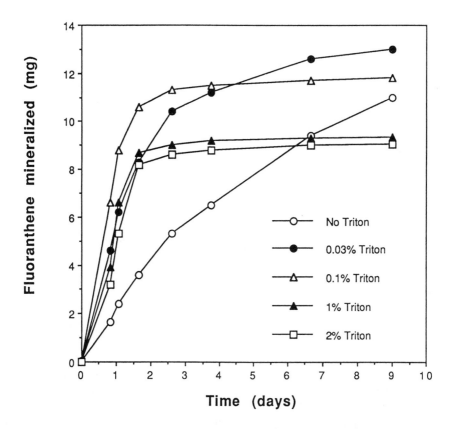

**FIGURE 2. Effect of different concentrations of Triton X-100 on the minerali-
zation of 20 mg of fluoranthene, expressed as mg equivalents, by strain
EPA 505 (inoculum size = 8 × 10^{10} cells/mL) in minimal salts medium.**

Along with the ability of EPA 505 to tolerate high surfactant concentrations, consideration also must be given to soil slurry systems where particle surfaces may decrease the aqueous concentration of a PAH compound due to sorption effects (Guerin & Boyd 1992), effectively reducing aqueous concentrations and thus decreasing the rate of biodegradation. However, the presence of soil particles in a solution can provide a higher solid-liquid contact surface area and, thereby, actually enhance the solid-liquid mass transfer. As shown in Figure 3, an aqueous suspension of soil particles (30 mg/mL) and fluoranthene crystals (20 mg) together resulted in greater mineralization rates than suspensions with only fluoranthene crystals. This would appear to be the effect of higher solid-liquid contact. Although increases in biomass or in the activity of the biomass as a result of exposure to soil particles may also explain the effect, this is unlikely because the biomass (8 × 10^{10} cells/mL) and the mineralization rates initially were high. Note that the effect of soil particles was equivalent to that of adding surfactant, a further indication of increased dissolution by either material.

TABLE 1. Mass balance of radioactivity (expressed as percent of total ^{14}C added) and fluoranthene (Fla) after 48 h incubation of EPA 505 cells (10^{10} cells/mL) in minimal salts medium with different concentrations of surfactant.

Triton	Recovery of ^{14}C	Soluble	Cell Mass[a]	Carbon Dioxide	Fla Residual
0.0%	81	4	54	42	4.73
0.005%	100	4	48	48	0.25
0.03%	95	9	28	63	0.15
0.1%	95	12	26	62	nd (< 5 µg)
2.0%	96	26	16	58	na
killed cells	100	< 1	99	< 1	30.03

(a) Radioactivity retained on 0.22-µm filter.
nd = not detected.
na = not analyzed (surfactant concentration too high for injection on GC).

Our initial results, and those of others (Zhang & Miller 1994, van Loosdrecht et al. 1987), strongly suggest that not enough attention has been paid to the important and useful characteristics of the microorganisms relative to improving bioavailability and biodegradation of PAHs. We know little about the dynamics of PAH movement between surfactant micelles and the bacterial cells themselves. We assume transport through the aqueous phase, but there may also be transport resulting from direct interaction between the micelle and the bacteria, particularly with cells that are surfactant tolerant and adhere to hydrophobic surfaces.

In addition, PAHs in contaminated field sites will likely be found in association with the oily (organic) contaminant phase, such as with creosote or coal tar residues. It is not clear how bioavailable the PAHs will be for biodegradation and how surfactants will affect the bioavailability. A recent study by Kohler et al. (1994) has shown that a PAH dissolved in an organic phase (heptamethylnonane) was in fact available to the aqueous phase and the bacteria contained therein. The importance of surfactants, and the bacteria capable of interacting with the PAH-containing micelles, must be studied further under these conditions.

SUMMARY AND CONCLUSIONS

Before surfactants can be used routinely to promote bioavailability and enhance bioremediation, considerable research is needed to more fully understand the mechanisms by which surfactants stimulate PAH bioremediation. Our approach has been to emphasize the characteristics of the microorganisms that are necessary to optimize surfactant-enhanced biodegradation of PAHs. We have, consequently, shown that the mineralization of ^{14}C-radiolabeled fluoranthene by *Sphingomonas paucimobilis*, strain EPA 505, can be enhanced by the presence of high concentrations

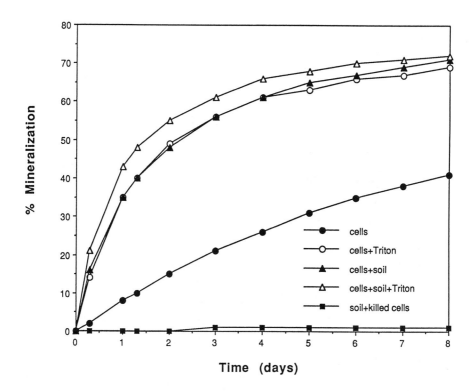

FIGURE 3. Effect of sterile soil (30 mg) and Triton X-100 on the mineralization of 20 mg fluoranthene by strain EPA 505 (inoculum size = 8 × 10^{10} cells/mL) in minimal salts medium.

of the surfactant Triton X-100. This organism, and others that are tolerant of high surfactant concentrations, will be further studied in an effort to determine how best to use surfactants effectively in the field for bioremediation.

REFERENCES

Cerniglia, C. E. 1993. "Biodegradation of polycyclic aromatic hydrocarbons." *Biodegradation* 3:351-368.

Edwards, D. A., R. G. Luthy, and Z. Liu. 1991. "Solubilization of polycyclic aromatic hydrocarbons in micellar nonionic surfactant solutions." *Environ. Sci. Technol.* 25:127-133.

Grosser, R. J., D. Warshawsky, and J. R. Vestal. 1991. "Indigenous and enhanced mineralization of pyrene, benzo[a]pyrene, and carbazole in soils." *Appl. Environ. Microbiol.* 57:3462-3469.

Guerin, W. F., and S. A. Boyd. 1992. "Differential bioavailability of soil-sorbed naphthalene to two bacterial species." *Appl. Environ. Microbiol.* 58:1142-1152.

Heitkamp, M. A., W. Franklin, and C. E. Cerniglia. 1988. "Microbial metabolism of polycyclic aromatic hydrocarbons: Isolation and characterization of a pyrene-degrading bacterium." *Appl. Environ. Microbiol.* 54:2549-2555.

Kohler, A., M. Schuttoff, D. Brynoik, and H-J. Knackmuss. 1994. "Enhanced biodegradation of phenanthrene in a biphasic culture system." *Biodegradation* 5:93-103.

Laha, S., and R. G. Luthy. 1992. "Effects of nonionic surfactants on the solubilization and mineralization of phenanthrene in soil-water systems." *Biotechnol. Bioeng.* 40:1367-1380.

Lin, Z., A. M. Jacobson, and R. G. Luthy. 1995. "Biodegradation of naphthalene in aqueous nonionic surfactant systems." *Appl. Environ. Microbiol.* 61:145-151.

Moore, M. N., D. R. Livingstone, and J. Widdows. 1989. "Hydrocarbons in marine mollusks: Biological effects and ecological consequences." pp. 291-328. In U. Varanasi (Ed.), *Metabolism of PAHs in the aquatic environment.* CRC Press, Inc., Boca Raton. Fl.

Mueller, J. G., P. J. Chapman, and P. H. Pritchard. 1989. "Action of a fluoranthene-utilizing bacterial community on polycyclic aromatic hydrocarbon components of creosote." *Appl. Environ. Microbiol.* 55:3085-3090.

Mueller, J. G., P. J. Chapman, E. O. Blattmann, and P. H. Pritchard. 1990. "Isolation and characterization of a fluoranthene-utilizing strain of *Pseudomonas paucimobilis.*" *Appl. Environ. Microbiol.* 56:1079-1086.

Mueller, J. G., S. E. Lantz, B. O. Blattmann, and P. J. Chapman. 1991. "Bench-scale evaluation of alternative biological treatment processes for the remediation of pentachlorophenol- and creosote-contaminated materials: Slurry-phase bioremediation." *Environ. Sci. Technol.* 25:1055-1061.

Mueller, J. G., S. E. Lantz, R. J. Colvin, D. Ross, D. P. Middaugh, and P. H. Pritchard. 1993. "Strategy using bioreactors and specially selected microorganisms for bioremediation of groundwater contaminated with creosote and pentachlorophenol." *Environ. Sci. Technol.* 27:691-698.

National Academy of Science (NAS). 1983. *Polycyclic aromatic hydrocarbons: Evaluation of sources and effects.* National Academy Press. Washington, DC.

Stucki, G., and M. Alexander. 1987. "Role of dissolution rate and solubility in biodegradation of aromatic compounds." *Appl. Environ. Microbiol.* 53:292-297.

Thomas, J. M., J. R. Yordy, J. A. Amador, and M. Alexander. 1986. "Rates of dissolution and biodegradation of water-insoluble organic compounds." *Appl. Environ. Microbiol.* 52:290-296.

Tiehm, A. 1994. "Degradation of polycyclic aromatic hydrocarbons in the presence of synthetic surfactants." *Appl. Environ. Microbiol.* 60:258-263.

U.S. Environmental Protection Agency. 1982. *Wood preservative pesticides: Creosote, pentachlorophenol, and the inorganic arsenical (wood uses).* Position Document 213. EPA 540/9-82/004. Environmental Protection Agency, Washington, DC.

van Loosdrecht, M. C., J. Lyklema, W. Norde, G. Schraa, and A.J.B. Zehender. 1987. "The role of bacterial cell wall hydrophobicity in adhesion." *Appl. Environ. Microbiol.* 53:1893-1897.

Volkerling, F., A. M. Breure, A. Sterkenburg, and J. G. van Andel. 1992. "Microbial degradation of polycyclic aromatic hydrocarbons: Effect of substrate availability on bacterial growth kinetics." *Appl. Microbiol. Biotechnol.* 36:548-552.

Walter, U., M. Beyer, J. Klein and H. J. Rehm. 1991. "Degradation of pyrene by *Rhodococcus* sp. UW1." *Appl. Microbiol. Biotechnol.* 34:671-676.

Weissenfels, W. D., M. Beyer, and J. Klein. 1990. "Degradation of phenanthrene, fluorene, and fluoranthene by pure bacterial cultures." *Appl. Microbiol. Biotechnol.* 34:528-535.

West, C. C., and J. H. Harwell. 1992. "Surfactants and subsurface bioremediation." *Environ. Sci. Technol.* 26:2324-2330.

Wodzinski, R. S., and J. E. Coyle. 1974. "Physical state of phenanthrene for utilization by bacteria." *Appl. Microbiol.* 27:1081-1084.

Zhang, Y., and R. M. Miller. 1994. "Effect of a *Pseudomonas* rhamnolipid biosurfactant on cell hydrophobicity and biodegradation of octadecane." *Appl. Environ. Microbiol.* 60:2101-2106.

Surfactant-Enhanced Bioremediation of PAH- and PCB-Contaminated Soils

Mriganka M. Ghosh, Ick Tae Yeom, Zhou Shi,
Chris D. Cox, and Kevin G. Robinson

ABSTRACT

The role of surfactants in the desorption of soil-bound polycyclic aromatic hydrocarbons (PAHs) and polychlorinated biphenyls (PCBs) was investigated. The solubilization of individual PAHs in an extract of a weathered, coal tar-contaminated soil containing a mixture of PAHs and other petroleum derivatives was found to be significantly less than that for pure compounds. Batch soil washing with Triton X-100 (a commercial, nonionic alkyl phenol ethoxylate) was found to increase the effective diffusion rate of PAHs from the contaminated soil by four orders of magnitude compared to that obtained by gas purging when the results were analyzed using a radial diffusion model. At concentrations of up to 24 times its critical micelle concentration (CMC), Triton X-100 did not seem to enhance hydrocarbon degradation in the coal tar-contaminated soil; however, the biosurfactant rhamnolipid R1, at a concentration of 50x CMC, increased the rate of mineralization of 4,4'-chlorinated biphenyl mobilized from a laboratory-contaminated soil by more than 60 times.

INTRODUCTION

Bioremediation has been frequently cited as an innovative technology to remediate certain organic compounds, notably petroleum hydrocarbons and PCBs (Cookson 1995). Bioremediation is anything but new; its mechanisms are deeply rooted in biological wastewater treatment. At a large number of contaminated sites, bioremediation is being used to remediate contaminated soil. Almost 78% of these are using ex situ bioremediation (U.S. Environmental Protection Agency 1994). PAHs, constituents of petroleum hydrocarbons, and PCBs are characterized by their low aqueous solubility and hydrophobicity. Thus, these hydrophobic organic compounds (HOCs) strongly bind to the soil matrix, primarily to the soil organic matter (Karickhoff 1984). The bioremediation of soil-bound HOCs appears to be limited by their slow rate of release (Ball & Roberts 1991). Commercial surfactants may dramatically increase the aqueous solubility of HOCs (Kile & Chiou 1989). In fact, surfactant washing with nonionic polyoxyethylene (POE) compounds forms the basis of in situ remediation of aquifers and ex situ treatment of excavated soils contaminated with PAHs (Aronstein et al. 1991, Edwards et al. 1991). Aggregates of micellar surfactant molecules create a hydrophobic, less polar core into which HOCs partition, thereby increasing their desorption from soils. Biosurfactants are produced in copious amounts by soil bacteria, *Pseudomonas, Rhodococcus*, and *Arthrobacter*, for example, under environmental stress (Ramana & Karanth 1989, Itoh & Suzuki 1972). Recent studies have demonstrated the effectiveness of biosurfactants produced by *Pseudomonas aeruginosa* in enhancing the biomineralization of sparingly soluble HOCs (Zhang & Miller 1992).

We report on a study of surfactant-mediated desorption and mineralization of PAHs and 4,4'-chlorobiphenyl (CB) from soils of different organic carbon contents (f_{oc}). In particular, the desorption of HOCs from a weathered, coal tar-contaminated soil obtained from a manufactured gas plant site (MGP) was studied; desorption using surfactant washing and gas purging were compared. Triton X-100, a nonionic branched alkyl phenol ethoxylate ($C_8PE_{9.5}$), was used. An attempt was made to interpret desorption in the light of adsorption theory developed in recent years. Rhamnolipid R1 was used to study the desorption and mineralization of 4,4'-CB from an artificially contaminated soil. To the extent that mineralization is constrained by bioavailability, the relationship between the period of weathering of contaminant hydrocarbons (contamination age) and bioremediation was qualitatively studied.

METHODOLOGY

Solubility enhancement of pure 4,4'-CB and PAHs (characterized in Table 1) was evaluated in batch tests where crystalline test compounds were mixed with various concentrations of surfactants in a mechanical shaker until equilibrium was achieved (<2 days). Samples were then centrifuged to separate any residual crystals, the aqueous phase was decanted, and the aqueous concentrations of test compounds were measured. The molar solubilization ratio (MSR), defined as the moles of HOC dissolved per mole of surfactant at dosages higher than CMC, and the coefficient K_m, describing equilibrium partitioning of HOC between the micellar and aqueous pseudophases, were used to characterize surfactant enhancement of solubilization (Edwards et al. 1991). The biosurfactant, rhamnolipid R1, was extracted from the supernatant of a *Pseudomonas aeruginosa* culture grown on glycerol and purified by silicic acid chromatography (Ramana & Karanth 1989). Using the method of Ramana and Karanth (1989), rhamnolipid R1 was characterized as a glycosylated anionic compound, having a CMC of 54 mg/L in 30 mM phosphate buffer (pH = 7), a molecular weight of 649 g/mol and a surface tension of 35.1 dyne/cm at the CMC. Similar characteristics of rhamnolipid R1 have been reported recently (Thangamani & Shreve 1994). The

TABLE 1. Characterization and solubilization of hydrophobic organic compounds (HOCs) by surfactants.

Surfactant	HOC	Molecular weight	Aqueous solubility (M)	log K_{ow}	CMC (M)	CMC (g/L)	MSR[a]	log K_m
Triton X-100		625			1.70e-04	0.110		
	Phenanthrene	178	7.20e-06	4.46			6.80e-02(1e-02)	5.61
	Anthracene	178	4.10e-07	4.45			2.99e-03(2.8e-3)	5.62
	Benzo(a)pyrene	252	1.51e-08	6.50			1.50e-02(2e-03)	7.56
	Pyrene	202	6.80e-07	4.90			3.50e-02(5.8e-03)	6.02
Rhamnolipid R1		649			8.32e-05	0.054		
	4,4'-CB	223	2.69e-07	5.58			3.79e-03	5.35
	Phenanthrene	178	7.20e-06	4.46			3.60e-02	5.5

(a)   Numbers in parentheses are for individual PAHs present as components of a PAH mixture obtained from MGP soil by soxhlet extraction using methylene chloride.

solubilization of PAHs in tar-contaminated soils was studied using whole, original soils and their extracts obtained by soxhlet extraction (Soxtec HT6, Itecator Co., Sweden) using boiling (110°C) methylene chloride according to the method suggested by Itecator Co. In the procedure, following extraction for 6 hours, the soil is rinsed for 1 hour with fresh methylene chloride, and the extract is purged with air for 2 minutes to remove the residual methylene chloride. The extraction scheme is designed to minimize volatilization losses. The extract was then dissolved in 5 mL of methylene chloride and diluted (1:100) with methanol. The PAHs in the extract were analyzed by HPLC. Four individual PAHs (phenanthrene, anthracene, benzo(a)pyrene, and pyrene) as well as total PAH were measured.

The mineralization of HOCs in the MGP soil in the presence of Triton X-100 was studied using an enriched consortium of microorganisms. The culture was developed using a mixture of soils obtained from an MGP site and from a coke-processing plant site, a superfund site in Tennessee. The soils were added to a flask containing a mixture of PAHs, known inducers for PAH biodegradation such as salicylic acid and succinic acid, and nutrients. After 3 weeks, an aliquot of the culture was transferred to another flask containing a mixture of PAHs, nutrients, and the MGP soil. The mixture was then incubated for 2 weeks. The presence of PAH-degrading microorganisms in the enriched culture was tested by the mineralization of ^{14}C-labeled PAHs (9-^{14}C-phenanthrene, 1-^{14}C-naphthalene, 4,5,9,10-^{14}C-pyrene and 7-^{14}C-benzo(a)pyrene). To estimate biodegradation, total CO_2 produced was collected in NaOH (0.5 N) traps and analyzed by a total carbon analyzer (Dohrmann Model 80). Also, the aqueous concentrations of three individual PAHs were measured as a function of time using HPLC. Augmentation

of biomineralization of ^{14}C-labeled 4,4'-CB desorbed from an artificially contaminated soil using rhamnolipid R1 was studied using a substrate-specific organism,
namely, *Alcaligenes eutrophus* (A5).

Additionally, experiments were conducted to determine the adsorption of
[3]H-Triton X-100 on two Tennessee soils of similar texture but different organic
carbon content (f_{OC}=0.011 and 0.022). First, 4 g of air-dried soil was suspended
in 10 mL phosphate buffer and wetted for 24 hours. Then 10 mL of radiolabeled
surfactant solutions of various concentrations were added to the soil suspensions
making up the volume of suspension to 20 mL. The samples were mixed for
24 hours in a mechanical shaker followed by centrifugation at 10,000 rpm for
50 minutes. Surfactant remaining in suspension was analyzed radiochemically.

RESULTS AND DISCUSSIONS

In the subCMC (CMC = 106 mg/L) range, the adsorption of ^{3}H-Triton X-100
on the two Tennessee soils was well described by the Freundlich isotherm (K =
7.86 and 26.3 (mg/g) (mg/mL)^n; n = 0.895 and 1.77; r^2 = 0.9868 and 0.9965, respectively, for soils with f_{OC} = 0.011 and 0.022) and was in agreement with that
reported in the literature (Liu et al. 1992). At the CMC, the adsorption of Triton
X-100 on soil was about 1 mg/g for both soils; above the CMC, adsorption was
found to be constant. Conceivably, sorbed surfactant could enhance the sorption
of contaminants via partitioning similar to that which occurs with natural organic
matter. However, applied at dosages above CMC, the same surfactant would
greatly enhance the solubilization of soil-bound contaminants.

Solubility enhancement of HOCs is reported in terms of the molar solubilization ratio (MSR), the number of moles of HOC solubilized per of mole of micellized surfactant, and by the micelle-water partitioning coefficient (K_m), defined
as the ratio of the contaminant mole fractions in the micellar (X_m) and aqueous
(X_a) pseudophases (Edwards et al. 1991, Diallo et al. 1994). The micelle-water
partitioning coefficient K_m of a PAH is given by:

$$K_m = X_m / X_a \tag{1}$$

where the mole fractions in micelle (X_m) and water (X_a), are given by equations 2
and 3, respectively:

$$X_m = \frac{MSR}{1 + MSR} \tag{2}$$

$$X_a = C_{aq} V_w \tag{3}$$

The molar solubility ratio (MSR) is experimentally determined. C_{aq} is the aqueous
solubility of pure PAH and V_w is the molar volume of water (0.018 L/mol). The

solubilization of test PAHs by Triton X-100 and of 4,4'-CB by rhamnolipid R1 are summarized in Table 1. For comparison, the results of solubilization of individual PAHs in the MGP soil extract are included in Table 1. In accordance with Raoult's law, the aqueous solubilities of individual PAHs in the MGP soil containing a multicomponent mixture were much less than those of pure compounds. The mole fractions of individual PAHs in micellar solution of the MGP soil extract were smaller than those in micellar solutions of pure PAHs. These are more realistic estimates of expected surfactant-solubilization of PAHs in tar-contaminated soils than those predicted from studies with pure compounds.

A radial diffusion model, similar to models used by other researchers (Wu & Gschwend 1986, Ball & Roberts 1991), was developed to compare the desorption of contaminants from soils by gas purging and surfactant washing. Figure 1 shows a comparison of desorption of anthracene and phenanthrene from the MGP soil using gas purging under closed-system aqueous-phase boundary conditions and washing with 5.5 g/L of Triton X-100 (~52x CMC); model simulations of the data are shown by solid lines. The particle radius was assumed to be

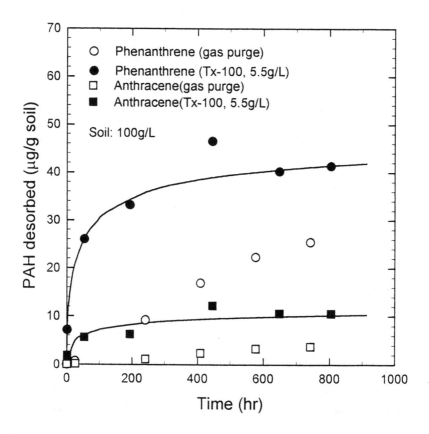

FIGURE 1. Desorption of PAHs by gas purging and washing with Triton X-100.

the characteristic length of the diffusional path and the only fitting parameter used in model calibration was the effective diffusivity, D_{eff}. The soil-water partitioning coefficient K_p in gas-purge experiments was estimated separately from the mass fractions of organic carbon in the soil and tar, average molecular weight of coal tar (230 to 780 g/mol), and the measured equilibrium concentrations of PAHs in soil (Cox et al. 1994). The partitioning coefficient K_p for surfactant washing experiments was estimated from PAH desorbed during the experiment.

The congruence of model with experimental results for the surfactant-washed soil was excellent, thus desorption of PAHs from the MGP soil appears to be diffusion-controlled. The model parameters for the two desorption methods are summarized in Table 2. The soil-water partitioning coefficients for the PAHs for the surfactant-washed soil were nearly 3 orders of magnitude smaller than those for surfactant-free soil-aqueous suspensions, as would be expected. The effective sorbent diffusivities for surfactant-washed soil were 4 orders of magnitude higher than those for aqueous soil suspensions. Clearly, surfactant plays a significant role in enhancing the effective diffusivity, given that the enhancements in solubility and soil-solution concentration gradients are accounted for in the model. How the facilitated transport is brought about by the surfactant is open to conjecture.

Preliminary results for the mineralization of contaminants in the weathered MGP soil in the presence of Triton X-100 are shown in Figure 2. The soil had 75% (w/w) organic matter ($f_{OC} = 0.62$) of which only 1.5% could be extracted by soxhlet extraction. The production of CO_2 was monitored as a measure of mineralization. To circumvent any toxicity caused by the surfactant, initial experiments were done at relatively low surfactant dosages. Triton X-100 at a dosage of up to 2.5 g/L (~24× CMC) did not seem to enhance the mineralization rate significantly.

TABLE 2. Parameters used in modeling desorption of PAHs at 25°C from MGP soil.

Parameter	Phenanthrene	Anthracene
Gas Purge[a]		
H (atm L mol^{-1})	0.0299	0.0282
K_p (L g^{-1})	55.9	58.5
D_{eff} (cm^2 s^{-1})[b]	$2.2 \times 10e\text{-}13$	$7.2 \times 10e\text{-}14$
Surfactant Wash (5.5 g L^{-1} Triton X-100)		
K_p (L g^{-1})	0.063	0.078
D_{eff} (cm^2 s^{-1})[b]	$3.2 \times 10e\text{-}9$	$1.4 \times 10e\text{-}9$

(a) Cox et al. 1994.
(b) Calculations are based on a porosity (n) of 0.3 and bulk density (ρ_b) of 1.72 g/cc for the MGP soil.

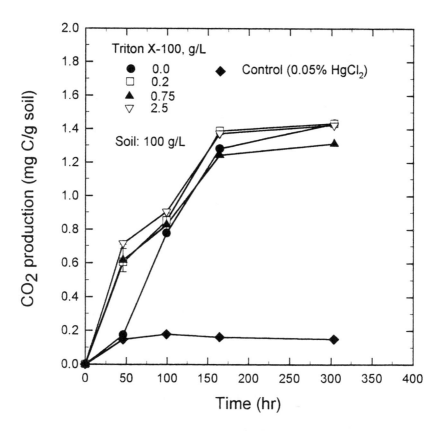

FIGURE 2. Mineralization of soil-bound contaminants in the presence of Triton X-100.

Further, Triton X-100 did not seem to inhibit mineralization. Also, micelle-incorporated PAHs were bioavailable. Typically, in weathered MGP soils, catabolic genes for the degradation of PAHs are fully expressed; bioremediation in such soils is more apt to be controlled by the mass transfer of contaminants and not by biokinetics. Mineralization experiments are now under way using a higher Triton X-100 dosage; both CO_2 production and aqueous concentration of representative PAHs are being measured as a function of time to study the progress of mineralization.

The results of adsorption, desorption, and mineralization of ^{14}C-4,4'-CB in an artificially contaminated Tennessee soil containing 1.8% organic carbon are shown in Figure 3a, 3b, and 3c, respectively. A PCB-specific organism, *Alcaligenes eutrophus* A5, was used to study the enhancement of mineralization of 4,4'-CB by the biosurfactant rhamnolipid R1. Almost 90% of the sorbed contaminant could be desorbed in three successive washings, each of 48-h duration, with rhamnolipid R1 at a dosage of 4,000 mg/L (~74x CMC); the recovery of the PCB at a surfactant dosage of 1,700 mg/L was about 60%). In 100 hours, about

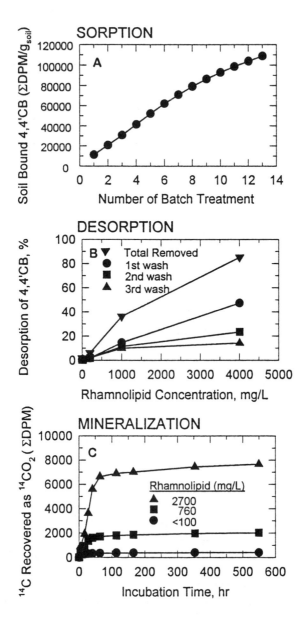

FIGURE 3. Sorption, desorption, and mineralization of 4,4′-chlorobiphenyl in a soil with 1.8% organic carbon using rhamnolipid R1.

6,000 dpm of radioactivity was ascribed to mineralization occurring in the presence of 2,700 mg/L of rhamnolipid R1. This represents 75% of the PCB that could be washed off the soil using 2,700 mg/L of the biosurfactant. The biosurfactant was effective in desorbing the PCB in the test soil. Further, it was not toxic to the test organism even at relatively high dosages.

ACKNOWLEDGMENTS

This study was supported by the U.S. Environmental Protection Agency through Grant No. R819168010 and by the Center for Environmental Biotechnology, University of Tennessee, Knoxville, TN. The paper has not been subjected to peer review by the sponsors and, thus, official endorsement of the results and conclusions should not be inferred. Technical assistance provided by Srinivas Sambangi, Wally Hunt, and Reed Stanley is gratefully acknowledged.

REFERENCES

Aronstein, B. N., Y. M. Caivallo, and M. Alexander. 1991. "Effect of Surfactants at Low Concentrations on the Desorption and Biodegradation of Sorbed Aromatic Compounds on Soil." *Environmental Science and Technology*, 25:1728-1731.

Ball, W. P., and P. V. Roberts. 1991. "Long-term sorption of halogenated organic compounds by aquifer material. 2. Intraparticle Diffusion." *Environmental Science Technology*, 25: 1237-1249.

Cookson, Jr., J. T. (1995). *Bioremediation Engineering: Design and Application*. McGraw-Hill, Inc., New York, NY. 524 pp.

Cox, C. D., J. R. Stanley, and M. M. Ghosh. 1994. "Desorption Rates of Polynuclear Aromatic Hydrocarbons from Soils." *Proceedings 67th Annual Conference*, Volume II, 49-60. Water Environment Federation, Chicago, Illinois, October 15-19, 1994.

Diallo, M. S., L. M. Abriola, and W. J. Weber, Jr. 1994. "Solubilization of Nonaqueous Phase Liquid Hydrocarbons in Micellar Solutions of Dodecyl Alcohol Ethoxylates." *Environmental Science and Technology*, 28:1829-1837.

Edwards, D. A., R. G. Luthy, and Z. Liu. 1991. "Solubilization of PAHs in Micellar Nonionic Surfactant Solutions." *Environmental Science and Technology*, 25:127-133.

Itoh, S., and T. Suzuki. 1972. "Effect of Rhamnolipids on Growth of *Pseudomonas aeruginosa* Mutant Deficient in *n*-Paraffin-Utilizing Ability." *Agricultural Biological Chemistry*, 36:2233-2235.

Karickhoff, S. W. 1984. "Organic Pollutant Sorption in Aquatic Systems." *Journal of Hydraulic Engineering, American Society of Civil Engineers*, 110: 707-735.

Kile, D. E., and C. T. Chiou. 1989. "Water Solubility Enhancement of DDT by Some Surfactants." *Environmental Science and Technology*, 23:832-838.

Liu, Z., D. A. Edwards, and R. G. Luthy. 1992. "Sorption of Nonionic Surfactants on Soil." *Water Research*, 26:1337-1345.

Ramana, K. V., and N. G. Karanth. 1989. "Factors Affecting Biosurfactant Production Using *Pseudomonas aeruginosa* CFTR-6 under Submerged Conditions." *Journal Chemical Technology and Biotechnology*, 45: 249-257.

Thangamani, S., and G. S. Shreve. 1994. "Effect of Anionic Biosurfactant on Hexadecane Partitioning in Multiphase Systems." *Environmental Science and Technology*, 28:1993-2000.

U.S. Environmental Protection Agency. 1994. *Bioremediation in the Field Search System (BFSS)*. EPA/540/R-94/511a. Washington, DC.

Wu, S. C., and P. M. Gschwend. 1986. "Sorption Kinetics of Hydrophobic Compounds to Natural Sediments and Soils." *Environmental Science Technology*, 20: 717-725.

Zhang, Y., and R. M. Miller. 1992. "Enhanced Octadecane Dispersion and Biodegradation by a *Pseudomonas* rhamnolipid Surfactant (Biosurfactant)." *Appl. Environ. Microbiol.*, 58: 3276-3282.

Surfactant Use With Nitrate-Based Bioremediation

Barbara H. Wilson, Stephen R. Hutchins,
and Candida C. West

ABSTRACT

This study presents results of an initial survey on the effect of six surfactants on the biodegradation of petroleum hydrocarbons in bioremediation applications using nitrate as the electron acceptor. Aquifer
material from Park City, Kansas, was used for the study. The three
anionic surfactants chosen were Steol CS-330, Dowfax 8390, and
sodium dodecylbenzene sulfonate (SDBS); the three nonionic surfactants were T-MAZ-60, Triton X-100, and Igepal CO-660. Both Steol CS-
330 and T-MAZ-60 biodegraded under denitrifying conditions. The
Steol inhibited biodegradation of benzene, toluene, ethylbenzene,
xylenes, and trimethylbenzenes (BTEXTMB). Only toluene was
rapidly degraded in the presence of T-MAZ-60. Biodegradation of all
compounds, including toluene, appears to be inhibited by Dowfax
8390 and SDBS. No biodegradation of Dowfax 8390 or SDBS was
observed. SDBS inhibited denitrification, but Dowfax 8390 did not.
For the microcosms containing Triton X-100 or Igepal CO-660, removal
of toluene, ethylbenzene, *m*-xylene, 1,3,5-TMB, and 1,2,4-TMB were
similar to their removals in the no-surfactant treatment. These two
surfactants did not biodegrade, did not inhibit biodegradation of the
alkylbenzenes, and did not inhibit denitrification. Further studies are
continuing with aquifer material from Eglin Air Force Base.

INTRODUCTION

Bioremediation of contaminated aquifer solids has been successful at many
sites with the addition of nitrate and nutrients to the groundwater. The
biodegradation of monocyclic aromatic hydrocarbons (Hutchins and Wilson
1994; Hutchins 1991; Kuhn et al. 1988; Major et al. 1988) and polycyclic aromatic
hydrocarbons (Mihelcic and Luthy 1988a,b; 1991) has been observed under
denitrifying conditions. However, bioremediation can be restricted by bioavailability. Bioavailability is limited by mass transfer of the compounds from the

bulk contamination into groundwater. Ideally, the use of surfactants would increase the rate of dissolution from the bulk contamination into groundwater, resulting in increased rates of microbial utilization of the contaminants (West and Harwell 1992). Studies on surfactant enhancement of biodegradation of aromatic hydrocarbons and lubricating-oil soil-slurry systems have shown increased biodegradation when surfactants were present (Aronstein and Alexander 1992; Aronstein et al. 1991; Rittmann and Johnson 1989). The apparent enhancement was due to increased desorption from soil into the aqueous phase and improved microorganism contact with the hydrocarbons. The basic goal of this study is to determine the feasibility of enhancement of nitrate-based bioremediation of aquifer solids contaminated with petroleum-derived products by the use of surfactants.

MATERIALS AND METHODS

The microcosms were constructed in an anaerobic glovebox using 50-mL serum bottles with 15 g soil, and amended with nitrate plus ammonium and phosphorus nutrients. The alkylbenzenes and the surfactants were added as separate aqueous solutions. The alkylbenzene solution contained benzene; toluene; ethylbenzene; o-, m-, and p-xylene; and 1,2,3-, 1,2,4-, and 1,3,5-trimethylbenzene (BTEXTMB). Total alkylbenzene concentrations were approximately 10 mg/L or less. Triton X-100, T-MAZ-60, Igepal CO-660, and SDBS were added at individual concentrations approximating 2 times the critical micelle concentration (CMC). Dowfax 8390 and Steol CS 330 were added at respective concentrations of 3,600 mg/L and 250 mg/L. Nitrate was added at 50 mg/L; ammonium and phosphorus were added at 10 mg/L each. Sodium azide and mercuric chloride were added as biocides to the control treatments at respective concentrations of 500 mg/L and 250 mg/L. In other studies (Hutchins 1993; 1991), these biocide concentrations have demonstrated their effectiveness in providing sterile, abiotic controls. After adding the above solutions, the microcosms had sterile water (½ Byrd's Mill Spring water + ½ distilled water) added so that no headspace would remain. The bottles were capped with a Teflon™-lined septa and incubated in a glovebox at 20°C. Triplicate microcosms were sacrificed at sampling times of 0, 1, 2, 4, and 8 weeks for measurement of benzene, alkylbenzenes, surfactant, NO_3, NO_2, NH_4, PO_4, and pH.

SAMPLING AND ANALYSIS

The surfactants were analyzed by high-performance liquid chromatography (HPLC) using reverse-phase chromatography developed at Robert S. Kerr Environmental Research Laboratory, Ada, Oklahoma (U.S. EPA 1994). The quantitation limit for the surfactants is 50 mg/L. The alkylbenzenes were analyzed by purge-and-trap gas chromatography as previously described (Hutchins 1993), except Antifoam A from Fluka was used in the samples containing surfactants to prevent foaming. The quantitation limit for the alkylbenzenes was 1 μg/L. The inorganic analyses were analyzed by EPA Methods 350.1 (ammonia nitrogen), 353.1 (nitrogen, nitrate-nitrite), 365.1 (phosphorous, all forms), and 150.1 (pH) (U.S. EPA 1983).

RESULTS

Results of the study are shown in Tables 1, 2, and 3. Results in the no-surfactant treatments showed that toluene, ethylbenzene, *p*-xylene, *m*-xylene, 1,3,5-TMB, and 1,2,4-TMB were degraded below 2 μg/L. The *o*-Xylene and 1,2,3-TMB were degraded to 39% and 87% of their original concentrations. No degradation of benzene was observed. These results are similar to previous studies of biodegradation of BTEXTMB under denitrifying conditions (Hutchins and Wilson 1994; Hutchins 1993; Hutchins 1991).

In the microcosms containing Steol CS-330, the initial surfactant concentrations were measured at approximately 190 mg/L in the living microcosms and 220 to 270 mg/L in the control microcosms. Two components (Component 1 and Component 2) were analyzed separately for Steol concentrations. Both components biodegraded under denitrifying conditions. The increased concentration of Component 1 at Day 56 may have been due to desorption. Biodegradation of BTEXTMB appeared to be inhibited with Steol present. In the no-surfactant treatment, toluene, ethylbenzene, *p*-xylene, *m*-xylene, 1,3,5-TMB, and 1,2,4-TMB were readily degraded. However, these same compounds were not biologically degraded in the presence of Steol.

T-MAZ-60 was added at concentrations of 200 mg/L. Concentrations of T-MAZ-60 decreased rapidly in both living and control samples (Table 3). T-MAZ-60 concentrations were 0 by week 1 in the controls and by week 2 in the living microcosms. T-MAZ-60 is an ethoxylated sorbitol fatty acid ester that is both readily biodegraded and hydrolyzed. Because no NO_3 demand was exerted in the controls, removal of T-MAZ-60 is most likely due to an abiotic process and not biodegradation. Toluene degradation was rapid in the living microcosms; ethylbenzene degraded to approximately 25% of its original concentration. No biodegradation of the other compounds was observed.

Triton X-100 was added at an initial concentration of 400 to 410 mg/L. Triton X-100 concentrations decreased 10% in the control samples by week 8. Removals of toluene, ethylbenzene, *m*-xylene, *o*-xylene, 1,3,5-TMB, and 1,2,4-TMB were similar to their removals in the no-surfactant treatment. No biodegradation of benzene occurred; *p*-xylene was degraded to 8.7% of its original concentration.

Igepal CO-660 was added at an initial concentration of 260 to 280 mg/L. At the end of 56 days, 78% and 85% of the Igepal were remaining in the living and

TABLE 1. Percentages of benzene and alkylbenzenes remaining after 56 days of incubation[a].

Compound	No Surfactant	No Surfactant Control	Steol CS 330	Steol CS 330 Control	T-MAZ-60	T-MAZ-60 Control	Triton X-100	Triton X-100 Control	Igepal CO-660	Igepal CO-660 Control	Dowfax 8390	Dowfax 8390 Control	SDBS[b]	SDBS Control
Benzene	112 ± 3.3	113 ± 18	103 ± 8.8	45 ± 4.0	103 ± 2.5	64 ± 7.2	65 ± 8.9	75 ± 7.7	80 ± 3.7	80 ± 4.4	133 ± 6.9	147 ± 5.0	139 ± 9.4	100 ± 16
Toluene	0.16 ± 0.04	102 ± 17	90 ± 7.7	41 ± 4.5	.21 ± 0.04	58.4 ± 7.0	0.07 ± 0.004	62 ± 6.7	0.06 ± 0.0	68 ± 5.7	121 ± 6.5	129 ± 6.7	121 ± 10	92 ± 16
Ethyl-benzene	0.13 ± 0.0	89 ± 14	56 ± 15	34 ± 2.0	25 ± 6.6	50 ± 5.8	0.12 ± 0.0	56 ± 7.2	0.12 ± 0.0	63 ± 7.6	103 ± 5.1	108 ± 9.2	100 ± 8.8	80 ± 15
p-Xylene	0.17 ± 0.02	84 ± 10	68 ± 8.4	38 ± 6.2	70 ± 2.8	49 ± 5.5	8.7 ± 3.4	58 ± 7.5	0.15 ± 0.0	65 ± 8.0	95 ± 3.9	99 ± 8.1	92 ± 7.2	75 ± 13
m-Xylene	0.11 ± 0.0	84 ± 11	75 ± 5.7	34 ± 2.5	68 ± 3.1	52 ± 5.8	0.37 ± 0.21	59 ± 7.6	0.16 ± 0.0	66 ± 7.7	94 ± 3.9	97 ± 6.8	91 ± 6.4	74 ± 12
o-Xylene	39 ± 3.3	90 ± 11	77 ± 5.1	33 ± 1.8	48 ± 0.68	55 ± 6.1	22 ± 8.0	60 ± 6.7	30 ± 5.0	66 ± 7.1	93 ± 4.6	93 ± 7.0	91 ± 7.6	75 ± 12
1,3,5-TMB[c]	0.21 ± 0.0	73 ± 10	63 ± 2.2	25 ± 1.6	63 ± 2.4	50 ± 5.3	2.7 ± .0	47 ± 6.1	0.27 ± 0.0	52 ± 6.4	66 ± 4.0	69 ± 5.6	64 ± 6.4	54 ± 9
1,2,4-TMB	0.19 ± 0.0	73 ± 9.4	66 ± 2.5	21 ± 4.6	65 ± 2.3	52 ± 5.3	0.59 ± 0.51	49 ± 6.1	0.23 ± 0.01	54 ± 6.8	64 ± 4.0	64 ± 5.3	58 ± 6.1	49 ± 9.1
1,2,3-TMB	87 ± 7.8	85 ± 9.2	72 ± 2.4	8.0 ± 1.5	69 ± 2.4	57 ± 6.1	43 ± 19	52 ± 6.7	55 ± 8.8	59 ± 7.1	61 ± 1.7	63 ± 5.4	57 ± 5.5	50 ± 8.7

(a) Percentages calculated from $C/C_o * 100 \pm$ standard deviation, where C values are the means of three replicate values at 56 days of incubation, and C_o values are means of three replicate no-surfactant samples at 0 days of incubation.

(b) SDBS = sodium dodecylbenzene sulfonate.

(c) TMB = trimethylbenzene.

TABLE 2. Concentrations of anionic surfactants (mg/L)[a].

Time (Days)	CS-330 Steol Component 1	CS-330 Steol Component 1 Control	CS-330 Steol Component 2	CS-330 Steol Component 2 Control	Dowfax 8390	Dowfax 8390 Control	SDBS[b]	SDBS Control
0	187 ± 1.6	270 ± 6.8	190 ± 3.0	217 ± 5.3	1720 ± 220	1450 ± 510	3690 ± 270	3510 ± 310
7	53 ± 4.0	225 ± 10	157 ± 10	198 ± 4.6	4080 ± 600	3170 ± 170	3780 ± 150	3400 ± 110
14	50 ± 0	232 ± 8.1	174 ± 7.0	239 ± 7.9	3860 ± 1930	3080 ± 80	3360 ± 110	3240 ± 260
28	50 ± 0	223 ± 90	127 ± 19	205 ± 103	3890 ± 270	2870 ± 190	3410 ± 200	3500 ± 390
56	50 ± 0	361 ± 46	99 ± 7.7	202 ± 12	3602 ± 350	3270 ± 38	3140 ± 200	2870 ± 330

(a) Concentrations are the means of three replicates ± standard deviations.
(b) Sodium dodecylbenzene sulfonate.

TABLE 3. Concentrations of nonionic surfactants (mg/L)[a].

Time (Days)	T-MAZ-60	T-MAZ-60 Control	Triton X-100	Triton X-100 Control	Igepal CO-660	Igepal CO-660 Control
0	240 ± 30	230 ± 36	410 ± 23	400 ± 46	280 ± 1	260 ± 22
7	63 ± 55	0 ± 0	340 ± 50	360 ± 20	160 ± 44	170 ± 26
14	0 ± 0	0 ± 0	350 ± 7	350 ± 17	250 ± 18	230 ± 7
28	0 ± 0	0 ± 0	360 ± 16	340 ± 34	200 ± 23	220 ± 8
56	0 ± 0	0 ± 0	500 ± 320	360 ± 40	220 ± 22	220 ± 21

(a) Concentrations are the means of three replicates ± standard deviations.

control microcosms, respectively. The fate of all BTEXTMB compounds was similar to their fates in the no-surfactant treatment.

The concentrations of Dowfax 8390 added to living and control microcosms were 3,200 to 4,000 mg/L. SDBS was added at an initial concentration of 3,500 to 3,700 mg/L. Biodegradation of all compounds, including toluene, appears to be inhibited by the Dowfax 8390 and SDBS. Concentrations of Dowfax 8390 at the end of 56 days compared to concentrations at 7 days were 88% in the living microcosms and 103% in the control microcosms. The cause of the decreased concentrations of Dowfax 8390 at Time 0 has not been adequately determined, but may be due to initial sorption followed by subsequent desorption. Concentrations of SDBS remaining at 56 days were 85% for the living microcosms and 82% for the control microcosms. SDBS inhibited denitrification, but the Dowfax 8390 did not (data not shown).

SUMMARY

Both Steol CS-330 and T-MAZ-60 biodegraded under denitrifying conditions. The Steol inhibited biodegradation of BTEX. Only toluene was rapidly degraded in the presence of T-MAZ-60. Concentrations of ethylbenzene, *p*-xylene, *m*-xylene, *o*-xylene, 1,3,5-TMB, 1,2,4-TMB, and 1,2,3-TMB slowly decreased in both living and control microcosms.

For the microcosms containing Triton X-100 or Igepal CO-660, removal of toluene, ethylbenzene, *m*-xylene, 1,3,5-TMB, and 1,2,4-TMB were similar to their removals in the no-surfactant treatment. These two surfactants did not biodegrade, did not inhibit biodegradation of TEXTMB, and did not inhibit denitrification of NO_3.

Abiotic removal of BTEXTMB in the controls is probably due to sorption. These compounds are resistant to hydrolysis under geochemical conditions found in groundwaters (Lyman et al. 1990). Biodegradation is doubtful because no removal of the available electron acceptor was measured in any of the control treatments. Removal of BTEXTMB was greater in controls containing the surfactants Steol CS-330, T-MAZ-60, Triton X-100, and Igepal Co-660 than in the no-surfactant controls. Incorporation of BTEXTMB into surfactant micelles is presumed to cause the increased removals.

A survey of the effect of six surfactants on denitrification of BTEXTMB components of fuel has been completed. The purpose of the study was to identify surfactants that were nonbiodegradable under denitrifying conditions and yet would not inhibit biodegradation of the alkylbenzenes or denitrification of NO_3. Two nonionic surfactants, Triton X-100 and Igepal CO-660, were identified for further study.

DISCLAIMER

Although the research described in this paper has been funded wholly or in part by the U.S. Environmental Protection Agency (U.S. EPA) and the U.S. Air Force (MIPR N92-08, AFDW/ACS-B, Bolling Air Force Base, and MIPR N92-65, AL/EQ-OL, Environmental Quality Directorate, Armstrong Laboratory, Tyndall Air Force Base), this article has not been subjected to U.S. EPA review and therefore does not necessarily reflect the views of the U.S. EPA. No official endorsement should be inferred.

REFERENCES

Aronstein, B. N., and M. Alexander. 1992. "Surfactants at low concentrations stimulate biodegradation of sorbed hydrocarbons in samples of aquifer sands and soil slurries." *Environ. Toxicol. Chem.* 11:1227-1233.

Aronstein, B. N., Y. M. Calvillo, and M. Alexander. 1991. "Effect of surfactants at low concentrations on the desorption and biodegradation of sorbed aromatic compounds in soil." *Environ. Sci. Technol.* 25:1728-1731.

Hutchins, S. R., and J. T. Wilson. 1994. "Nitrate-based bioremediation of petroleum-contaminated aquifer at Park City, Kansas: Site characterization and treatability study." In R. E. Hinchee, B. C. Alleman, R. E. Hoeppel, and R. N. Miller (Eds.), *Hydrocarbon Bioremediation*, pp. 80-92. Lewis Publishers, Boca Raton, FL.

Hutchins, S. R. 1993. "Biotransformation and mineralization of alkylbenzenes under denitrifying conditions." *Environ. Toxicol. Chem.* 12:1413-1423.

Hutchins, S. R. 1991. "Optimizing BTEX biodegradation under denitrifying conditions." *Environ. Toxicol. Chem.* 10:1437-1448.

Kuhn, E. P., J. Zeyer, P. Eicher, and R. P. Schwarzenbach. 1988. "Anaerobic degradation of alkylated benzenes in denitrifying laboratory aquifer columns." *Appl. Environ. Microbiol.* 54(2):490-496.

Lyman, W. J., W. F. Reehl, and D. H. Rosenblatt. 1990. *Handbook of Chemical Property Estimation Methods.* American Chemical Society, Washington, DC.

Major, D. W., C. I. Barker, and J. F. Barker. 1988. "Biotransformation of benzene by denitrification in aquifer sand." *Ground Water.* 26(1):8-14.

Mihelcic, J. R. and R. G. Luthy. 1991. "Sorption and microbial degradation of naphthalene in soil-water suspensions under denitrification conditions." *Environ. Sci. Technol.* 25:169-177.

Mihelcic, J. R., and R. G. Luthy. 1988a. "Degradation of polycyclic aromatic hydrocarbon compounds under various redox conditions in soil-water systems." *Appl. Environ. Microbiol.* 54(5):1182-1187.

Mihelcic, J. R. and R. G. Luthy. 1988b. "Microbial degradation of acenaphthene and naphthalene under denitrification conditions in soil-water systems." *Appl. Environ. Microbiol.* 54(5):1188-1198.

Rittmann, B. E. and N. M. Johnson. 1989. "Rapid biological clean-up of soils contaminated with lubricating oil." *Wat. Sci. Tech.* 21:209-219.

U.S. Environmental Protection Agency (U.S. EPA). 1983. *Methods for Chemical Analysis of Water and Wastes.* U.S. EPA-600/4-79-020.

U.S. Environmental Protection Agency (U.S. EPA). 1994. Standard Operating Procedure for Quantitative Analysis of T-MAZ-60 (RSKSOP 165); Standard Operating Procedure for Quantitative Analysis of Steol CS-330 (RSKSOP 167); Standard Operating Procedure for Quantitative Analysis of Triton X-100 (RSKSOP 169); Standard Operating Procedure for Quantitative Analysis of Igepal CO-660 (RSKSOP 170); Standard Operating Procedure for Quantitative Analysis of Dowfax 8390 (RSKSOP 171); Standard Operating Procedure for Quantitative Analysis of Sodium Dodecylbenzene Sulfonate (RSKSOP 172). Robert S. Kerr Environmental Research Laboratory. Ada, OK.

West, C. C., and J. H. Harwell. 1992. "Surfactants and subsurface remediation." *Environ. Sci. Technol.* 26(12):2324-2330.

Bioavailability Enhancement by Addition of Surfactant and Surfactant-Like Compounds

Janet M. Strong-Gunderson and Anthony V. Palumbo

ABSTRACT

The bioavailability and microbial degradation of contaminant compounds (e.g., toluene and naphthalene) were enhanced by adding synthetic surfactants, biosurfactants, and nutrients with surfactant–like properties. In addition to enhanced contaminant degradation, these surfactant compounds have the potential to change the availability of natural organic matter (NOM), and thus may affect overall site bioremediation. Two bacterial bioreporter strains that are induced by toluene or naphthalene were used to directly measure contaminant bioavailability. A cell-free biosurfactant product, Tween-80, and an oleophilic fertilizer were added to aqueous suspensions and soil slurries containing toluene or naphthalene. The addition of these surfactant compounds at or below the critical micelle concentration (CMC) enhanced bioavailability as measured by increased levels of bioluminescence. Bioluminescence data were coupled with gas chromatographic analyses. The addition of Tween-80 increased not only the bioavailability of the contaminants but also, in a separate assay, the bioavailability of recalcitrant NOM. The enhanced NOM bioavailability was inferred from measurements of biomass by optical density increases and plate counts. Thus, adding surfactant compounds for enhanced contaminant degradation has the potential to introduce additional competition for nutrients and microbial metabolism, a significant area of concern for in situ site remediation.

INTRODUCTION

The remediation of contaminated sites by microorganisms is dependent on the bioavailability of the contaminants. Because biodegradation takes place in the liquid phase, contaminants present as nonaqueous-phase liquids (NAPLs) or sorbed to the soil matrix are not available for microbial degradation. Thus, rates of degradation are limited by mass-transfer problems even when other parameters are optimized. Mass-transfer problems can be reduced by adding surface-active agents, i.e., surfactants. Surfactants increase solubilization rates and thus make a greater fraction of the contaminants amenable to biodegradation.

Synthetic surfactants have been used in soil-washing technologies for enhanced removal of sorbed contaminant from the subsurface (Abdul et al. 1990). However, a significant problem with the high surfactant concentrations used in soil washing is their toxicity. Furthermore, surfactants used above the CMC may sequester the contaminant in the micelle, thus making the compound unavailable for microbial degradation. The use of biosurfactants has been proposed, in part, to overcome toxicity. Several reports have shown that biosurfactants can increase degradation of NAPLs (Francy et al. 1991; Zhang and Miller 1992). There is apparently no information on the effects of either synthetic surfactants or biosurfactants on NOM. Limited data indicate that a labile fraction of NOM associated with higher molecular weight inhibits the bioavailability of NOM (Meyer et al. 1987).

The goal of this study was to increase the bioavailability of contaminants sorbed to a soil matrix through the use of surfactants. Unless otherwise specified, we are using the generic term "surfactant" to encompass three categories of bioavailability-enhancing compounds: (1) surfactants, (2) biosurfactants, and (3) fertilizer with surfactant properties. These experiments were specifically designed to measure the effects of surfactants on toluene and naphthalene bioavailability in aqueous and soil-based systems. Surfactant concentrations were tested at or below the aqueous CMC. Furthermore, we investigated the effects that surfactants have on the bioavailability of recalcitrant natural organic matter found in deep subsurface groundwaters.

EXPERIMENTAL PROCEDURES AND MATERIALS

Bacteria and Contaminant Bioavailability Assay

Two bacterial strains containing the transcriptional fusion between a catabolic pathway and genes encoding light production were used in the bioavailability assays. The toluene bioreporter strain can assess toluene bioavailability (G. Sayler, University of Tennessee) and the naphthalene strain can assess naphthalene bioavailability (R. Burlage, Oak Ridge National Laboratory; Burlage et al. 1990). The biosensors were grown in Luria-Bertani (LB) medium (Maniatis et al. 1982), which was supplemented with 50 mg/L kanamycin at room temperature at an optical density (OD) of ca. 2.0. The cultures were diluted with Nate, a minimal salts medium (Little et al. 1988), to a standard working optical density of 1.0 at 600 nm using a Gilford spectrophotometer (Strong-Gunderson et al. 1994; Summers et al. 1995).

The bioavailability assay was performed as described previously (Strong-Gunderson et al. 1994). Briefly, equal volumes of the bioreporter suspension and treatment supernatant were combined in 15-mL EPA vials, capped with Teflon™-lined septa. At 1-h intervals, supernatant aliquots were taken and luminescence was quantified using an ATP Photometer (Turner Designs, California). This instrument assigns light readings (light units, LU) that are dimensionless units. These readings were compared to a positive control strain that was induced with a known amount of toluene or naphthalene.

In the soil systems, toluene or naphthalene (0.125 ppm) was added and allowed to equilibrate for 24 h on a rotary shaker. After 24 h, the surfactant compounds were added at the aqueous CMC (ca. 0.03%) and allowed to equilibrate for an additional 24 h before bioavailability was measured.

Surfactants

The surfactants tested were Tween 80 (Sigma Chemical), a biosurfactant (produced by *Pseudomonas aeruginosa*, ATCC#9721), and an oleophilic fertilizer (Inipol EAP 22, Elf Atochem North America, Philadelphia, Pennsylvania). The aqueous CMC was determined using a DuNouy ring tensiometer (CSC Scientific Company, Inc., Fairfax, Virginia). Sample aliquots (15 mL) were transferred to small petri dishes and ring measurements were obtained, in triplicate, within 5 min to minimize dynamic surface effects. The surface tension was measured for all surfactants at various aqueous concentrations to determine the CMC.

Analytical Methods

Gas chromatographic (GC) analyses for toluene were run on a 30-μL (gas-tight syringe) headspace sample, using a Perkin-Elmer GC equipped with a packed AT1000 column (column 200°C, injector 200°C, detector 250°C) and a flame ionization detector. Tests on all samples were performed in duplicate.

NOM Bioavailability Assay

The NOM bioavailability assay followed a previously developed procedure for low-level carbon values (Strong-Gunderson et al. 1995). Briefly, bacteria were suspended in Nate and held overnight (starvation period) to minimize any stored nutrients. A standard concentration of NOM (2 ppm) was added and supplemented with inorganic nutrients to ensure that organic carbon was the only limiting constituent. These bacterial suspensions were inoculated into a 96-well microtiter plate, and bacterial growth on the organic carbon source was measured by an increase in optical densities (OD) at 24-h intervals. Previous experiments verified that an increase in OD from 0.02 to 0.05 corresponds to an increase in cell number of ca. one order of magnitude. ODs of controls (bacteria and no organic carbon) were compared to the OD of bacteria plus various organic carbon sources (e.g., Tween, natural organic carbon). Each value was calculated as the mean of eight replicates.

In the surfactant/NOM test system, the Tween was added at 2 mg carbon/L to the groundwater that was composed of previously identified/recalcitrant

NOM at a concentration of 2 mg carbon/L (final organic carbon concentration = 4 mg carbon/L). The solution was inoculated into the 96-well plates and incubated at room temperature for 72 h. Optical densities (590 nm) were recorded at 24-h intervals.

RESULTS AND DISCUSSION

The addition of surfactant compounds to an aqueous system has a substantial effect on enhancing toluene and naphthalene bioavailability. Our previous experiments have shown that luminescence (i.e., contaminant bioavailability) can be increased following the addition of surfactants (Strong-Gunderson et al. 1994). Current analytical experiments for toluene, a very volatile compound, determined that less contaminant was in the headspace (Figure 1) in the presence of the surfactant and no cells. This reduction in headspace concentration may be responsible for increased aqueous bioluminescence. The surfactant affected the partitioning and decreased the headspace concentration. The surfactant compounds alone (negative controls) had no effect on bioluminescence (<10 LU).

The light emitted from the bioreporter assay containing soil plus toluene was only ca. 1,800 LU. When surfactants were added, the LU did not notably

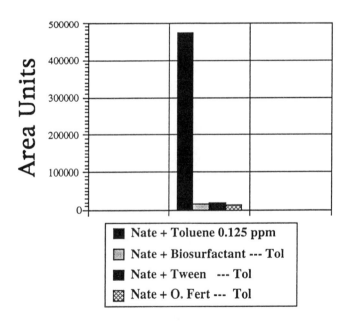

FIGURE 1. GC analyses of the toluene showed less of the compound in the headspace if the surfactant compounds were added to the solution. The area unit of ca. 500,000 corresponds to 0.125 ppm toluene. Nate = minimal salts solution, tol = toluene, n = 2 SD within bar symbol.

increase for toluene (2,000 LU) (data not shown) implying that the addition of surfactants below the CMC had limited effect on increasing toluene bioavailability. However, when the surfactant compounds were added before the toluene, there was a significantly higher level of luminescence (Figure 2A). The

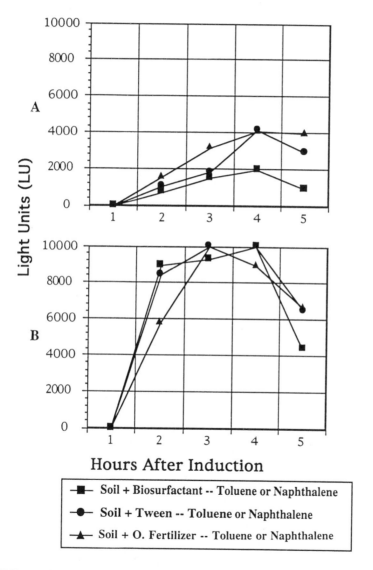

FIGURE 2. Surfactants were added to soil slurries and subsequently contained with 0.125 ppm (A) toluene or (B) naphthalene. The addition of surfactant compounds before contaminant addition resulted in a substantial increase in luminescence. Mean data points n = 2, SD 10 to 500 LU omitted for clarity.

surfactants added prior to contaminant addition may have associated with the binding sites on the soil, thus preventing contaminant sorption. However, we used surfactants at such low concentrations (aqueous phase did not show a decrease in surface tension) that the soil binding sites should not have been saturated.

Naphthalene results were significantly different from toluene experiments. Soil plus naphthalene induced the bioreporter to ca. 2,000 LU and increased to ca. 4,000 LU when surfactant was added after naphthalene. There was a very significant increase in luminescence to >10,000 LU (off scale) when the surfactants were added before the naphthalene. Again, the surfactant compounds alone do not cause induction of this bioreporter (<100 LU; data not shown).

To our knowledge, no report has examined the interaction between surfactants and NOM. The bioavailability of deep subsurface organic carbon is often low because of the carbon type, and it may not support microbial growth even when all other parameters (oxygen and inorganic nutrients) are adequate (Strong-Gunderson et al. 1995). However, there are indications in the literature that some labile NOM may be coupled or associated with recalcitrant NOM (Meyer et al. 1987), and surfactant addition could make this fraction more available. In control experiments, neither the Tween (as the sole carbon source) nor the NOM supported microbial growth (Figure 3A, 3B). However, the combination of these two resulted in an increased optical density from 0.02 to 0.04 for select isolates (Figure 3C). The addition of Tween to the NOM solution increased the bioavailability of the NOM. This increased NOM bioavailability suggests potential competition for or interaction with microbial metabolism between the contaminant and NOM/Tween.

No investigation of the surfactant-contaminant complex for in situ remediation would be complete without addressing the potential "interference" between surfactants and natural organic carbon. Enhanced bioavailability of NOM may interfere with bioremediation (e.g., by microbial utilization of inorganic nutrients in NOM metabolism, which could plug stimulated growth on NOM) if conditions are not controlled and potential interactions are not investigated. Thus, the use of surfactants has the potential to increase the bioavailability not only of contaminants but also of recalcitrant NOM.

ACKNOWLEDGMENTS

We thank S. Carroll for excellent technical assistance in all phases of this research. We also appreciate the assistance of T. Davidson, T. Mehlhorn, B. Summers, and B. Clark. We thank R. S. Burlage (Oak Ridge National Laboratory) for the naphthalene bioreporter and B. Applegate and G. Sayler (University of Tennessee, Knoxville) for the toluene bioreporter.

The surfactant-enhanced bioavailability research was sponsored by the In-Situ Remediation Technology Development Program of the Office of Technology Development (J. S. Walker), U.S. Department of Energy (U.S. DOE).

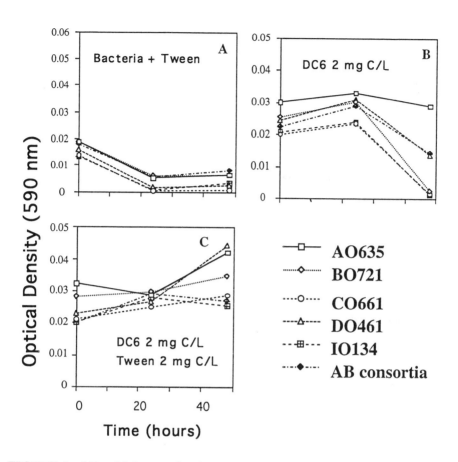

FIGURE 3. Microbial growth (changes in optical density) for (A) NOM, (B) Tween, and (C) NOM plus Tween. All carbon concentrations are 2 ppm. Only the combination of NOM and Tween supported microbial growth. Bacteria AO635, BO721, CO661, and DO461 were isolated from Savannah River Laboratory and AB consortium is AO635 and BO72 combined. Identities currently unknown.

The natural organic carbon bioavailability research was sponsored by the Subsurface Science Program of the Office of Health and Environmental Research (F. J. Wobber), U. S. DOE. ORNL is managed by Martin Marietta Energy Systems, Inc., under contract DE-AC05-84OR21400 with the U.S. DOE.

REFERENCES

Abdul, A. S., T. L. Gibson, and D. N. Rai. 1990. "Selection of Surfactants for the Removal of Petroleum Products from Shallow Sandy Aquifers." *Ground Water 28*, 920-926.

Burlage, R. S., G. S. Sayler, and F. Larimer. 1990. "Monitoring of Naphthalene Catabolism by Bioluminescence with *nah-lux* Transcriptional Fusions." *J. Bacteriol. 172*, 4749-4757.

Francy, D. S., J. M. Thomas, R. L. Raymond, and C. H. Ward. 1991. "Emulsification of Hydrocarbons by Subsurface Bacteria." *J. Industr. Microbiol. 8*, 237-246.

King, J. M. H., P. M. DiGrazia, B. Applegate, R. Burlage, J. Sanseverino, P. Dunbar, F. Larimer, F., and G. S. Sayler. 1991. "Rapid, Sensitive Bioluminescent Reporter Technology for Naphthalene Exposure and Biodegradation." *Science* 249:778-781.

Jardine, P. M., N. L. Weber, and J. F. McCarthy. 1989. "Mechanisms of Dissolved Organic Carbon Adsorption on Soil." *Soil Sci. Soc. of America J. 53*, 1378-1385.

Little, C. D., A. V. Palumbo, S. E. Herbes, M. E. Lindstrom, R. L. Tyndall, and P. J. Gilmer. 1988. "Trichloroethylene Biodegradation by Pure Cultures of a Methane-Oxidizing Bacterium." *Appl. Environ. Microbiol. 54*, 951-956.

Maniatis, T., E. F. Fritsch, and J. Sambrook. 1982. *Molecular Cloning: A Laboratory Manual.* Cold Spring Harbor Laboratory, Cold Spring Harbor, NY.

McCarthy, J. M., A. V. Palumbo, and J. M. Strong-Gunderson. 1994. "The Significance of Interactions of Humic Substances and Organisms in the Environment." In *Humic Substances in the Global Environment, Proceedings of the Sixth Meeting of the International Humic Substances Society.* Elsevier Publishers, Amsterdam. In press.

Meyer, J. L., R. T. Edwards, and R. Risley. 1987. *Micro. Ecol. 13*: 12-29.

Strong-Gunderson, J. M., A. V. Palumbo, and J. McCarthy. 1995. "Simple Method of Determining the Microbiological Bioavailability of Natural Organic Matter." Submitted.

Strong-Gunderson, J. M., B. Applegate, G. Sayler, and A. V. Palumbo. 1994. "Biosurfactants and Increased Bioavailability of Sorbed Organic Contaminants: Measurements Using a Biosensor." *Proceed. Inst. Gas Technol.* In press.

Summers, B., A. V. Palumbo, and J. M. Strong-Gunderson. 1995. "Biosurfactant Production: Stimulation by Contaminant Exposure." Submitted.

Zhang, Y., and R. M. Miller. 1992. "Enhanced Octadecane Dispersion and Biodegradation by a *Pseudomonas* Rhamnolipid Surfactant (Biosurfactant)." *Appl. Environ. Microbiol. 58*, 3276-3282.

Increased Mineral Oil Bioavailability in Slurries by Monovalent Cation-Induced Dispersion

Hubert de Jonge and J. M. Verstraten

ABSTRACT

Bioavailability of apolar contaminants is an important limiting factor for microbial reclamation of polluted soils. This paper describes a laboratory study of the relation between microaggregate stability and bioavailability of mineral oil in soil-water slurries. The stability of microaggregates in slurries is regulated by the valence and surface affinity of the cations in the system, and by the complexing anion $P_2O_7^{4-}$ (metaphosphate). A silt loam, contaminated with a weathered gas oil (3,000 mg-kg^{-1}), was collected from an oil refinery site. Degradation rates were monitored in small-scale incubations at solid:liquid ratios of 1:5 (w/w). The solution contained Ca, Na, or K as the dominant cation. The levels of nutrients (NH_4, PO_4) and metaphosphate were varied. Biodegradation rates increased with the sequence Ca < K < Na as the dominant cation in solution. Addition of NaCl increased the biodegradation rate by 70% (no ammonium added) and 30% (NH_4Cl 10 mM added) when compared to a $CaCl_2$ treatment. Measurements of the particle size distribution in the slurry showed that an increase in the finer fractions qualitatively correlated with enhanced biodegradation. This is a strong indication that dispersion of the microaggregates increased bioavailability of the contaminant.

INTRODUCTION

Bioavailability of organic contaminants is an important limiting factor for biodegradation rates in soils (Rijnaarts et al. 1990, Scow 1993). Contaminants are not directly available to bacteria if present in pores smaller than approximately 0.2 μm. Mass transfer within these pores limits the biodegradation rate, because intraparticle diffusion rates of hydrophobic compounds are low due to sorption to internal pore surfaces (Wu and Gschwend 1986, Yiacoumi and Tien 1994). Until now, efforts to increase bioavailability are based mainly on addition of surfactants, thereby increasing the solubility of the compounds (Rouse et al. 1994).

An alternative approach is to decrease the particle size, thereby reducing the diffusion distance of the contaminant. Pores smaller then $0.2 \mu m$ are found mainly in microaggregates ($<50 \mu m$) stabilized by clay-clay and clay-organic matter complexes. An important mechanism regulating dispersion (collapse) of colloidal particles is expansion/compression of the diffuse double layer (DDL) around negatively charged clay surfaces, as influenced by cation type and ionic strength (van Olphen 1963). According to the Schulze-Hardy rule, dispersive strength decreases with the charge of the cation. A secondary effect is that, given a certain charge of the cation, dispersive strength increases with decreasing affinity for the surface. As a result, the following order in dispersive strength is found given a constant ionic strength: $Ca < Mg << NH_4 < K < Na < Li$. The sequence of cations with equal valence is known as a lyotropic series. A second mechanism leading to dispersion of aggregates is specific complex formation of anions with the positively charged edges of clay particles, thereby breaking edge-to-surface bonds. Metaphosphate ($P_2O_7^{4-}$) therefore is a strong dispersant (van Olphen 1963).

Normally Ca and Mg are present in large amounts in soil-water systems and therefore dominate as counter-ions adsorbed to the clay-surface. Considering the above, aggregate stability can be influenced by addition of monovalent cations and possibly by addition of metaphosphate. Strong dispersion is expected to lead to reduction of the diffusive path length and, consequently to increased biodegradation rates. The purpose of the study described in this paper was to establish the effect of addition of monovalent cations (K, Na) and metaphosphate on the biodegradation rate of mineral oil in slurries. Treatments were incubated both with and without addition of nutrients (ammonium, orthophosphate).

MATERIALS AND METHODS

A silt loam (8% clay, 2.6% organic matter) contaminated with gas oil was sampled from an oil refinery site at Pernis, The Netherlands. In earlier lysimeter experiments (Freijer 1994), the hydrocarbon concentration was reduced from 10,000 to 3,000 $mg \cdot kg^{-1}$. Hence, at the start of experiments the gas oil was strongly weathered, and volatile components ($<C10$) were no longer present. Incubations were carried out in closed jars (400 cm^3) equipped with a septum. The slurries consisted of 10 g soil and 50 mL solution. The treatments that were performed are presented in Table 1, showing that the charge contribution of the cations was constant for treatments 2 through 5, which had no ammonium addition, and 6 through 12, having a combined treatment of a salt and NH_4Cl. Three replicates for each treatments were incubated at $25 \pm 2°C$, using a rotary shaker (90 rpm).

TABLE 1. Initial salt concentrations and biodegradation rates of the various incubation treatments.

No.	Salt Added	concentration [mM]	NH_4Cl [mM]	rate[a] [mmol kg^{-1}]
1	NaN$_3$ control	15.00	0	—
2	CaCl$_2$	5.00	0	53 ± 6
3	KCl	10.00	0	74 ± 1
4	NaCl	10.00	0	89 ± 2
5	Na$_4$P$_2$O$_7$	2.50	0	92 ± 10
6	CaCl$_2$	5.00	10.0	257 ± 28
7	KCl	10.00	10.0	281 ± 67
8	NaCl	10.00	10.0	325 ± 53
9	Na$_4$P$_2$O$_7$	2.50	10.0	231 ± 3
10	K$_2$HPO$_4$/KH$_2$PO$_4$	5.00	10.0	250 ± 20
11	Na$_2$HPO$_4$/NaH$_2$PO$_4$	5.00	10.0	331 ± 34
12	—	0.00	20.0	252 ± 28

(a) Based on oxygen consumption after 1 week of incubation, corrected for sterilized treatment, using relation shown in Figure 1 for conversion to oil degradation. Error based on three replicates.

Degradation rates were monitored by measuring oxygen concentrations in the headspace and were validated by measuring residual oil concentrations of three treatments at the end of the experiment. Oxygen concentrations were quantified by gas chromatography (molsieve 5 Å, 2-m (¹/₈-in.) SS column, isotherm 50°C, hot wire detection). Mineral oil was extracted (original soil and treatments 1, 2, and 4) twice with an acetone-pentane mixture, using octane (Aldrich, 29,698-8) as an internal standard. After removal of acetone, Florisil™ (Merck 12518) was added to remove extracted humic compounds. The oil concentration in the extract was measured by gas chromatography (DB-1 capillary column, 60°C, 4°/min to 320°C, flame ionization detection).

The following analyses were performed to relate biodegradation rates to physicochemical parameters of the slurries: first, the chemical composition of the solution was measured (treatments 6 through 11) after 3 weeks of incubation. After filtration over a 0.2-μm filter, the following compounds were measured: Ca, Mg, K, Na, Fe, Mn (atomic absorption spectrophotometer, Perkin

Elmer 5000); NH_4, NO_3, NO_2, Cl, ortho-P, SO_4, dissolved organic carbon (Autoanalyzer, Breda Scientific); pH; and alkalinity. Saturation indices of the solution to various salts and minerals were calculated using the speciation program MINEQL (Schecher and McAvoy 1991). Second, the particle size distribution in the slurries was measured after 3 weeks of incubation (treatments 6 through 9) using a Microscan II™ (Quantachrome). Third, identical slurries as in Table 1 were prepared for measurement of particle size distribution after 72 hours of incubation. Finally, slurries were prepared for measurement of pH in solution during incubation.

RESULTS AND DISCUSSION

Table 1 gives an overview of the initial salt concentrations and initial oxygen consumption rates of the various incubation treatments. The relation between oxygen consumption and residual oil content for treatments receiving no ammonium is shown in Figure 1. Oxygen consumption was strongly influenced by the absence or presence of nitrogen, which is in accordance with earlier literature data (Dibble and Bartha 1979). However, oxygen consumption is related to nitrification of ammonium in the slurries, and does not necessarily reflect differences in oil degradation. However, because the trend in the CO_2 production very closely resembled the O_2 consumption kinetics (data not shown), oxygen consumption was considered a good relative measure for biodegradation rates in subsets receiving the same amount of nitrogen.

The main objective of this study was to establish the influence of cationic composition on biodegradation rates. Table 1 and Figure 2 show the effect of the applied cation for treatments having the same N-concentration at the start of the experiment. Addition of NaCl gives a 70% increase in the initial rate when compared to Ca, for the treatments receiving no nitrogen. The increase is about 30% in incubations receiving combined treatments with ammonium. The degradation rates in the K-treatment gradually decrease to lower levels than in the Ca treatment (Figure 2B). Treatments receiving a Na-phosphate buffer show higher rates than K-phosphate treatments. The sequence of the initial rates (Ca < K < Na) is consistent with the expectation that increased dispersion will lead to higher rates, as it fits in the Schulze-Hardy rule and the lyotropic series of monovalent cations. If NH_4 is introduced as well, the relative effect is smaller because ammonium itself is an efficient dispersant. However, it is not possible on the basis of these results to explain the lower rates in the K treatments in the later phase of the incubation.

Addition of orthophosphate (PO_4, treatments 10 and 11) did not enhance biodegradation. The effect metaphosphate addition had on the biodegradation rates was not consistent; Treatment 5 shows a higher rate than do 2 through 4 (see Figure 2B), whereas the combined ammonium and metaphosphate addition (treatment 9) did not lead to increased rates when compared to treatments

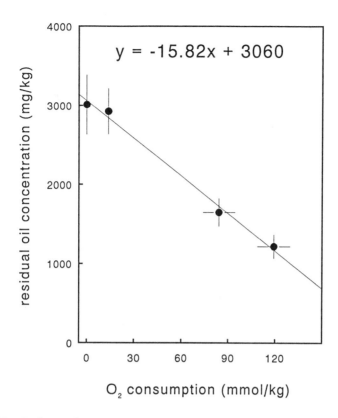

FIGURE 1. Relation between total oxygen consumption with residual oil concentration at the end of the experiment. Error bars represent one standard deviation.

6 through 8. Higher rates were expected due to the strong dispersive strength of metaphosphate.

Figure 3 shows that the degree of dispersion in the slurries qualitatively correlates with the trends of the biodegradation rates. Clearly, monovalent cations treatments have a higher fraction of finer particles compared to Ca treatments. The size distributions after 3 min of ultrasonic treatment gave an interesting result: in all monovalent cation treatments the finer fraction increased, but the Ca treatment showed no significant change. This finding implies that the ionic Ca-complexes are relatively strong and are not affected by high energy input by ultrasonic treatment. After 3 weeks of incubation, the $CaCl_2$ treatments were still less dispersed than the NaCl and KCl treatments, as in the combined treatments. A striking result is that the $Na_4P_2O_7$ treatment is the least dispersed after 3 weeks of incubation. This was confirmed by visual observation, clearly showing flocculation in the slurry, and it coincides with the lower observed biodegradation rate in this treatment.

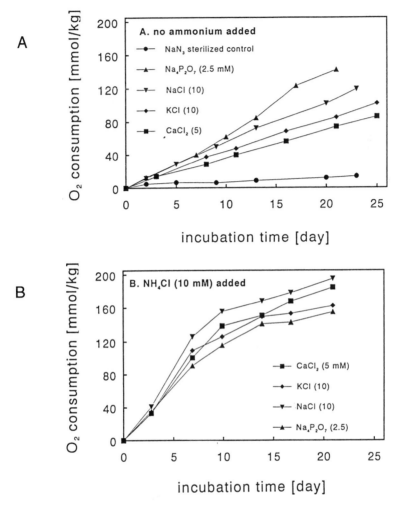

FIGURE 2. Cumulative oxygen consumption in the incubated slurries. See Table 1 for initial salt concentrations. A: no nitrogen added. B: 10 mM NH₄Cl added at the start of the experiment.

Analysis of the solution chemistry partly explains the complex response to addition of orthophosphate and metaphosphate. First, a pH decrease from 7.2 initially to 6.2 ± 0.2 was observed in incubations receiving a combined treatment of nitrogen and phosphate, whereas the pH was stable in treatments receiving no phosphate (7.6 ± 0.2). Second, ammonium concentrations in solution decreased from 10,000 μM (start of the experiment) to <146 μM after 3 weeks of incubation. In the same period, nitrate concentrations increased from 231 μM (3 days of incubation) to 8,940 μM (3 weeks of incubation). Therefore, it can be concluded that almost all of the added ammonium was nitrified,

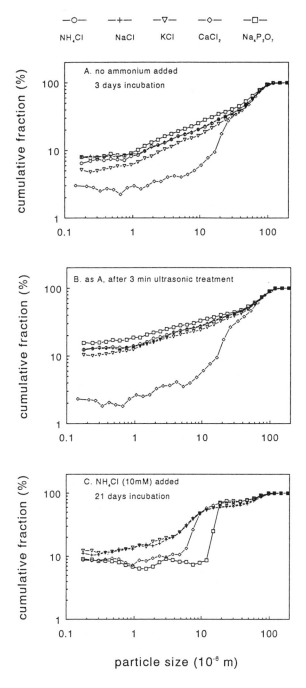

FIGURE 3. Particle size distributions in slurries (particles < 150 μm). A: after 3 days of incubation, no nitrogen added. B: as A, but after 3 minutes ultrasonic treatment. C: after 3 weeks of incubation, 10 mM NH₄Cl added as N-fertilizer.

and approximately 10% was incorporated in biomass. Third, only 13% of the added phosphate was recovered from solution after 3 weeks of incubation (treatments 9 through 11), and a lower amount of Ca was recovered in these treatments. Calculations with the speciation program MINEQL confirmed that these treatments were oversaturated with respect to $CaHPO_4$, whereas the other treatments were at equilibrium with $CaCO_3$.

Considering the above, the decrease in pH was caused by the combined effect of nitrification and precipitation of $CaHPO_4$, resulting in less favorable growth conditions for the microbial population. This process also occurred in the treatment receiving metaphosphate, because this compound is not stable in the soil and gradually transforms to orthophosphate. Because the dissolved organic carbon (DOC) concentration was considerably higher in the metaphosphate treatment, the authors believe that the flocculation was the result of newly formed clay-organic matter complexes. A strong argument for this explanation is the pH decrease, probably leading to precipitation of organic acids.

CONCLUSIONS

The following conclusions can be drawn from the present study. Biodegradation rates in slurries were affected by the dispersive strength of the dominant cation in solution. The rates increased with the sequence Ca < K < Na as the dominant cation, both with and without addition of ammonium. This sequence qualitatively correlated with the degree of dispersion in the slurries. Therefore, the increase probably is caused by a higher bioavailability of the contaminant. Addition of ammonium has a twofold positive effect on biodegradation, as both a fertilizer and a dispersant. However, ammonium can be nitrified in the course of a reclamation process. Because Na is the most effective dispersing cation, its addition to bioreactors will improve efficiency by maintaining a stable, dispersed slurry. Addition of metaphosphate is recommended only if pH conditions remain slightly alkaline.

ACKNOWLEDGMENTS

The authors would like to thank Joke Westerveld, Leen de Lange, and Bert de Leeuw for technical assistance, and Jan Dopheide for the analysis with MINEQL.

REFERENCES

Dibble, J. T., and R. Bartha. 1979. "Effect of Environmental Parameters on the Biodegradation of Oil Sludge." *Applied Environmental Microbiology* 37: 729-739.

Freijer, J. I. 1994. "Mineralization of Hydrocarbons and Gas Dynamics in Oil Contaminated Soils. Experiments and Modeling." Ph.D. thesis, University of Amsterdam, Amsterdam, The Netherlands, ISBN 90-6787-038-2.

Rijnaarts, H. H. M., A. Bachmann, J. C. Jumelet, and A.J.B. Zehnder. 1990. "Effect of Desorption and Intra-Particle Diffusion on the Aerobic Biomineralization of α-Hexachlorocyclohexane in a Contaminated Calcareous Soil." *Environmental Science and Technology* 24(9): 1349-1354.

Rouse, J. D., D. A. Sabatini, J. H. Suflita, and J. H. Harwell. 1994. "Influence of Surfactants on Microbial Degradation of Organic Compounds." *Critical Reviews in Environmental Science and Technology* 24(4): 325-370.

Schecher, W. D., and D. C. McAvoy. 1991. *MINEQL: A Chemical Equilibrium Program for Personal Computers. User Manual Version 2.1.* The Proctor and Gamble Co., Cincinnati, OH.

Scow, K. M. 1993. "Effect of Sorption-Desorption and Diffusion Processes on the Kinetics of Biodegradation of Organic Chemicals in Soils." In D. M. Linn (Ed.), *Sorption and Degradation of Pesticides and Organic Chemicals in Soils, SSSA Special Publication 32,* pp. 73-114. SSSA, Madison, WI.

van Olphen, H. 1963. *An Introduction to Clay Colloid Chemistry for Clay Technologists, Geologists and Soil Scientists.* Wiley Interscience, New York, NY.

Wu, S. and P. M. Gschwend. 1986. "Sorption Kinetics of Hydrophobic Organic Compounds to Natural Sediments." *Environmental Science and Technology* 20: 717-725.

Yiacoumi, S., and C. Tien. 1994. "A Model of Organic Solute Uptake from Aqueous Solutions by Soils." *Water Resources Research* 30(2): 571-580.

Surfactant Influence on PAH Biodegradation in a Creosote-Contaminated Soil

Louise Deschênes, Pierre Lafrance,
Jean-Pierre Villeneuve, and Réjean Samson

ABSTRACT ▄▄▄▄▄▄▄▄▄▄▄▄▄▄▄▄▄▄▄▄▄▄▄▄▄▄▄

This study consisted of assessing the biodegradation of 13 of the 16 U.S. Environmental Protection Agency (EPA) priority polycyclic aromatic hydrocarbons (PAHs) in a creosote-contaminated soil, using both biological and chemical surfactants. The assumption was that surfactants may enhance the mobilization of the hydrophobic PAHs, and possibly their biodegradation. The rhamnolipid biosurfactants were produced by *Pseudomonas aeruginosa* UG2. The chemical surfactant was sodium dodecyl sulfate (SDS). Over a period of 45 weeks, PAHs were periodically extracted from soil and quantified by gas chromatography/mass spectrometry (GC/MS). Results showed that, at three studied concentrations, surfactant addition did not enhance PAH biodegradation in the creosote-contaminated soil. Furthermore, for the four-ring PAHs, surfactant presence seemed harmful to the biodegradation process, the residual concentrations of each studied PAH decreasing more slowly than those found in the untreated soil. Moreover, this effect increased as a function of surfactant concentration. The negative effect was less evident with biosurfactants than for the chemical surfactant. The high-molecular-weight PAHs (more than four rings) were not degraded by the indigenous microorganisms. For the PAHs in general, the higher the molecular weight, the more recalcitrant was the contaminant. It is suggested that the surfactants were used as a preferential substrate by the indigenous microflora, which may have interfered with the biodegradation of the PAHs.

INTRODUCTION

In recent years, there has been great concern about remediating soil contaminated with polycyclic aromatic hydrocarbons (PAHs) because of their toxic, carcinogenic, and mutagenic potential. Among decontamination technologies, biological treatment constitutes an attractive alternative to other costly solutions, such as incineration for the removal of most PAHs (Wilson and Jones 1993). In weathered contaminated soils, high-molecular-weight PAHs are strongly sorbed to soil and are difficult to degrade in solid-phase remediation treatment. To obtain a better performance, it is important to improve the biodegradation rate and the removal efficiency of these hydrophobic contaminants. The addition of surface active agents (surfactants) is known to favor the mass transfer rates of contaminants from solid to liquid phases. Surfactants were shown to be effective in mobilization of PAHs in soil (Liu et al. 1991), but few studies have been performed to evaluate their real effect on the biodegradation of hydrophobic contaminants. In soil systems spiked with contaminants, recent work (Jain et al. 1992; Aronstein et al. 1991; Aronstein and Alexander 1993) suggests that low concentrations of surfactants may be useful for in situ bioremediation of sites contaminated with hydrophobic pollutants such as PAHs. Even if the results obtained are potentially applicable in the field, few data are available on the effects of surfactants in an actual contaminated soil. In most studies, radiolabeled or freshly added compounds were used rather than soils containing weathered contaminants. This is of particular importance because, in reconstituted systems, contaminants are not expected to exhibit the same bioavailability as in a weathered soil in which the contaminants are strongly sorbed for a long period of time.

The purpose of this study was to verify the effect of surfactants at low concentrations on the biodegradation of sorbed PAHs in a weathered contaminated soil from a wood-preserving plant. The experiment involved use of both a chemical and a biological surfactant. *Pseudomonas aeruginosa* UG2 rhamnolipids were chosen because the strain was isolated from a hydrocarbon-contaminated soil; in addition, these biosurfactants have been well characterized and were shown to increase the removal of PAHs from soil (Van Dyke et al. 1993; Scheibenbogen et al. 1994). Moreover, glycolipids constitute one of the most studied types of biosurfactants (Cooper and Zajic 1980). The chemical anionic surfactant sodium dodecyl sulfate (SDS) has been employed in soil-washing techniques (Clarke et al. 1992; Dipak et al. 1994). Low concentrations of surfactant were used because studies on the subject (Jain et al. 1992; Aronstein et al. 1991; Aronstein and Alexander 1992, 1993) have already shown the potential of surface active agents (*P. aeruginosa* UG2 biosurfactants and other chemical surfactants) used at low concentrations (10 and 100 g/g) to promote the biodegradation of hydrophobic hydrocarbons in soil.

MATERIALS AND METHODS

Soil Description

The weathered creosote-contaminated soil used for all experiments was collected in a wood-preserving site at Delson, Québec (Canada). Samples were collected near the creosote autoclaving plant at a 5- to 30-cm depth. At this location, the soil had been contaminated for several years with PAHs. The soil was classified as a sandy loam (13% clay, 16% silt, 71% sand). The soil organic matter was 6.35%, and the pH of the soil was 7.5. Table 1 lists the soil contaminants and their concentrations, which show a high level of contamination. In fact, PAH concentrations (about 3,042 mg/kg of total PAHs) were above the Quebec criterion C of 200 mg/kg. No evidence of contamination by inorganic wood preservatives (Cu, Cr, As) was found.

Surfactants

Pseudomonas aeruginosa UG2 was obtained from J. T. Trevors (University of Guelph, Guelph, Ontario, Canada). This strain was first isolated from an oil-contaminated soil (MacElwee et al. 1990). Rhamnolipids were produced and then partially purified according to the method of Van Dyke et al. (1993). Reagent-grade SDS (Sigma, St. Louis, Missouri, USA) was used as received. The properties of the studied surfactants are shown in Table 2.

Biodegradation Experiment

To evaluate the effect of surfactants on PAH biodegradation in the creosote-contaminated soil, 24 microcosms were prepared. Each microcosm was made by adding 350 g of contaminated soil in 1-L glass jars closed with Teflon™ covers.

TABLE 1. Chemical characterization of the creosote-contaminated soil.

Contaminant	Concentration (mg/kg)	Quebec criterion C[a] (mg/kg)
naphthalene	114	50
acenaphthylene	17	100
acenaphthene	247	100
fluorene	181	100
phenanthrene	700	50
anthracene	126	100
fluoranthene	705	100
pyrene	485	100
benzo(a)anthracene	128	10
chrysene	127	10
benzo(b)fluoranthene	80	10
benzo(k)fluoranthene	48	10
benzo(a)pyrene	46	10
indeno(1,2,3-cd)pyrene	24	10
dibenzo(a,h)anthracene	3	10
benzo(ghi)perylene	13	10
PAHs (total)	3,042	200
PCP	103	5
mineral oil and greases	4,843	5,000

(a) MENVIQ (1988).

TABLE 2. Properties of the studied surfactants.

Surfactants	Structure	Ionic state	Minimal surface tension (mN/m)	CMC[a] (g/L)	Reference
rhamnolipids	see MacElwee et al. 1990	anionic	31.4	0.03	Van Dyke et al. 1993
SDS	CH_3-$(CH_2)_{11}$-$SO_4^-Na^+$	anionic	38-39	2.3	Clarke et al. 1992

(a) Critical micellar concentration.

Seven different treatments were performed in triplicate: (1) soil without surfactant; (2) soil amended with 10 μg/g SDS; (3) soil amended with 100 μg/g SDS; (4) soil amended with 500 μg/g SDS; and (5), (6), and (7) soil amended with *P. aeruginosa* UG2 biosurfactants at the same three concentrations (10, 100, and 500 μg/g). Abiotic controls were made by addition of 0.02% sodium azide. The soil moisture was adjusted and maintained at 80% of field capacity with sterile mineral salts medium (MSM) in which the appropriate amount of the desired surfactant was dissolved. The treated soil was mixed weekly to enhance oxygen availability, and every 2 weeks previously selected amounts of surfactants were added in an appropriate MSM volume. The jars were incubated in the dark for 45 weeks at 20°C to prevent photodegradation of the PAHs.

In a previous study, it had been shown that SDS was completely biodegraded within 10 days by the indigenous microflora of creosote-contaminated soil (Deschênes et al. 1995), Therefore, to ensure that the surfactants remained effective during the treatment period (45 weeks) and that no accumulation occurred, surfactants were added every 2 weeks. Soil samples were taken periodically for complete chemical analysis. PAHs were extracted with dichloromethane and quantified by gas chromatographyl/mass spectroscopy (GC/MS). The naphthalene, acenaphthylene, and acenaphthene PAHs are not discussed in the present paper. Because of their rapid biodegradation rate, it would be difficult to determine their biodegradation rate with the analytical methods used here.

RESULTS AND DISCUSSION

Results, expressed as a percentage of the abiotic controls, are presented in Table 3. For all treatments, almost all the fluorene and phenanthrene was degraded within 11 weeks of incubation (result not shown). In fact, values of around 90 to 98% of biodegradation were obtained at 11 weeks and maintained for the 45 weeks of treatment. A maximum biodegradation of 86% was attained for the anthracene, the PAH that exhibits the lowest aqueous solubility (0.073 mg/L, compared to 1.98 and 1.29 mg/L for fluorene and phenanthrene, respectively [Dzombak and Luthy 1984]). At the end of the treatment, no effect

TABLE 3. Percentage of removal of PAHs by biodegradation after 45 weeks.

PAHs		MSM with biosurfactants (µg/g)			MSM with SDS (µg/g)		
	MSM	10	100	500	10	100	500
fluorene	93	94	93	94	94	94	92
phenanthrene	98	98	97	98	98	98	98
anthracene	88	84	84	84	86	86	82
fluoranthene	94	93	87	75	94	77	43
pyrene	94	94	90	66	95	81	36
benzo(a)anthracene	90	86	83	68	90	77	35
chrysene	80	82	69	46	82	54	19
benzo(b)fluoranthene	2	13	14	4	2	5	5
benzo(k)fluoranthene	−1	−14	2	12	−6	12	4
benzo(a)pyrene	8	13	−7	−9	12	−9	−2
indeno(1,2,3-cd)pyrene	−5	−4	−10	−13	−4	−8	−7
dibenzo(a,h)anthracene	0	10	0	−29	5	−4	−6
benzo(ghi)perylene	−18	−13	−10	−22	−1	−10	17

from either of the two surfactants on biodegradation of the three-ring PAHs was detected. The biodegradation of PAHs decreased as the number of rings increased. The biodegradation of the four-ring PAHs (fluoranthene, pyrene, benzo(a)anthracene, and chrysene) was slower than that of the three-ring PAHs. Similar results have been obtained in the literature (Park et al. 1990). After 45 weeks of treatment without addition of surfactants, the four-ring PAHs were almost completely biodegraded (80 to 94% biodegradation).

Results showed that addition of low concentrations (10 µg/g) of either of the two studied surfactants did not affect biodegradation of four-ring PAHs, when compared to treatments without surfactant addition. On the contrary, the addition of 100 and 500 µg/g of either surfactant inhibited biodegradation, especially for four-ring PAHs. For example, 94% of the fluoranthene was biodegraded in the presence of 10 µg/g of SDS, compared to only 77 and 43% in the presence of 100 and 500 µg/g SDS, respectively. The higher the surfactant concentration, the slower was the PAH biodegradation. For example, 87 and 75% of the fluoranthene was biodegraded in the presence of 100 and 500 µg/g biosurfactants, respectively. At the same concentration, addition of biosurfactants inhibited the biodegradation of four-ring PAHs, but not three-ring PAHs. The inhibition was less significant than that produced by addition of SDS. The same patterns were obtained for the other four-ring PAHs.

PAHs that contained more than four rings were not biodegraded by the indigenous microflora of the contaminated soil, and values of near 0% biodegradation were obtained. Negative values are explained by the fact that the PAH concentration in the abiotic control was lower than the one obtained in the biotic treatment. The large variations obtained can be explained by the small concentrations of the high-molecular-weight PAHs in the contaminated soil (see Table 1).

Even though some studies have shown that low concentration of surfactants may promote the biodegradation of hydrophobic compounds in soil (Jain et al. 1992; Aronstein et al. 1991; Aronstein and Alexander 1992, 1993). The studies have shown that surfactants in reconstituted systems inhibit biodegradation. In fact, Tiehm (1994) showed that the biodegradation rate of phenanthrene was reduced by the presence of different concentrations of SDS in an aqueous-phase system. The author suggested that SDS was preferred as growth substrate. In our case, it was shown that the indigenous microflora of creosote-contaminated soil quickly biodegraded SDS at the studied surfactant concentrations (Deschênes et al. 1995). Moreover, in the literature it was shown that the *P. aeruginosa* UG2 biosurfactants can be biodegraded by soil microorganisms (Providenti 1994). It was therefore suggested that the studied surfactants were used as preferential substrates by the indigenous microflora. This substrate competition was assumed to slow down the biodegradation of the four-ring PAHs.

Other factors may be responsible for this inhibition of PAH biodegradation. Among these, the possible toxicity of surfactants or their biodegradation intermediates for the microorganisms is an important hypothesis to consider. For rhamnolipids, research is also needed to verify that they could accumulate to high concentrations in the studied soil. Finally, complex interactions exist between surfactants, microorganisms, and hydrophobic contaminants; research is needed to better understand the effect of surfactants used in environmental biotechnologies.

PERSPECTIVES FOR SITE RESTORATION

Optimum efficiency is the first criterion for the selection of a surfactant to offer a viable surfactant-based remediation process (West and Harwell 1992). This assumes nontoxicity and biodegradability of the surfactant. Results from this study indicate that the addition of SDS or *P. aeruginosa* UG2 biosurfactants are not very effective in improving aboveground biotreatment of creosote-contaminated soil. At high concentrations their addition slows down PAH biodegradation. On the other hand, results in the literature (Deschênes et al. 1995) indicated that SDS was easily mineralized by the indigenous microflora. In practice, this observation implies that an optimal addition sequence, as well as an optimal concentration, must be determined prior to treatment. This is particularly true if an aboveground heap pile or a bioreactor is to be used to remediate soil.

In the case of in situ bioremediation, these results could be advantageous because persistence of surfactant in the aquifer is an important concern. The high biodegradability of surfactant suggests that no residual surfactant will persist after the treatment. However, as with aboveground treatment, the surfactant should be injected at a determined rate and concentration to maintain high efficiency throughout the treatment. Overloading the aquifer with surfactants would be detrimental to biodegradation, because surfactants can be used as a preferential substrate by the indigenous microflora. This would result in significant inhibition of biodegradation of the more recalcitrant contaminants. It is also possible that the surfactant or its biodegradation intermediates could be toxic to indigenous microflora. Bioavailability of contaminants contained in micelles should be considered if surfactants are expected to be used at a concentration above the effective critical micelle concentration. Also, the use of surfactants may contribute to the spreading of contaminants in the aquifer if biotreatment is carried out in situ. In such a case, effective control of the transport of solubilized hydrocarbons is needed to prevent groundwater contamination. Therefore, to develop effective bioremediation techniques, research is needed on the behavior and ultimate fate of surfactants.

REFERENCES

Aronstein, B. N., and M. Alexander. 1993. "Effect of a non-ionic surfactant added to the soil surface on the biodegradation of aromatic hydrocarbons within the soil." *Appl. Microbiol. Biotechnol. 39*: 386-390.

Aronstein, B. N., and M. Alexander. 1992. "Surfactants at low concentrations stimulate the biodegradation of sorbed hydrocarbons in samples of aquifer sands and soil slurries." *Environ. Toxicol. Chem. 11*: 1227-1233.

Aronstein, B. N., Y. M. Calvillo, and M. Alexander. 1991. "Effect of surfactants at low concentrations on the desorption and biodegradation of sorbed aromatic compounds in soil." *Environ. Sci. Technol. 25*: 1728-1731.

Clarke, A. N., R. D. Mutch Jr., D. J. Wilson, and K. H. Oma. 1992. "Design and implementation of pilot scale surfactant washing/flushing technologies including surfactant reuse." *Wat. Sci. Tech. 26*: 127-135.

Cooper, D. G., and J. E. Zajic. 1980. "Surface-active compounds from microorganisms." *Adv. Appl. Microbiol. 26*: 229-253.

Deschênes, L., P. Lafrance, J. P. Vileneuve, and R. Samson. 1995. "The effect of an anionic surfactant on the mobilization and the biodegradation of PAHs in a creosote contaminated soil." *Hydro. Sci. J.* (in press).

Dipak, R., L. M. Minwen, and W. Guang-te. 1994. "Modeling of anthracene removal from soil columns by surfactant." *J. Environ. Sci. Health A-29*: 197-213.

Dzombak, D., and R. G. Luthy. 1984. "Estimating adsorption of polycyclic aromatic hydrocarbons on soils." *Soil Science 137*: 292-308.

Jain, D. K., H. Lee, and J. T. Trevors. 1992. "Effect of addition of *Pseudomonas aeruginosa* UG2 inocula or biosurfactants on biodegradation of selected hydrocarbons in soil." *J. Ind. Microbiol. 10*: 87-93.

Liu, Z., S. Laha, and R. G. Luthy. 1991. "Surfactant solubilization of polycyclic aromatic hydrocarbon compounds in soil-water suspensions." *Wat. Sci. Tech. 23*: 475-485.

MacElwee, C. G., H. Lee, and J. T. Trevors. 1990. "Production of extracellular emulsifying agent by *Pseudomonas aeruginosa* UG1." *J. Ind. Microbiol. 5*: 25-32.

MENVIQ. 1988. *Guide Standard de Caractérisation de Terrains Contaminés.* Direction des substances dangereuses, Ministère de l'Environnement du Québec, Québec, Canada, report QEN/SD-2.

Park, K. S., R. C. Sims, and R. R. Dupont. 1990. "Transformation of PAHs in soil systems." *J. Environ. Eng., 116*: 632-641.

Providenti, M. A. 1994. "Effect of *Pseudomonas aeruginosa* UG2 inocula and biosurfactants on phenanthrene mineralization by *Pseudomonas* sp. UG14." Master's Thesis, University of Guelph, Guelph, Ontario, Canada.

Scheibenbogen, K., R. G. Zytner, H. Lee, and J. T. Trevors. 1994. "Enhanced removal of selected hydrocarbons from soil by *Pseudomonas aeruginosa* UG2 biosurfactants and some chemical surfactants." *J. Chem. Tech. Biotechnol. 59*: 53-59.

Tiehm, A. 1994. "Degradation of polycyclic aromatic hydrocarbons in the presence of synthetic surfactants." *Appl. Environ. Microbiol. 60*: 258-263.

Van Dyke, M. I, P. Couture, M. Brauer, H. Lee, and J. T. Trevors. 1993. "*Pseudomonas aeruginosa*-UG2 rhamnolipid biosurfactants: structural characterization and their use in removing hydrophobic-compounds from soil." *Can. J. Microbiol. 39*: 1071-1078.

West, C. C., and J. H. Harwell. 1992. "Surfactants and subsurface remediation." *Environ. Sci. Technol. 26*: 2324-2330.

Wilson, S. C., and K. C. Jones. 1993. "Bioremediation of soil contaminated with polynuclear aromatic hydrocarbons (PAHs): a review." *Environ. Pollut. 80*: 229-249.

Biodegradation Kinetics of Phenanthrene Solubilized in Surfactant Micelles

Stefan J. Grimberg and Michael D. Aitken

ABSTRACT

The biodegradation of phenanthrene solubilized in surfactant micelles was studied using a simple, well-defined laboratory system. The system was designed to evaluate whether phenanthrene present in micelles of the nonionic surfactant Tergitol NP-10 was available to a phenanthrene-degrading bacterium. Results indicate that micellized phenanthrene is essentially unavailable to the microorganism, so that only the phenanthrene present in the aqueous phase is degraded. A modified Michaelis-Menten equation was developed to quantify the effects of surfactant concentration on phenanthrene uptake rates. Experimental data were described well with this equation.

INTRODUCTION

Many of the pollutants found in contaminated soil are poorly soluble in water, and therefore exist in nonaqueous-phase liquids as trapped residuals in pore spaces, or partitioned into hydrophobic domains such as soil organic matter. The limited availability of these compounds in the aqueous phase can limit their rates of biodegradation (Volkering et al. 1992, Weissenfels et al. 1992). The use of surfactants has been proposed to help solubilize hydrophobic chemicals, and in some cases coupling surfactants to biodegradation processes has been suggested (Laha and Luthy 1991; Aronstein and Alexander 1992; Abramowicz et al. 1993; Bury and Miller 1993). Results of laboratory experiments on biodegradation of hydrophobic compounds in the presence of surfactants have been contradictory, and range from virtually complete inhibition of biodegradation (Laha and Luthy 1991, Abramowicz et al. 1993) to stimulation of biodegradation (Aronstein and Alexander 1992, Bury and Miller 1993).

Most previous work on surfactant-aided bioremediation has been conducted in complex matrixes, from which it is difficult to establish cause-and-effect relationships. To better understand the influences of surfactants on biodegradation of hydrophobic chemicals, we have undertaken a series of experiments in well-defined laboratory systems. The model system described in this report involves phenanthrene as a model hydrophobic pollutant, a well-characterized bacterium, and a nonionic surfactant that does not affect the metabolic activity of the microorganism and is not biodegradable over short time scales. The purpose of this work was to evaluate the bioavailability of phenanthrene solubilized in surfactant micelles.

BIODEGRADATION MODEL

In the presence of a surfactant at concentrations above its critical micelle concentration (CMC), hydrophobic compounds such as phenanthrene will partition between the surfactant micelles and the aqueous phase (Edwards et al. 1991). For many combinations of pollutant, surfactant, and microorganisms, it is possible that the pollutant partitioned into micelles is not available directly for biodegradation. To test this hypothesis, it is necessary to measure rates of biodegradation in the presence of surfactant, and to correlate those rates with the aqueous-phase solute concentration. In micellar systems, however, it is not possible to distinguish between solute in the aqueous phase and solute in micelles using conventional measurement techniques; instead, only the total liquid-phase concentration can be measured easily. It is therefore necessary to derive a correlation between the aqueous-phase concentration and the total liquid-phase concentration using measurable parameters.

In a saturated solution, the amount of solute in micelles is related to the amount of surfactant present in micellar form and the solubilization capacity (SC) of the micelles. The solubilization capacity, which also has been referred to as the molar solubilization ratio (Edwards et al. 1991), represents the mass of solute in micelles per unit mass of surfactant above the CMC. Therefore, the total mass of solute in micelles at saturation is given by:

$$M_m = SC(C_{surf} - CMC) \bullet V \tag{1}$$

where C_{surf} is the surfactant concentration and V is the total liquid volume. At saturation, the solute concentration in the aqueous phase is simply equal to its aqueous solubility, $C_{a,sat}$. Assuming that the volume occupied by micelles is negligible compared to the total liquid volume, a mass balance on solute at saturation gives:

$$C_{T,sat} = C_{a,sat} + \frac{M_m}{V} = C_{a,sat} + SC(C_{surf} - CMC) \tag{2}$$

where $C_{T,sat}$ is the total liquid-phase solute concentration at saturation. If an equilibrium distribution of solute exists between the micellar and aqueous phases, the aqueous-phase solute concentration, C_a, at any total liquid-phase concentration, C_T, is given by:

$$Ca = CT \cdot \left[\frac{C_{a,sat}}{C_{a,sat} + SC\,(C_{surf} - CMC)} \right] = CT \cdot \frac{C_{a,sat}}{C_{T,sat}} \qquad (3)$$

Finally, if it is assumed that only solute present in the aqueous phase is available for biodegradation, and that the solute equilibrates rapidly between the micellar and aqueous phases, Eq. 3 can be used to derive a modified Michaelis-Menten equation incorporating the effects of surfactant on specific solute degradation rates:

$$\frac{q}{q_{max}} = \frac{C_T/C_{T,sat}}{K_S/C_{a,sat} + C_T/C_{T,sat}} \qquad (4)$$

where q is the specific solute uptake rate and K_S is the half-saturation coefficient. Eq. 4 is normalized by the total liquid-phase saturation concentration, which permits direct comparison of kinetic data obtained in the presence and absence of surfactant. The effects of surfactant concentration can be quantified by substituting the definition of $C_{T,sat}$ shown in Eq. 2.

MATERIALS AND METHODS

Microorganism

Pseudomonas stutzeri P-16 was isolated from soil contaminated with creosote, and has been well characterized with respect to phenanthrene biodegradation (Stringfellow and Aitken 1994a, Stringfellow and Aitken 1994b). In the absence of surfactants the organism is known to degrade phenanthrene by Michaelis-Menten kinetics and appears to degrade phenanthrene only in the liquid phase (Stringfellow and Aitken 1994b). P-16 was grown in media containing peptone (Stringfellow and Aitken 1994b) or in tap-water buffer (TWB; Stringfellow and Aitken 1994b) in a flask containing a monolithic, cylindrical block of crystalline phenanthrene (Grimberg et al. 1994).

Surfactant

The nonionic surfactant Tergitol NP-10 (Sigma Chemical Co., St. Louis, Missouri), a nonylphenol ethoxylate, was used without further purification. The biodegradability of Tergitol NP-10 by P-16, as well as its effect on the growth of P-16, was tested by growing the organism in peptone medium containing 500 mg/L of the surfactant. Growth was followed by measuring optical density at 420 nm, and was compared to growth in peptone medium without surfactant. Aliquots were removed from the growth medium periodically for measurement of the surfactant concentration by high-pressure liquid chromatography (HPLC) as described below.

The potential effect of Tergitol NP-10 on expression of phenanthrene metabolism was determined by growing P-16 on solid phenanthrene in TWB as described above, except 500 mg/L of the surfactant was added to the medium. The culture was harvested by centrifugation, resuspended in TWB, then centrifuged again and resuspended in TWB to a final absorbance at 420 nm of 0.1. Initial rates of phenanthrene uptake were measured as a function of phenanthrene concentration as described below, and were compared to initial rates determined for a control culture grown in the absence of surfactant.

The CMC and solubilization capacity of Tergitol NP-10 were measured using procedures described elsewhere (Grimberg et al. 1995).

Phenanthrene Uptake Kinetics

Experiments were conducted on micellized phenanthrene only, so that no excess (crystalline) phenanthrene was present. For phenanthrene concentrations below 1 mg/L, a spectrophotometric method described in detail elsewhere (Stringfellow and Aitken 1994b) was used. The initial phenanthrene uptake rate at each concentration was determined by following decreases in absorbance at 250 nm. Specific phenanthrene uptake rates were calculated by normalizing the measured rate by the biomass concentration (Stringfellow and Aitken 1994b). The presence of surfactant did not interfere with the spectrophotometric method.

Uptake rates in the presence of surfactant at phenanthrene concentrations greater than 1 mg/L were determined with an HPLC technique. Saturated solutions of phenanthrene were prepared by adding an excess of phenanthrene to a solution containing a known concentration of Tergitol NP-10 dissolved in TWB. The mixture was stirred for 20 h, then filter sterilized through a 0.2-μm polycarbonate filter. Solutions containing phenanthrene at concentrations below saturation were prepared by diluting the saturated solution with presterilized surfactant solution. To initiate the rate assay, 1 mL of washed culture grown on solid phenanthrene in TWB was added to 19 mL of the phenanthrene/surfactant solution. Phenanthrene concentration was monitored with time by withdrawing 1 mL aliquots, adding formaldehyde to a 10% final concentration to terminate microbial activity, and measuring phenanthrene by HPLC. Nine different initial phenanthrene concentrations were tested in this experiment. To estimate rate coefficients, the integrated form of Eq. 4 was fit to the combined concentration vs. time data by nonlinear regression using SYSTAT (SYSTAT, Inc., Evanston, Illinois). Parameter values used to simulate rates as a function of surfactant concentration were $K_S = 0.2$ mg/L (Stringfellow and Aitken 1994b), CMC = 82 mg/L, and SC = 0.029 g phenanthrene/g Tergitol NP-10 (Grimberg et al. 1995).

HPLC

Both phenanthrene and Tergitol NP-10 were analyzed using a Waters 600E HPLC system, a Waters 470 fluorescence detector, and a C_{18} reverse-phase column (25 cm × 1.5 mm; Supelco, Inc., Bellefonte, Pennsylvania). The mobile phase (1 mL/min) was 70:30 (v/v) acetonitrile in water for 0 to 6 min, increased linearly to 100% acetonitrile over 6 to 8 min, and maintained at 100% acetonitrile through 18 min. Phenanthrene was analyzed at an excitation wavelength of 259 nm and emission wavelength of 370 nm, while Tergitol NP-10 was analyzed at 225 nm and 295 nm, respectively.

RESULTS AND DISCUSSION

Tergitol NP-10 did not influence the growth of *Pseudomonas stutzeri* P-16 in a peptone medium and was not able to grow on Tergitol as a sole carbon source (data not shown). To evaluate possible indirect effects of Tergitol NP-10 on phenanthrene metabolism, P-16 was grown on phenanthrene as sole carbon source both in the presence and absence of the surfactant. Cells from each medium were harvested and washed to remove the surfactant and residual phenanthrene, then specific rates of phenanthrene uptake were determined for each culture as a function of phenanthrene concentration. The results shown in Figure 1 indicate that growth in the presence of surfactant had no influence on the ability of P-16 to metabolize phenanthrene.

Phenanthrene-uptake rates also were measured over a range of phenanthrene concentrations in the presence of 500 mg/L Tergitol NP-10. Michaelis-Menten kinetic coefficients (q_{max} and K_S) were determined by fitting the integrated form of Eq. 4 to the combined concentration vs. time data for all initial phenanthrene concentrations. A curve generated from the fitted parameters is shown on Figure 1. The close overlap of this curve with the experimental data on phenanthrene-uptake rates in the absence of surfactant (data points) indicates that the modified Michaelis-Menten equation accurately describes the effects of surfactant on biodegradation of phenanthrene solubilized in micelles. At 500 mg/L Tergitol NP-10, the saturation concentration of phenanthrene is approximately 12 mg/L, more than ten times the solubility in water alone. However, it appears that only the phenanthrene actually present in the aqueous phase is available for biodegradation. If all of the micellized phenanthrene were available, then maximum specific rates of phenanthrene uptake would have been observed over nearly the entire range of phenanthrene concentrations tested.

Specific rates of phenanthrene uptake were also measured for constant total phenanthrene concentration (0.78 mg/L) in the presence of varying concentrations of Tergitol NP-10. Results are shown in Figure 2. Figure 2 also includes a line representing the simulated effect of surfactant if only phenanthrene present in the aqueous phase is available for biodegradation; the simulation was conducted using independently measured parameters. Below about 250 mg/L Tergitol NP-10, the data closely match the simulation, but at higher surfactant concentration the simulation slightly underestimates phenanthrene uptake rates.

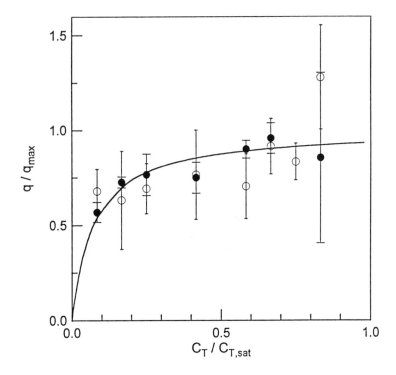

FIGURE 1. Specific rate of phenanthrene uptake (normalized by the maximum rate) as a function of phenanthrene concentration (normalized by the saturation concentration) in the absence of surfactant (data points) or in the presence of 500 mg/L Tergitol NP-10 (solid line). For the experiments performed in the absence of surfactant, data are from P-16 grown in the absence of surfactant (O) or in the presence of 500 mg/L Tergitol NP-10 (●). Error bars represent the standard deviation of triplicate measurements. The solid line represents the use of Eq. 4 and kinetic parameters fit to the experimental data obtained when phenanthrene uptake was measured in the presence of the surfactant.

The combined results of experiments illustrated in Figures 1 and 2 indicate that phenanthrene solubilized in micelles of the nonionic surfactant Tergitol NP-10 is largely unavailable to *Pseudomonas stutzeri* P-16. A modified kinetic model based on rapid equilibration of phenanthrene between micelles and the aqueous phase described the effects of surfactant concentration reasonably well. These findings, therefore, might serve as a foundation from which to construct models of the influences of surfactants on biodegradation in more complex situations. Such situations include systems in which nonaqueous-phase contaminant is present, the use of surfactants that are themselves biodegraded, or combinations of surfactants and microorganisms that lead to direct interactions between microbial cells and surfactant micelles.

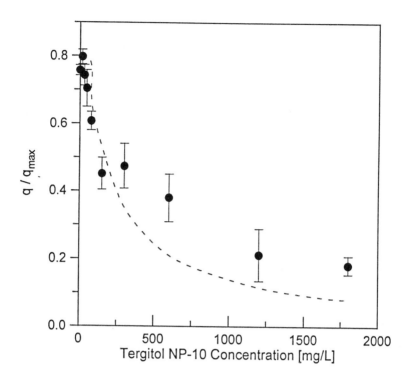

FIGURE 2. Specific rate of phenanthrene uptake as a function of surfactant concentration for a total liquid-phase phenanthrene concentration of 0.78 mg/L. Error bars represent the standard deviation of triplicate measurements. The dashed line represents the simulated effect of surfactant concentration using Eq. 4 and independently measured parameters.

ACKNOWLEDGMENTS

This work was supported by the U.S. Geological Survey (grant # 14-08-001-G2103) and the University of North Carolina Water Resources Research Institute (grants # 20162 and 70138).

REFERENCES

Abramowicz, D.A., M.J. Brennan, H.M. van Dort, and E.L. Gallagher 1993. "Factors Influencing the Rate of Polychlorinated Biphenyl Dechlorination in Hudson River Sediments." *Environmental Science and Technology* 27: 1125-1131.

Aronstein, B.N., and M. Alexander. 1992. "Surfactants at Low Concentrations Stimulate Biodegradation of Sorbed Hydrocarbons in Samples of Aquifer Sands and Soil Slurries." *Environmental Toxicology and Chemistry* 11: 1227-1233.

Bury, S.J., and C.A. Miller. 1993. "Effect of Micellar Solubilization on Biodegradation Rates of Hydrocarbons." *Environmental Science and Technology* 27: 104-110.

Edwards, D.A., R.G. Luthy, and Z. Liu. 1991. "Solubilization of Polycyclic Aromatic Hydrocarbons in Micellar Nonionic Surfactant Solutions." *Environmental Science and Technology* 25: 127-133.

Grimberg, S.J., M.D. Aitken, and W.T. Stringfellow. 1994. "The Influence of a Surfactant on the Rate of Phenanthrene Mass Transfer into Water." *Water Science and Technology* 30: 23-30.

Grimberg, S.J., J. Nagel, and M.D. Aitken. 1995. "Kinetics of Phenanthrene Dissolution into Water in the Presence of Nonionic Surfactants." *Environmental Science and Technology,* in press.

Laha, S., and R.G. Luthy. 1991. "Inhibition of Phenanthrene Mineralization by Nonionic Surfactants in Soil-Water Systems." *Environmental Science and Technology* 25: 1920-1930.

Stringfellow, W.T., and M.D. Aitken. 1994a. "Kinetics of Phenanthrene Degradation by Soil Isolates." In R.E. Hinchee, D.B. Anderson, F.B. Metting, Jr., and G.D. Sayles (Eds.), *Applied Biotechnology for Site Remediation,* pp. 310-314. Lewis Publishers, Boca Raton, FL.

Stringfellow, W.T. and M.D. Aitken. 1994b. "Comparative Physiology of Phenanthrene Degradation by Two Dissimilar Pseudomonads Isolated from a Creosote-Contaminated Soil." *Canadian Journal of Microbiology* 40: 432-438.

Volkering, F., A.M. Breure, A. Sterkenburg, and J.G. van Andel. 1992. "Microbial Degradation of Polycyclic Aromatic Hydrocarbons: Effect of Substrate Availability on Bacterial Growth Kinetics." *Applied Microbiology and Biotechnology* 36: 548-552.

Weissenfels, W.D., H.-J. Klewer, and J. Langhoff. 1992. "Adsorption of Polycyclic Aromatic Hydrocarbons (PAHs) by Soil Particles: Influence on Biodegradability and Biotoxicity." *Applied Microbiology and Biotechnology* 36: 689-696.

Detoxification of Aromatic Pollutants by Fungal Enzymes

Jean-Marc Bollag and Jerzy Dec

ABSTRACT

Fungal enzymes, such as laccase, peroxidase, and tyrosinase, play a prominent role in catalyzing the transformation of various aromatic compounds in the environment. The enzyme-mediated oxidative coupling reaction results in covalent binding of chlorinated phenols and anilines to soil organic matter or polymerization of the substrates in aquatic systems. Both of these processes are accompanied by a detoxification effect. Therefore, it has been postulated that they be exploited for the treatment of polluted soil and water. The mechanism and efficiency of oxidative coupling in pollutant removal were studied by incubation of chlorinated phenols and anilines with various humic substances or soil and analysis of the reaction products by chromatography and mass and ^{13}C nuclear magnetic resonance (NMR) spectrometry. The decontamination effect could be enhanced by optimization of the reaction conditions and immobilization of enzymes on solid materials. The results obtained strongly support the concept of using enzymes for control of environmental pollution.

INTRODUCTION

Enzyme-mediated binding of xenobiotic pollutants to soil is considered a potential detoxification technique (Bollag 1992). It has been postulated that aromatic carbons of the contaminant molecules involved in binding become constituent units of soil humus and do not represent toxic residues. Initially, the validity of this concept was questioned by reports of the possible release of bound xenobiotics from humic substances. However, recent studies and a closer examination of the previous data revealed that the observed release was very limited, and the idea of using binding for decontamination purposes remained an alternative.

For general acceptance of using binding of pollutants to organic matter, it is necessary to obtain a thorough understanding of the binding mechanism and the chemical nature of the bonds between the reacting substances.

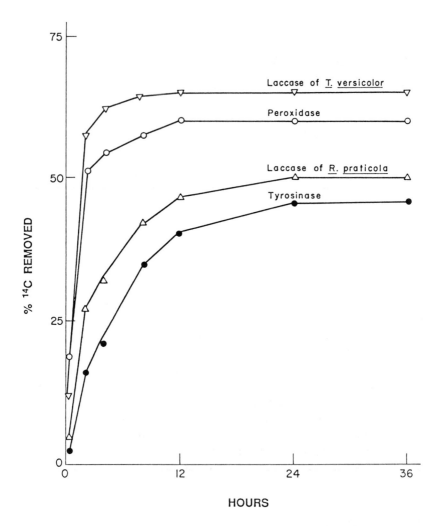

FIGURE 1. Removal of ^{14}C -2,4-dichlorophenol through binding to fulvic acid and polymerization in the presence of oxidoreductases.

Generally, xenobiotics may bind to soil either through various adsorption mechanisms or through covalent binding to humic substances. We have primarily investigated covalent binding because it constitutes the strongest type of interaction between xenobiotics and humus, and therefore it is considered most desirable from an environmental point of view (Bollag 1992). In particular, we studied the oxidative coupling reaction believed to result in covalent binding of chlorinated phenols and anilines to soil organic matter or polymerization of these compounds in aquatic systems. The catalysts used for stimulation of these oxidative coupling processes were various microbial enzymes, such as laccases, peroxidases, and tyrosinases.

EVIDENCE FOR COVALENT BINDING

Environmental concerns made it essential to determine with certainty whether covalent bonds are formed during the enzymatic binding reactions. Three methods were used to investigate this question. The first approach involved model coupling between selected chlorinated substrates, such as 4-chloroaniline or 2,4-dichlorophenol, and humic acid constituents, such as guaiacol or syringic acid, in the presence of different oxidoreductases. The reaction products were analyzed by various chromatographic techniques and mass spectrometry. Model compounds may not adequately simulate humus; therefore, in the second method, ^{13}C-labeled 2,4-dichlorophenol was enzymatically incorporated into natural humic acid and examined by ^{13}C NMR spectrometry. The third approach relied on monitoring release of chloride ions during binding of chlorinated substrates to humic acid.

As indicated by mass spectra, the reaction mediated by laccase from the fungus *Rhizoctonia praticola* resulted in cross-coupling between chlorinated phenols or anilines and representative humus constituents (Bollag 1992). The hybrid products from reactions of 2,4-dichlorophenol with oricinol, vanillin, vanillic acid, and syringic acid ranged from dimers to pentamers. In a study with 4-chloroaniline and guaiacol, the same aminoquinone, carbazole, and iminoquinone structures were always formed without regard to the type of enzyme (horseradish peroxidase, tyrosinase, or laccases from *R. praticola* and *Trametes versicolor*).

The formation of hybrid oligomers in these model reactions constituted an indication that halogenated substrates can couple covalently to humus in the terrestrial environment. Apparently, the role of enzymes is to oxidize the substrates to free radicals or quinones; the subsequent coupling of these oxidation products to humus probably occurs without further involvement of the catalyst.

The application of ^{13}C NMR provided direct evidence for covalent binding of ^{13}C-labeled 2,4-dichlorophenol to humic acid (Hatcher et al. 1993). After incubation without enzyme, only the labeled carbons of the free, unbound pollutant were detected both in the presence and in the absence of humic acid. However, when the labeled compound and humic acid were incubated in the presence of peroxidase, the resulting ^{13}C NMR spectrum displayed a large dispersion of signals representing various covalent interactions of 2,4-dichlorophenol with the humic polymer. Based on the position of these signals on the spectrum, the interactions could be identified as carbon-carbon binding, carbon-oxygen binding via an ether linkage, and carbon-oxygen binding via an ester linkage.

Some ^{13}C NMR signals suggested interactions at chlorinated sites of the 2,4-dichlorophenol molecule, indicating removal of the chlorine atom. Such a dehalogenation reaction was confirmed through direct measurement of the release of chloride ions into the reaction mixture during enzyme-mediated binding of chlorophenols to humic acid (Dec and Bollag 1994a). Since the dehalogenation could not be explained by adsorption mechanisms such as van

der Waal's forces, H-bonding, ligand exchange, etc., it was concluded that covalent binding was involved, and that dehalogenation was a consequence of the
free-radical mechanism of oxidative coupling.

DECONTAMINATION REACTIONS

Binding Reactions

The polymerization or incorporation of chlorinated substrates into soil
organic matter is expected to decrease their toxicity because the covalently
bound contaminants are no longer accessible to living organisms. To examine
this assumption, 2,4-dichlorophenol and 4-chloro-2-methylphenol were incubated in a fungal growth medium of *R. praticola* in the presence or absence of a
humic acid constituent (syringic acid) and a laccase (Bollag et al. 1988). The
growth of *R. praticola* was inhibited through the addition of chlorophenols in
the absence of enzyme and syringic acid. However, addition of laccase and
syringic acid caused the formation of various nontoxic, cross-coupling products, resulting in elimination of the inhibition effect.

The efficiency of binding in the removal of pollutants was demonstrated in
a study of Sarkar et al. (1988), who incubated ^{14}C-labeled 2,4-dichlorophenol
with fulvic acid in the presence of various oxidoreductases. As a result of the
enzyme-mediated reactions, about 65% of the initial radioactivity was found to
be firmly incorporated into fulvic acid (Figure 1); only insignificant amounts of
bound material were released during extensive washing with water and
organic solvents. No incorporation was observed in control samples that contained boiled enzymes. The rate of incorporation with active enzymes depended on pH, temperature, and amount of enzyme.

Shannon and Bartha (1988) demonstrated enzymatic immobilization of
[14]C-labeled 2,4-dichlorophenol in soil treated with horseradish peroxidase and
H_2O_2. The soil was placed in glass columns and leached with distilled water to
model the prevention of groundwater pollution. As a result of this treatment,
the amount of radioactivity leached from the soil was reduced to about 10% of
that leached from the nontreated control. No remobilization of the bound pollutant was detected during a 4-week incubation period.

Polymerization Reactions

In the absence of humic acid, the free radicals and quinoid products generated during incubation of phenolic and anilinic pollutants with oxidoreductases can couple to each other, resulting in the formation of water-insoluble
oligomers that may be easily removed by sedimentation and/or filtration.
Klibanov et al. (1980) proposed to exploit this phenomenon for decontamination of aqueous solutions. The mechanism of polymerization was studied by
Minard et al. (1981) who isolated and identified several phenolic and quinoid
oligomers resulting from incubation of 2,4-dichlorophenol with *R. praticola* lac-

case. In a study by Dec and Bollag (1990), horseradish peroxidase, tyrosinase, and laccases from *R. praticola* and *T. versicolor* were incubated with different phenols under various reaction conditions. The laccases were capable of removing 2,4-dichlorophenol if added at concentrations of up to 1,600 mg/L (Table 1). The removal of phenols depended on the chemical structure and concentration of the substrate, pH of the reaction mixture, activity of the enzyme, incubation time, and temperature. The enzymes retained their activity throughout broad ranges of pH (3 to 10) and temperature (5 to 55°C). The removal of halogenated phenols decreased with increasing number of chlorines and increasing molecular weight of the substituent. Mass spectra and gel permeation chromatography indicated that precipitates formed during polymerization of 2,4-dichlorophenol constituted a mixture of oligomers with molecular weights of up to 800.

Decontamination of Nonreactive Pollutants

A problem associated with the potential use of the enzyme-mediated oxidative coupling for decontamination is the relative inertness of some pollutants to enzymatic action. However, substrates that react readily with the enzyme may enhance the removal of those phenols and anilines that are less reactive. Klibanov et al. (1980) demonstrated that only about 50% of *o*-aminophenol was removed in the presence of peroxidase alone, whereas the

TABLE 1. Transformation of catechol and various chlorophenols by different enzymes.[a]

Compound	Substrate remaining (%)			
	Horseradish peroxidase	*R. praticola* laccase	*T. versicolor* laccase	Tyrosinase
Catechol	55.4 ± 1.2	0	0	0
4-Chlorophenol	17.0 ± 0.4	49.0 ± 0.8	34.4 ± 1.1	0
2,4-Dichlorophenol	0	35.3 ± 1.0	9.1 ± 0.4	58.8 ± 1.2
2,4,5-Trichlorophenol	24.0 ± 0.5	89.6 ± 1.4	62.5 ± 1.1	95.5 ± 1.8
Pentachlorophenol	60.0 ± 1.3	100.0 ± 1.3	90.0 ± 1.8	100.0 ± 1.6

Source: Dec and Bollag (1990).

(a) The substrate (0.1 m$\underline{M}$) was incubated in citrate-phosphate buffer at pH 5.5 for horseradish peroxidase (0.1 purpurogallin unit/mL, 0.1 m$\underline{M}$ H_2O_2), at pH 6.8 for *Rhizoctonia praticola* laccase (1 DMP unit/mL), at pH 4.0 for *Trametes versicolor* laccase (1 DMP unit/mL) and at pH 7.0 for tyrosinase (20 L-tyrosine units/mL). The reaction mixtures were incubated at 27°C for 15 h.

removal efficiency increased to about 90% with the addition of 1-naphthol, *p*-phenylphenol, or 2,3-dimethylphenol. The authors suggested that the improved rate of mixed polymer formation was due to an increase in the overall yield of free radicals in the presence of reactive substrates. According to Berry and Boyd (1985), the reason for the enhancement is a secondary chemical reaction between less reactive compounds and reaction products formed during the enzymatic oxidation of easily removed substrates. This concept is in agreement with experimental data. In studies by Roper et al. (1995), for instance, ferulic acid, a reactive constituent of humus, was incubated with peroxidase; the reaction products were filtered to exclude the enzyme and compounds of a molecular weight greater than 500. Incubation of the resulting enzyme-free filtrate with a solution of unreacted 2,4,5-trichlorophenol caused precipitation and a 72% removal of the halogenated pollutant. The observed phenomenon may be of particular advantage in the treatment of wastewater containing heterogeneous pollution. Mixing two streams of wastewaters—one polluted with inert chemicals, another containing reactive substrates—may be a useful strategy for enhancing contaminant removal.

EFFECTIVE USE OF ENZYMES

Enzymatic treatment as a decontamination technique may be limited by rapid deactivation of the applied enzymes under severe environmental and industrial conditions. Alberti and Klibanov (1981) proposed the use of crude enzymes to compensate for the losses in enzymatic activity. Other authors demonstrated that further improvement can be achieved through optimization of reaction conditions and application of various additives, such as gelatin or polyethylene glycol, to prevent losses of enzymatic activity caused by adsorption of enzyme molecules on polymers (Nakamoto and Machida 1992). Use of plant materials containing oxidoreductases was proposed as an additional method of enzymatic treatment (Dec and Bollag 1994b). This approach is discussed in detail in a separate chapter of this volume.

The efficiency of treatment can also be enhanced through the use of enzymes bound to solid supports (Bollag 1992). The immobilization of laccase on mineral supports increased the thermostability of the enzyme, its resistance to degradation by proteases, and its half-life. For example, laccase immobilized on kaolinite or soil removed 2,4-dichlorophenol as efficiently as did the free enzyme, but retained its activity for a much longer time period. Furthermore, the immobilized enzyme could be recovered from the reaction solution and reused to treat additional polluted water, with minimal loss of activity.

CONCLUSIONS

The purpose of this study was to investigate the possible use of enzymes for environmental cleanup. Use of various oxidoreductases appears to be effective in removing pollutants from contaminated water and soil. The removal of pollutants can be greatly enhanced by the use of immobilized enzymes and coprecipitation of nonreactive contaminants.

REFERENCES

Alberti, B. N., and A. M. Klibanov. 1981. "Enzymatic removal of dissolved aromatics from industrial aqueous effluents." *Biotechn. Bioeng. Symp. 11*: 373-379.

Berry, D. F., and S. A. Boyd. 1985. "Reaction rates of phenolic humus constituents and anilines during cross-coupling." *Soil Biol. Biochem. 17*: 631-636.

Bollag, J.-M. 1992. "Decontamination of soil with enzymes: An in situ method using phenolic and anilinic compounds." *Environ. Sci Technol. 26*: 1876-1881.

Bollag, J.-M., K. L. Shuttleworth, and D. H. Anderson. 1988. "Laccase-mediated detoxification of phenolic compounds." *Appl. Environ. Microbiol. 54*: 3086-3091.

Dec, J., and J.-M. Bollag. 1990. "Detoxification of substituted phenols by oxidoreductive enzymes through polymerization reactions." *Arch. Environ. Contam. Toxicol. 19*: 543-550.

Dec, J., and J.-M. Bollag. 1994a. "Dehalogenation of chlorinated phenols during oxidative coupling." *Environ. Sci. Technol. 28*: 484-490.

Dec, J., and J.-M. Bollag. 1994b. "Use of plant material for the decontamination of water polluted with phenols." *Biotechnol. Bioeng. 44*: 1132-1139.

Hatcher, P. G., J.M. Bortiatynski, R. D. Minard, J. Dec, and J.-M. Bollag. 1993. "Use of high-resolution ^{13}C NMR to examine the enzymatic covalent binding of ^{13}C-labeled 2,4-dichlorophenol to humic substances." *Environ. Sci. Technol. 27*: 2098-2103.

Klibanov, A. M., B. N. Alberti, E. D. Moris, and L. M. Felshin. 1980. "Enzymatic removal of toxic phenols and anilines from waste waters." *J. Appl. Biochem. 2*: 414-421.

Minard, R. D., S.-Y. Liu, and J.-M. Bollag. 1981. "Oligomers and quinones from 2,4-dichlorophenol." *J. Agric. Food Chem. 29*: 250-253.

Nakamoto, S., and N. Machida. 1992. "Phenol removal from aqueous solutions by peroxidase-catalyzed reaction using additives." *Wat. Res. 26*: 49-54.

Roper, J. C., J. M. Sarkar, J. Dec, and J.-M. Bollag. 1995. " Enhanced enzymatic removal of chlorophenols in the presence of co-substrates." *Water Res.*, in press.

Sarkar, J. M., R. L. Malcolm, and J.-M. Bollag. 1988. "Enzymatic coupling of 2,4-dichlorophenol to stream fulvic acid in the presence of oxidoreductases." *Soil Sci. Soc. Amer. J. 52*: 688-694.

Shannon, M. J. R., and R. Bartha. 1988. "Immobilization of leachable toxic soil pollutants by using oxidative enzymes." *Appl. Environ. Microbiol. 54*: 1719-1723.

Biodegradation of Naphthalene From Nonaqueous-Phase Liquids

Subhasis Ghoshal, Anuradha Ramaswami,
and Richard G. Luthy

ABSTRACT

Dissolution of polycyclic aromatic hydrocarbons (PAHs) from a non-aqueous-phase liquid (NAPL) to the aqueous phase renders these compounds bioavailable to microorganisms. Subsequent biodegradation of organic phase PAH then results in a depletion of PAH from the NAPL. This study focuses on identifying the rate-controlling processes affecting naphthalene biomineralization from a complex multicomponent NAPL, coal tar, and a simple two-component NAPL. A simplified dissolution degradation model is presented to identify quantitative criteria to assess whether mass transfer or biokinetic limitations control the overall rate of biotransformation of PAH compounds. Results show that the rate of mass transfer may control the overall rate of biotransformation in certain systems (e.g., when coal tar is present as a globule). Mass transfer does not limit biodegradation in slurry systems when coal tar is distributed in the micropores of a large number of small microporous silica particles. The end points of naphthalene degradation from the NAPLs have been evaluated, and results suggest that depletion of a significant mass of naphthalene from the NAPL phase is possible.

INTRODUCTION

Prior to the widespread use of natural gas, manufactured gas plants (MGPs) supplied gaseous fuel derived from coal, coke, and/or oil. A major by-product of manufactured gas processes was coal which today is often associated with subsurface contamination at former MGP sites (Luthy et al. 1994). Coal tars are denser-than-water nonaqueous-phase liquids (DNAPLs), and are primarily composed of polycyclic aromatic hydrocarbons (PAHs). Naphthalene, a two-ring aromatic hydrocarbon, is often the most abundant PAH compound in coal tars. When coal tar is discharged into the subsurface, significant amounts of coal tar may remain entrapped in the soil/aquifer matrix. The slow dissolution of organic solutes from such entrapped coal tar can cause long-term groundwater contamination problems (Mercer and Cohen 1990).

Although the biodegradation of two- and three-ring PAH compounds sorbed onto soils or in aqueous solution have been extensively studied (Guerin and Boyd 1992; Woodzinski and Bertolini 1972), there have been few studies on biodegradation of the relatively soluble PAH compounds from NAPLs. The work reported here focuses on quantitative evaluation of rate-limiting processes in slurry systems, which govern biodegradation of naphthalene from coal tar, and from a NAPL simpler in composition than coal tar. Independent measurements of rates of naphthalene dissolution and biodegradation have been made to identify the rate-limiting process. The biodegradation end points for naphthalene from the two NAPLs have also been determined.

CONCEPTUAL MODEL DEVELOPMENT

Physicochemical phenomena related to dissolution and mass transfer of PAH compounds from the organic phase to the bulk aqueous phase, and bio-kinetic phenomena pertaining to the intrinsic microbial degradation rates, may influence the overall rate of biotransformation of PAH compounds released from NAPLs. The dissolution-degradation model describes the effect of mass transfer on biodegradation in slurry batch systems. As a working hypothesis, it was assumed that only naphthalene in the aqueous phase is available to the microorganisms; however, the hypothesis was not verified in the experiments. The dynamic change in aqueous-phase PAH concentration, C_t [M/L^3], is described as:

$$\frac{\partial C_{(t)}}{\partial t} = k_l a \left[C_{eq\,(t)} - C_{(t)} \right] - k_{bio}\, C_{(t)} \tag{1}$$

where the first term on the right-hand side represents the rate of input of PAH to water due to dissolution from the NAPL, and the second term represents the rate of removal of bulk aqueous phase PAH due to biodegradation. The biodegradation rate is expressed by a pseudo-first-order biokinetic coefficient, k_{bio} [$1/T$], derived from the Monod or Michaelis-Menten models for conditions of low substrate concentration and stable microbial populations. The dissolution rate is represented by a lumped mass-transfer rate coefficient, $k_l a$ [$1/T$], which incorporates the specific surface area for mass transfer, a [L^2/L^3], and a linear driving-force term that represents the departure of the aqueous concentration, $C_{(t)}$, from the equilibrium concentration, $C_{eq(t)}$. The aqueous-phase concentration of a PAH compound in equilibrium with the NAPL is predicted by Raoult's law as:

$$C_{eq(t)} = x_{t,PAH} * C_{pure\,liquid}^{PAH} * \gamma_{NAPL} \tag{2}$$

where $C_{eq(t)}$ is the equilibrium aqueous-phase concentration of a NAPL-derived PAH compound; $X_{t,PAH}$ is the mole fraction of the PAH compound in the NAPL at any time, t; $C_{pure\ liquid}^{PAH}$ is the solubility of the pure subcooled liquid compound in water; and γ_{NAPL} is the activity coefficient of the PAH compound in the NAPL (γ_{NAPL} for naphthalene in the coal tar was found to be close to unity). If a significant amount of the remaining fraction of the NAPL is essentially insoluble, as in coal tar, the PAH mole fraction in the NAPL and the equilibrium aqueous-phase PAH concentration should decrease as cumulatively increasing quantities of the PAH compound are solubilized and then degraded by microbes. The equilibrium concentration at any time, $C_{eq(t)}$, is calculated from Equation 2 with the knowledge of the fraction of PAH depleted.

The overall rate of biotransformation of a PAH compound such as naphthalene is controlled by the slower of either mass-transfer or degradation processes. It can be shown that when $k_l a >> k_{bio}$ (i.e., when mass transfer occurs much faster than biodegradation), the aqueous-phase concentration of naphthalene in the reactor is close to equilibrium such that $C_{(t)} \simeq C_{eq(t)}$. Assuming zero growth of microorganisms during the initial phase of the test, the rate of mineralization of naphthalene is given as:

$$\frac{\partial C_{CO_2}}{\partial t} = k_{bio}\, C_{(t)}\, F_{min} \simeq k_{bio}\, C_{eq\,(t)}\, F_{min} \quad \text{for} \quad k_{bio} << k_l a \qquad (3)$$

where the left-hand side of Equation 3 represents the increment of PAH mass converted to CO_2 per unit time, per unit aqueous volume in the reactor and F_{min} is the mass of naphthalene mineralized per unit mass of naphthalene degraded. An average value of F_{min} equal to 0.73 has been determined from biodegradation tests with naphthalene dissolved in heptamethylnonane (Ghoshal et al. 1995). When $k_{bio} << k_l a$, the initial slope of the biomineralization curve is directly related to the biokinetic rate constant, k_{bio}. Conversely, when $k_l a << k_{bio}$ (i.e., when mass transfer occurs much slower than biodegradation), analysis of Equation 1 shows that the aqueous-phase naphthalene concentration in the reactor is small compared to the equilibrium concentration, and the slope of the biomineralization curve is given by:

$$\frac{\partial C_{CO_2}}{\partial t} = k_l a\, C_{eq\,(t)}\, F_{min} \quad \text{for} \quad k_l a << k_{bio} \qquad (4)$$

Equation 4 represents a system in which the overall biotransformation rate is controlled by mass transfer processes. For such a condition, the slope of the biomineralization curve is directly related to the mass transfer rate coefficient, $k_l a$.

EXPERIMENTAL TECHNIQUES

The coal tars used in these experiments were obtained from former MGP sites located in Stroudsburg, Pennsylvania, and Baltimore, Maryland. The coal tars were primarily a mixture of PAHs with naphthalene as the most abundant PAH compound (2.2% w/w for the Stroudsburg tar, and 10% w/w for the Baltimore tar). Radiolabeled ^{14}C naphthalene dissolved in a small volume of methanol was added to the coal tars. Addition of the radiolabeled naphthalene did not alter the mole fraction of naphthalene in the coal tar. The radiolabeled coal tar was imbibed into microporous silica beads (mean diameter $\simeq 250$ μm, average pore diameter $= 140$ Å, PQ Corporation, Pennsylvania) to model coal tar entrapped in microporous media. The coal-tar-containing microporous media was used for the mass transfer and biodegradation experiments. Mass-transfer and biodegradation experiments were also conducted with single globules of coal tar containing radiolabeled naphthalene. The globules of coal tar were produced by carefully expressing the coal tar through a syringe needle immersed in water. A two-component NAPL was prepared by dissolving naphthalene crystals in a liquid matrix of 2,2,4,4,6,8,8-heptamethylnonane (Sigma Chemical Co., Missouri), and radiolabeled naphthalene was added to the mixture. Two-component NAPLs with naphthalene mole fractions approximately equal to that for the two coal tars were used for biodegradation experiments.

Mass-transfer tests were carried out in batch and flowthrough systems. In batch tests, the coal tar, present either as globules or imbibed within the silica beads, was contacted with water and the aqueous-phase naphthalene concentration was measured at equilibrium. The dissolution of naphthalene from coal tar was studied in gently stirred flowthrough reactors. Descriptions of these experiments have been published elsewhere (Luthy et al. 1993; Ramaswami 1994).

Biomineralization experiments were carried out under aerobic conditions in 250-mL biometer flasks fitted with a side tube. The side tube and the flask were sealed with neoprene stoppers and the biometers were shaken in a gyratory shaker. In slurry tests, 5 g of silica beads was imbibed with 1 mL of coal tar spiked with radiolabeled ^{14}C naphthalene and then was added to each biometer along with 50 mL of autoclaved nutrient medium. Tests with coal-tar globules were performed with a single globule of 0.7 mL coal tar and 50 mL of nutrient medium. Biometers used for the two-component NAPL had a stoppered glass vial fused to the inside bottom of the flask. The vial contained 1.25 mL of the heptamethylnonane-naphthalene mixture, which was less dense than water and remained on the water surface. Slots at the bottom of the glass vials allowed the bulk aqueous phase in the flask to be continuous with the aqueous phase in the glass vial.

Biometers were inoculated with an actively growing bacterial culture containing approximately 10^8 naphthalene-degrading organisms. Microbial concentrations were not measured during the experiment, but it is expected that the high initial concentration of microorganisms would result in stable microbial concentrations during the initial phase. Radiolabeled CO_2 from biomineralization of naphthalene was trapped in the NaOH contained in the side arms of the biometers. The NaOH was sampled periodically, and the activity of the

$^{14}CO_2$ in the NaOH was measured with a Beckman LS 5000TD scintillation counter.

RESULTS AND DISCUSSION

Batch tests were conducted to measure the initial aqueous-phase solubility of PAH compounds from coal tar. Results from these tests matched well with the theoretical predictions from Raoult's law, indicating an initial equilibrium aqueous-phase naphthalene solubility of 3.8 mg/L for the Stroudsburg tar and 18.5 mg/L for the Baltimore tar. For the system where coal tar was imbibed into microporous silica, the lumped mass-transfer coefficients derived from measured aqueous concentrations decreased with time from about 6,000/day to 500/day. The variation in mass-transfer rates with time, observed upon "aging" coal tar in water, was believed to be a result of interfacial film formation in tar-water systems (Luthy et al. 1993). A mass-transfer rate coefficient equal to 1.8/day was obtained from flowthrough tests with a 0.7-mL coal-tar globule.

Naphthalene mineralization profiles were obtained by plotting the fraction of ^{14}C naphthalene mineralized with time. Mineralization profiles for systems with coal tar imbibed in microporous media, and with 0.7-mL coal-tar globules suspended in water are displayed in Figure 1a and 1b. The system, comprised of small microporous silica particles, exhibits a significantly greater initial biomineralization rate (apparent initial mineralization rate constant k_{app} = 34/day for the Stroudsburg tar and 14/day for the Baltimore tar) compared to that containing the coal-tar globule (1 to 2/day).

For small media, the measured mass transfer rate coefficients of 500 to 6,000/day were significantly greater than the biokinetic rate constants of 1 to 25/day reported in the literature (Guerin and Boyd 1992; Mihelcic and Luthy 1991), indicating that the biotreatment system composed of small particles in a coal-tar slurry system is limited by biokinetic factors during the initial period of the test. The apparent initial mineralization rate constant, k_{app}, is thus the first-order initial biokinetic rate constant, k_{bio}, for these systems and is within the range of values reported in the literature. The naphthalene mineralization profiles obtained from the two-component NAPL test were similar to those obtained with the coal-tar/silica-bead systems, and the measured mass-transfer rate coefficient for naphthalene from the NAPL phase was greater than the initial mineralization rate constant for naphthalene.

For biomineralization of naphthalene from a 0.7-mL coal-tar globule (Figure 1b), the initial apparent mineralization rate constant, k_{app}, of 1 to 2/day was much smaller than the k_{bio} determined with the biokinetic controlled system of coal-tar containing silica beads. This indicated that the biotransformation of naphthalene from large coal-tar blobs was mass-transfer limited. This also suggested that the value of the mass-transfer coefficient should be related to the initial slope of the experimental biomineralization data in Figure 1b through Equation 4. When Equation 4 was applied to the initial slope of the

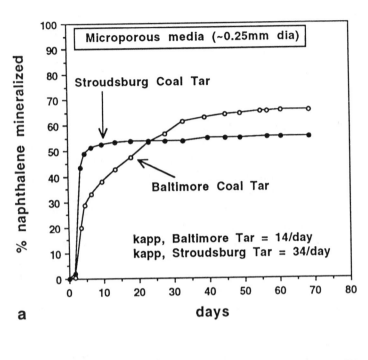

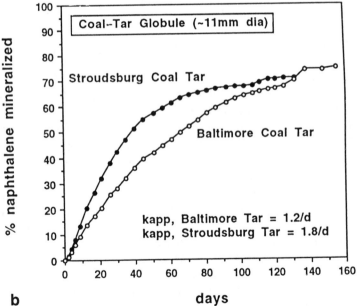

FIGURE 1. Naphthalene mineralization profiles. (a) Slurry system with coal tar imbibed in 0.25-mm-diameter microporous silica beads. Mineralization in this system is biokinetic limited. (b) System with a single 11-mm-diameter coal-tar globule. Mineralization in this system is mass-transfer limited.

mineralization curve in Figure 1b, mass-transfer coefficient ($k_l a$) values of approximately 1.8/day for the Stroudsburg tar and 1.2/day for the Baltimore tar were estimated. These estimated values of $k_l a$ match fairly well the measured value of $k_l a \simeq 1.8$/day obtained in the companion flowthrough mass-transfer test.

During the biodegradation test, on day 50, the aqueous concentrations of naphthalene in the coal-tar-globule systems were measured, and concentrations significantly less than the equilibrium concentrations were obtained, as expected. In the systems evaluated here, the initial rate of naphthalene mineralization did not exceed the rate of naphthalene mass transfer. However, other studies have reported rates of naphthalene mineralization to be higher than the rates of phenanthrene and naphthalene mass transfer in experiments with naphthalene or phenanthrene dissolved in NAPLs such as heptamethyl-nonane or diethylhexylphthalate (Efroymson and Alexander 1994; Ortega-Calvo and Alexander 1994). Those studies have attributed microorganisms at the interface to be utilizing PAH compounds directly from the NAPL.

At the conclusion of both mass-transfer and biodegradation experiments, the residual radioactivity in the NAPL phase and in the supernatant was measured to check for mass balance of the radiolabeled naphthalene. Recovery of ^{14}C from the systems ranged from 82% to 94%, with an average of 84%. Endpoints of naphthalene degradation were assessed by measuring the naphthalene concentration in the NAPL phase when mineralization activity in the biometers had ceased. In the case of the two-component NAPL, approximately 0.5% of the naphthalene initially present in the naphthalene/heptamethyl-nonane mixture remained in the NAPL phase. In the experiments with coal-tar globules, only 7% of the initial mass of naphthalene remained in the coal-tar globule. These results suggest that significant depletion of naphthalene from NAPLs may be possible in slurry-treatment systems.

In conclusion, the mineralization patterns and the end points of naphthalene degradation from the tars from the two different sites and from the synthesized two-component NAPL were comparable. Also, the simple dissolution-degradation model described the general behavior of the NAPL-water systems studied here.

ACKNOWLEDGMENTS

This study was supported by Baltimore Gas and Electric Co., Baltimore, Maryland, and by Texaco Inc. Research and Development, Beacon, New York.

REFERENCES

Efroymson, R. A., and M. Alexander. 1994. "Role of Partitioning in Biodegradation of Phenanthrene Dissolved in Nonaqueous-Phase Liquids." *Environ. Sci. Technol.* 28(9):1172-1179.

Ghoshal, S., A. Ramaswami, and R. G. Luthy. 1995. "Biodegradation of Naphthalene From Coal Tar in Mixed Batch Systems." Submitted for publication. Carnegie Mellon University, Pittsburgh, PA.

Guerin, W. F., and A. S. Boyd. 1992. "Differential Bioavailability of Soil-Sorbed Naphthalene to Two Bacterial Species." *Appl. Environ. Microbiol.* 58(4):1142-1152.

Luthy, R. G., A. Ramaswami, S. Ghoshal, and W. Merkel. 1993. "Interfacial Films in Coal Tar Nonaqueous-Phase Liquid-Water Systems." *Environ. Sci. Technol.* 27(13):2914-2918.

Luthy, R. G., D. A. Dzombak, C. A. Peters, S. B. Roy, A. Ramaswami, D. V. Nakles, and B. R. Nott. 1994. "Remediating Tar-Contaminated Soils at Manufactured Gas Plant Sites." *Environ. Sci. Technol.* 28(3):266A-276A.

Mercer, J. W., and R. M. Cohen. 1990. "A Review of Immiscible Fluids in the Subsurface: Properties, Models, Characterization and Remediation." *J. Contam. Hydrol.* 6:107-168.

Mihelcic, J. R., and R. G. Luthy. 1991. "Sorption and Microbial Degradation of Naphthalene in Soil-Water Suspensions Under Denitrification Conditions." *Environ. Sci. Technol.* 25(1): 169-177.

Ortega-Calvo, J., and M. Alexander. 1994. "Role of Bacterial Attachment and Spontaneous Partitioning in Biodegradation of Naphthalene Initially Present in Nonaqueous-Phase Liquids." *Appl. Microbiol.* 60(7):2643-2646.

Ramaswami, A. 1994. Ph.D. Thesis. Carnegie Mellon University, Pittsburgh, PA.

Woodzinski, R. S., and D. Bertolini. 1972. "Physical State in which Naphthalene and Bibenzyl Are Utilized by Bacteria." *Appl. Microbiol.* 23(6):1077-1081

Induction of PAH Degradation in a Phenanthrene-Degrading Pseudomonad

William T. Stringfellow, Shu-Hwa Chen, and Michael D. Aitken

ABSTRACT

Recent evidence suggests that different polycychic aromatic hydro-carbon (PAH) substrates are metabolized by common enzymes in PAH-degrading bacteria, implying that inducers for low-molecular-weight PAH degradation may coinduce for the metabolism of higher-molecular-weight compounds. We have tested this hypothesis with a well-characterized PAH-degrading bacterium, *Pseudomonas saccharophila* P-15. Growth of P-15 on salicylate, a metabolite of phenanthrene degradation, and a known inducer for naphthalene degradation, induced the metabolism of both substrates. Several potential inducers were then tested for their effects on metabolism of the four-ring compounds pyrene and fluoranthene, neither of which is a growth substrate for P-15, but both of which can be metabolized by this organism. Incubation of P-15 in the presence of phenanthrene or salicylate induced the metabolism of pyrene and fluoranthene in resting-cell assays. Catechol, another intermediate of naphthalene and phenanthrene degradation, did not induce the metabolism of either compound and interfered with the inducing effect of salicylate. These results have implications for strategies designed to maintain PAH degradation in contaminated environments, particularly for compounds that are degraded slowly or are degraded only by nongrowth metabolism.

INTRODUCTION

The biodegradability of PAHs has been well established, but relatively little is known about the regulation of bacterial PAH metabolism for substrates other than naphthalene. Experience with bioremediation of PAH-contaminated soils has indicated that the removal of compounds with four or more rings is often significantly lower than the removal of low-molecular-weight PAH (Wilson and Jones 1993). Compounds with four rings serve as growth substrates for only a limited number of bacteria, and no organism has been observed to grow on a PAH with five or more rings. For such nongrowth substrates, it is clear that, to sustain biodegradation activity after the growth substrates have been depleted, strategies must be developed to permit growth of PAH degraders and simultaneously to induce for PAH metabolism. One such strategy is to add a known inducing substrate as a carbon source to stimulate both selective growth and induction of the desired metabolic activity (Ogunseitan et al. 1991, Colbert et al. 1993).

Recent evidence (Denome et al. 1993, Menn et al. 1993, Sanseverino et al. 1993, Kiyohara et al. 1994, Stringfellow and Aitken 1995) indicates that multiple PAHs are metabolized by common enzymes in PAH-degrading bacteria. Extension of this concept to metabolic regulation suggests that induction of known pathways for low-molecular-weight PAH degradation may coinduce the degradation of high-molecular-weight compounds. We have tested this hypothesis by studying the coinduction of naphthalene, phenanthrene, fluoranthene, and pyrene metabolism by *Pseudomonas saccharophila* P-15 (identified tentatively as a *Pseudomonas vesicularis* in a preliminary report [Stringfellow and Aitken 1994a]). This organism was isolated from a creosote-contaminated soil by enrichment on phenanthrene as sole carbon source, and has been well characterized with respect to phenanthrene biodegradation (Stringfellow and Aitken 1994a, Stringfellow and Aitken, 1994b). P-15 metabolizes phenanthrene by a pathway that proceeds through 1-hydroxy-2-naphthoic acid, salicylate, and catechol (Stringfellow and Aitken 1994b). It is able to grow on naphthalene and phenanthrene, but not on pyrene or fluoranthene. It is, however, able to metabolize both of these four-ring substrates. Therefore, we evaluated whether fluorene and pyrene metabolism could be induced in the presence of phenanthrene or intermediates of phenanthrene metabolism.

MATERIALS AND METHODS

To determine the effects of growth on naphthalene on phenanthrene metabolism, and the effects of growth on salicylate on naphthalene or phenanthrene metabolism, P-15 was grown on either substrate as sole carbon source in tap-water buffer (TWB) (Stringfellow and Aitken 1994b). Naphthalene or salicylate was dissolved in methanol, the methanol solution (8 g/L) was added to sterile test tubes, and the methanol was evaporated under a hood. Next, 20 mL of TWB was added to each tube to give a final substrate concentration of 40 mg/L, and the tubes were inoculated with 0.05 mL of a culture suspension. Cells were harvested and washed in TWB, and oxidation of naphthalene or phenanthrene was determined by respirometry as described previously (Stringfellow and Aitken 1994b, Stringfellow and Aitken 1995).

For experiments on induction of pyrene and fluoranthene metabolism, P-15 was grown in TWB containing peptone, harvested by centrifugation, washed, and resuspended in TWB using sterile procedures as described previously

(Stringfellow and Aitken 1994b); then 5 mL of the washed culture was placed in replicate sterile culture flasks (50 mL). Each flask was then supplemented with 10 mL of TWB and 5 mL of sterile distilled water. Killed controls were prepared by adding 5 mL of formalin instead of the distilled water. Pyrene and fluoranthene were added together to each flask in 100 μL of methanol. Each potential inducer (phenanthrene, salicylate, or catechol) was added to triplicate flasks by injecting 100 μL of a methanol solution. Final concentrations of salicylate, catechol, or phenanthrene were 100 μM, while the concentrations of pyrene and fluoranthene were at their aqueous solubility limits (approximately 0.6 and 1.0 μM, respectively).

Triplicate flasks containing pyrene and fluoranthene at these concentrations also were prepared without any inducing substrate, to serve as uninduced controls. Flasks were incubated for 20 h, then 1-mL aliquots were removed and placed in 0.5 mL of formalin to terminate activity. Residual pyrene and fluoranthene were measured by high-pressure liquid chromatography (HPLC) with a Waters 600E HPLC system and Waters 470 fluorescence detector. Isocratic analysis was performed with a C_{18} column (25 cm $\times$ 4.6 mm, 5-μm particle size; Supelco, Inc., Bellefonte, Pennsylvania) eluted with 70:30 (v/v) acetonitrile in water. Pyrene and fluoranthene were detected at their optimum excitation and emission wavelengths (Risner and Conner 1991). Detection limits were 0.6 and 1.0 nM, respectively.

PAH disappearance in controls was shown to be caused by sorption to the glassware and to microbial cells as follows. Three uninoculated flasks were prepared containing 0.6 μM pyrene in 20 mL distilled water, 15 mL TWB plus 5 mL distilled water, or 15 mL TWB plus 5 mL formalin. These three flasks and an inoculated but killed control were incubated for 20 h as described above, then half of the liquid volume (10 mL) was removed and replaced with 10 mL of acetonitrile. Pyrene concentration was determined by HPLC as described above, then corrected to the concentration in the original volume.

RESULTS AND DISCUSSION

Growth of *Pseudomonas saccharophila* P-15 on peptone plus phenanthrene has been shown to induce its ability to oxidize naphthalene, 1-methylnaphthalene, 2-methylnaphthalene, and fluorene in resting-cell assays (Stringfellow and Aitken 1995). In the current study, growth of P-15 on naphthalene as sole carbon source induced the ability to oxidize phenanthrene, and growth on salicylate induced the ability to oxidize both naphthalene and phenanthrene (Table 1). Thus, it appears that naphthalene and phenanthrene metabolism is regulated by a common inducer, which is likely to be salicylate. Salicylate is a known inducer of naphthalene degradation in *Pseudomonas putida* (Menn et al. 1993). These results support previous work from which it was concluded that naphthalene and phenanthrene are metabolized by common enzymes in P-15 (Stringfellow and Aitken 1995).

TABLE 1.   Oxidation of naphthalene and phenanthrene after growth on inducer substrates.

Growth Substrate	SOUR[a] on Test Substrate	
	Naphthalene	Phenanthrene
Naphthalene	ND[b]	1.5
Salicylate	3.0	4.3

(a) Specific oxygen uptake rate, mg oxygen min^{-1} (g cells)$^{-1}$. SOUR on phenanthrene for uninduced culture is 0.1 (Stringfellow and Aitken 1994b).
(b) ND = not determined.

Because it appears that P-15 degrades several low-molecular-weight PAHs by a common pathway (Stringfellow and Aitken 1995), we were interested in determining whether induction of low-molecular-weight PAH degradation would coinduce for the metabolism of the four-ring compounds pyrene and fluoranthene. P-15 was grown in peptone medium, then washed-cell suspensions were used to measure pyrene and fluoranthene disappearance in resting-cell assays; the resting cell assay mixture contained both of the four-ring PAHs, plus a potential inducer. Mixtures with no inducer served as uninduced controls, and inoculated medium containing formaldehyde served as a killed control.

A small amount of pyrene was removed from solution in uninoculated controls over a 20-h incubation period in the resting-cell assay, and more extensive removal was observed in the killed controls, as shown in Table 2. The addition of acetonitrile to these controls led to essentially complete recovery of the pyrene initially added to the mixtures (Table 2), presumably by desorbing pyrene that may have sorbed to the glass or to the biomass. Reversible sorption to biomass appears to be more significant than sorption to glass. Similar sorptive processes occurred in experiments with fluoranthene (not shown).

When pyrene and fluoranthene were incubated together with resting cells, no removal of pyrene (relative to killed controls) occurred in the absence of an inducer substrate over a 20-h incubation period (Table 3). Phenanthrene and salicylate induced pyrene metabolism, whereas catechol did not induce pyrene degradation and actually repressed the inducing effect of salicylate (Table 3). Similar results were obtained for fluoranthene, except approximately 60% removal of fluoranthene occurred in the absence of inducing substrates. The presence of a low constitutive level of PAH-degrading enzymes could be responsible for fluoranthene degradation in the absence of inducers. Because PAHs in mixtures have been shown to be competitive substrates for P-15

TABLE 2. Recovery of pyrene in controls in resting-cell experiment.

Sample	Added Pyrene Remaining in Solution (%)[a]
Formaldehyde-killed control	35 ± 1.0
Formaldehyde-killed + acetonitrile	95
Uninoculated controls	87 ± 2.1
Uninoculated + acetonitrile	103 ± 5.7

(a) Data represent mean ± standard deviation of triplicates, except for formaldehyde-killed control amended with acetonitrile (which was a single measurement).

TABLE 3. Effect of potential inducers on removal of pyrene and fluoranthene by resting cells.

Potential Inducer	Compound Remaining as % of Killed Control[a]	
	Pyrene	Fluoranthene
None	103 ± 1.5	41 ± 2.7
Phenanthrene	28 ± 1.3	0.0 ± 0.0
Salicylate	55 ± 2.9	5.0 ± 0.8
Catechol	122 ± 5.5	77 ± 4.6
Catechol + Salicylate	101 ± 0.9	97 ± 1.0
Formaldehyde-killed control	100 ± 7.0	100 ± 6.5

(a) Data represent mean ± standard deviation of triplicates.

(Stringfellow and Aitken 1995), it is possible that fluoranthene competitively inhibits pyrene metabolism in the presence of such constitutive activity. The observed differences in the extent of pyrene and fluoranthene removal in uninduced (as well as induced) conditions might also be a result of faster rates of fluoranthene degradation relative to pyrene degradation.

The apparent repressive effects of catechol on pyrene and fluoranthene metabolism is an interesting finding. Catechol is not toxic to P-15, and in fact is oxidized rapidly by resting cells of P-15 grown on phenanthrene (Stringfellow and Aitken 1994b). It is possible that catechol exerts a catabolite-repression phenomenon with respect to PAH metabolism by P-15, but further experiments are required to elucidate the role of catechol as repressor.

CONCLUSIONS

Results from this study clearly indicate that degradation of high-molecular-weight PAH (pyrene and fluoranthene) is coinduced when the pathway for low-molecular-weight PAH degradation is induced by either phenanthrene or salicylate. This is an important observation, because P-15 cannot grow on either pyrene or fluoranthene. If such an observation can be extended to other PAH-degrading bacteria, there are direct implications of this work for strategies to sustain high-molecular-weight PAH biodegradation in the field. Salicylate, for example, is a water-soluble substrate that could select for growth of PAH degraders and also induce PAH metabolism. However, not all bacteria oxidize phenanthrene through salicylate (Cerniglia and Heitkamp 1989), and inducers for the alternative pathways are unknown. The addition of salicylate to soil has been shown to stimulate PAH-degrading bacteria (Ogunseitan et al. 1991), but extensive studies on contaminated soil have not been conducted. Limited bioavailability of poorly soluble compounds clearly is a confounding factor that might still limit the ultimate degradation of high-molecular-weight PAH in contaminated soils, but it is just as clear that such degradation cannot be sustained in the absence of growth substrates.

ACKNOWLEDGMENTS

This work was supported by the National Institute of Environmental Health Sciences (grant #P42ES05948) under its Superfund Basic Research program.

REFERENCES

Cerniglia, C.E., and M.A. Heitkamp 1989. "Microbial Degradation of Polycyclic Aromatic Hydrocarbons (PAH) in the Aquatic Environment." In U. Varanasi (Ed.), *Metabolism of Polycyclic Aromatic Hydrocarbons in the Aquatic Environment*, pp. 41-68. CRC Press, Boca Raton, FL.

Colbert, S.F., T. Isakeit, M. Ferri, A.R. Weinhold, M. Hendson, and M.N. Schroth. 1993. "Use of an Exotic Carbon Source to Selectively Increase Metabolic Activity and Growth of *Pseudomonas putida* in Soil." *Appl. Environ. Microbiol. 59*: 2056-2063.

Denome, S.A., D.C. Stanley, E.S. Olson, and K.D. Young. 1993. "Metabolism of Dibenzothiophene and Naphthalene in *Pseudomonas* Strains: Complete DNA Sequence of an Upper Naphthalene Catabolic Pathway." *J. Bacteriol. 175*: 6890-6901.

Kiyohara, H., S. Torigoe, N. Kaida, T. Asaki, T. Iida, H. Hayashi, and N. Takizawa. 1994. "Cloning and Characterization of a Chromosomal Gene Cluster, *pah*, That Encodes the Upper Pathway for Phenanthrene and Naphthalene Utilization by *Pseudomonas putida* OUS82." *J. Bacteriol. 176*: 2439-2443.

Menn, F.-M., B.M. Applegate, and G.S. Sayler. 1993. "NAH Plasmid-Mediated Catabolism of Anthracene and Phenanthrene to Naphthoic Acids." *Appl. Environ. Microbiol. 59*: 1938-1942.

Ogunseitan, O.A., I.L. Delgado, Y.-L. Tsai, and B.H. Olson. 1991. "Effect of 2-Hydroxybenzoate on the Maintenance of Naphthalene-Degrading Pseudomonads in Seeded and Unseeded Soil." *Appl. Environ. Microbiol. 57:* 2873-2879.

Risner, C.H. and J.M. Conner. 1991. "The Quantification of 4- to 6-Ring Polynuclear Aromatic Hydrocarbons in Indoor Air Samples by High-Performance Liquid Chromatography." *J. Liquid Chromatog. 14:* 437-463.

Sanseverino, J., B.M. Applegate, J.M.H. King, and G.S. Sayler. 1993. "Plasmid-Mediated Mineralization of Naphthalene, Phenanthrene, and Anthracene." *Appl. Environ. Microbiol. 59:* 1931-1937.

Stringfellow, W.T., and M.D. Aitken. 1994a. "Kinetics of Phenanthrene Degradation by Soil Isolates." In R.E. Hinchee, D.B. Anderson, F.B. Metting, Jr. and G.D. Sayles (Eds.), *Applied Biotechnology for Site Remediation,* pp. 310-314. Lewis Publishers, Boca Raton, FL.

Stringfellow, W.T., and M.D. Aitken. 1994b. "Comparative Physiology of Phenanthrene Degradation by Two Dissimilar Pseudomonads Isolated From a Creosote-Contaminated Soil." *Can. J. Microbiol. 40:* 432-438.

Stringfellow, W.T., and M.D. Aitken. 1995. "Competitive Metabolism of Naphthalene, Methylnaphthalenes, and Fluorene by Phenanthrene Degrading Pseudomonads." *Appl. Environ. Microbiol. 61:* 357-362

Wilson, S.C., and K.C. Jones. 1993. "Bioremediation of Soil Contaminated With Polynuclear Aromatic Hydrocarbons (PAHs): A Review." *Environ. Pollut. 81:* 229-249.

Bioremediation Potential of Crude Oil Spilled on Soil

Sara J. McMillen, Adolpho G. Requejo,
Gary N. Young, Pam S. Davis,
Peter D. Cook, Jill M. Kerr, and Nancy R. Gray

ABSTRACT

Spills sometimes occur during routine operations associated with exploration and production (E&P) of crude oil. These spills at E&P sites typically are small, less than 1 acre (0.4 ha), and the spill sites may be in remote locations. As a result, bioremediation often represents a cost-effective alternative to other cleanup technologies. The goal of this study was to determine the potential for biodegrading a range of crude oil types and determining the effect of process variables such as soil texture and soil salinity. Crude oils evaluated ranged in American Petroleum Institute (API) gravity from 14° to 45°. The extent of biodegradation was calculated from oxygen uptake data and the total extractable material (TEM) concentration. Based on the data collected, a simple model was developed for predicting the bioremediation potential of a range of crude oil types. Biodegradation rates were significantly lower in sandy soils. Soil salinities greater than approximately 40 mmhos/cm adversely impacted soil microbial activity and biodegradation rate.

INTRODUCTION

Crude oils vary widely in composition depending on factors such as source bed type and generation temperatures (Hunt 1979). Biodegradation rates for crude oils will vary due to differences in composition, as reflected by hydrocarbon class distribution (saturates, aromatics, and polars), and the amount of *n*-alkanes versus branched and cyclic alkanes within the saturated hydrocarbon class (Cook et al. 1974). Other factors that influence hydrocarbon biodegradation rates include soil texture (Song et al. 1974), moisture content, nutrient addition, temperature (Dibble and Bartha 1979), and salinity (Haines et al. 1994). Various studies have addressed the bioremediation of crude oils and crude oil wastes. Most of these focused on the repeated applications of crude oil wastes to soil in landfarming operations (Loehr et al. 1992). Bioremediation using composting and land treatment for E&P oily soils and wastes has also been employed successfully by some groups (Fyock et al. 1991, Zimmerman and Robert 1991, and McMillen and Gray 1994). However, data do not exist that will allow one to predict whether bioremediation will be successful for the range of crude oil types and environmental conditions that may exist at E&P sites.

EXPERIMENTAL PROCEDURES

Electrolytic respirometry was used to compare the biodegradability of crude oils and to determine the effect of increased soil salinity. Respirometry flasks contained 50 g of soil. A negative control contained 2 wt% mercuric chloride, a positive control contained 0.5% glucose, and a background soil contained nutrients but no added carbon source. All determinations were conducted in triplicate. For comparative biodegradability studies, identical experimental conditions were used so that any differences in oxygen uptake were strictly a function of crude oil type. The conditions were as follows: (1) loam soil, (2) nutrients to yield a 100:5:1 ratio of carbon to nitrogen and phosphorus (C:N:P), (3) moisture content adjusted to 70% of the water-holding capacity (WHC), and (4) crude oil mixed into the soil to give 0.5 wt% on a dry weight basis. The crude oils were "topped" at 40°C to remove volatiles before being mixed into the soil. (Topping removes hydrocarbons below C12, in order to minimize the effect of volatilization losses on the results.) For salinity studies, crude oil #3 was used, and all variables were held constant except the electrical conductivity (EC). Sodium chloride was added to a loam soil to vary the EC of the soil from 0.3 to 41.7 mmhos/cm.

Microcosms were used to validate the biodegradation model and evaluate the effects of soil texture. For the soil texture experiments, three soil types were used: sand, clay, and loam. Each soil texture microcosm contained a homogeneous mixture of 2 kg of soil and 4 wt% crude oil #3. For the model validation experiments, five different crude oils were added to a loam soil to give 3.4 wt% oil. Nutrients were added to yield a C:N:P of 100:2:1. Moisture was maintained at 60 to 80% of the WHC, and microcosms were incubated at room temperature. Duplicate microcosms were constructed for each set of test conditions, and composite samples were taken initially and over time. TEM measurements were made in quadruplicate, and the results presented are the mean of the quadruplicate extractions and duplicate microcosms.

The TEM concentrations were obtained by tecator-solvent extraction with methylene chloride:methanol (9:1). The extract was dried at 40°C with a nitrogen stream to a constant weight. This method yields the C12 to C15+ boiling point fraction. Class distributions were characterized by asphaltene precipitation followed by alumina-silica gel chromatography [for saturates, aromatics, and nitrogen-sulfur-oxygen compounds (NSOs)] (McMillen and Gray 1994).

GC was used to obtain saturate and aromatic hydrocarbon "fingerprints" and to quantify both resolved and unresolved complex mixtures of these fractions (Farrington et al. 1979). The fractions were also analyzed by gas chromatography/mass spectroscopy (GC/MS) to determine concentrations of specific polycyclic aromatic hydrocarbons (PAHs) (Gray et al. 1993).

API gravity is the API measurement of specific gravity of a crude oil at 16°C (the reported value is converted to an integer). The percent sulfur in crude oils was determined by high-temperature combustion using a LECO SC-132 instrument. Soil texture, pH, EC, and WHC were determined by American Society for Agronomy Standard Methods (Page et al. 1982).

RESULTS

A range of crude oil types were selected for the study based upon their origin and compositional characteristics. Some of the characteristics of each oil are summarized in Table 1. A total of 77 parameters for each oil were investigated, including the concentration of 39 PAHs, 9 isoprenoids, and 18 *n*-alkanes. Other characteristics examined were API gravity, class distribution, percent sulfur, and quantitative amounts of resolved and unresolved saturates and aromatics detectable by GC.

TEM biodegradation losses for 4 of the crude oils were compared with those losses measured by electrolytic respirometry. The respirometry data, as

TABLE 1. Major compositional characteristics of 17 crude oils.

Crude Oil	API gravity°	% Sulfur	% Saturates	% n-Alkanes	% Aromatics	% NSOs	% Asphaltenes	Sat/Aro Ratio
1	45	0.05	76	46	19	1	4	4.00
2	34	0.12	74	24	22	3	1	3.36
3	38	0.08	74	47	24	1	1	3.08
4	38	2.12	58	43	38	4	0	1.53
5	40	0.49	68	27	29	1	2	2.34
6	34	0.11	70	43	27	2	1	2.59
7	34	1.67	39	18	31	12	18	1.26
8	30	1.50	51	17	38	5	6	1.34
9	23	0.90	54	22	29	9	8	1.86
10	38	0.21	68	35	26	2	4	2.62
11	30	0.47	70	16	25	4	1	2.80
12	22	3.03	35	13	34	12	19	1.03
13	29	0.18	67	5	26	4	3	2.56
14	29	0.60	53	14	39	6	2	1.36
15	14	3.88	19	19	49	15	17	0.39
16	19	0.10	47	4	39	11	3	1.21
17	15	2.34	29	9	46	14	11	0.63

cumulative milligrams of oxygen consumed, were divided by the stoichiomet-
ric amounts of oxygen required to degrade hydrocarbons in order to calculate
approximate percent biodegradation losses. Figure 1a compares the calculated
losses based on respirometry with the TEM results. The data indicate that
respirometry results are an adequate approximation of TEM losses.

Respirometry was then used to compare the biodegradability of the 17 selected crude oils, and the results are shown in Figure 1b. The calculated losses
ranged from 10 to 63% in 4 weeks. These percent losses would be observed
only under the specific conditions of this study, and the rate of loss may vary
under different conditions such as higher initial oil concentration.

A multivariate correlation analysis (all statistical analyses were conducted
with STAT-GRAPHICS®) of the biodegradation losses and compositional characteristics was conducted to provide a preliminary view of relationships
between composition and biodegradability. The correlation coefficients must
range between +1 and –1, with positive correlations indicating that the variables increase in concert. The results shown in Figure 2a reveal that the
amount of *n*-alkanes, total saturates, total resolved and unresolved saturates
and aromatics detected by GC/FID, and the ratio of saturates to aromatics correlate positively with biodegradation losses. Negative correlations were
obtained for the total aromatics, NSOs, asphaltenes, and sulfur content.
Correlation analysis also revealed that API gravity correlates with the compositional characteristics in the same manner as biodegradation losses. This
follows from the fact that API gravity is inversely proportional to specific gravity. The components of crude oil with the lowest specific gravity are saturated
and aromatic hydrocarbons, whereas those with the highest specific gravity are
NSOs and asphaltenes, hence the observed correlations. Therefore, API gravity can be used as a surrogate for the other measured properties in a regression
model to predict biodegradability of other oils.

Regression analysis revealed that API gravity can be used to predict
biodegradability, where API gravity^2 * 0.03358 = percent biodegraded. The R-
square adjusted value for this model is 0.965. The R-square value is the amount
of variance accounted for by the model, 96.5% in this case. Figure 2b illustrates
the losses plotted versus API gravity. Further regression analysis indicated that
additional oil parameters could be included in the equation to produce a
slightly more accurate predictive model. However, that model has an only
slightly improved R-square adjusted value of 0.980.

The model was validated with the microcosm studies using five different
API gravity crude oil (15°, 22°, 29°, 34°, and 38°). The results shown in Figure 3
illustrate the predicted range of loss calculated for each oil and the observed
loss of total extractables after 4 months. The actual biodegradation losses
observed either fell within the predicted range (3 of 5 oils) or was no more the
±15% of the mean predicted value.

Biodegradation losses in a sandy soil were lower than in loam or clay
soils, as illustrated in Figure 4. Also, the lag time was longer in the sandy soil,
as evidenced by the 5% loss observed after 4 weeks. Sandy soils have both

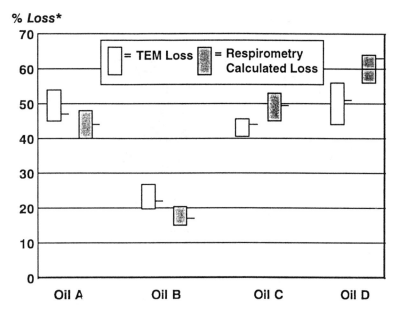

FIGURE 1a. Comparison of TEM losses with calculated respirometry losses.

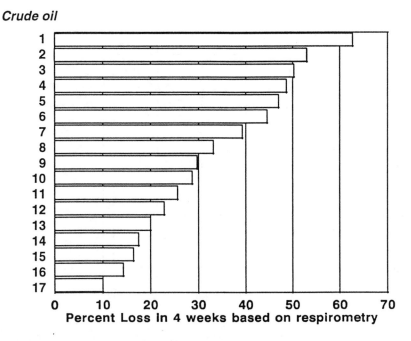

FIGURE 1b. Biodegradability of crude oils.

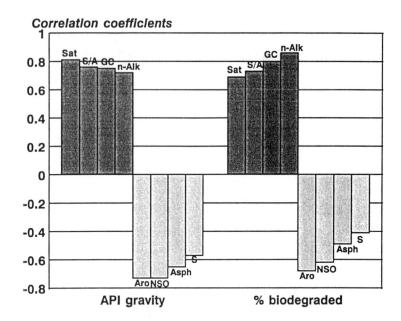

FIGURE 2a. Correlation coefficients for saturates (Sat), the saturate to aromatic ratio (S/A), the total resolved and unresolved hydrocarbons detected by GC (GC), the *n*-alkanes (*n*-Alk), aromatics (Aro), asphaltenes (Asph), and the % sulfur (S).

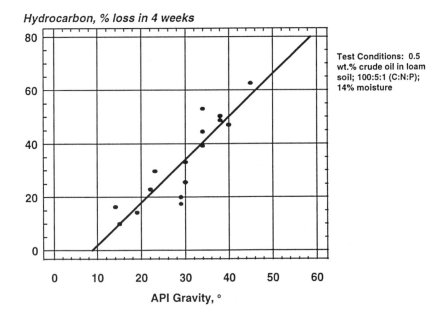

FIGURE 2b. Hydrocarbon losses observed for different API gravity crude oils.

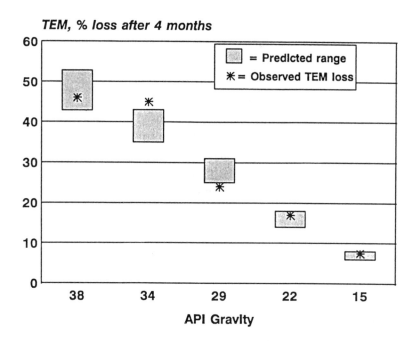

FIGURE 3. Validation of API gravity predictive model.

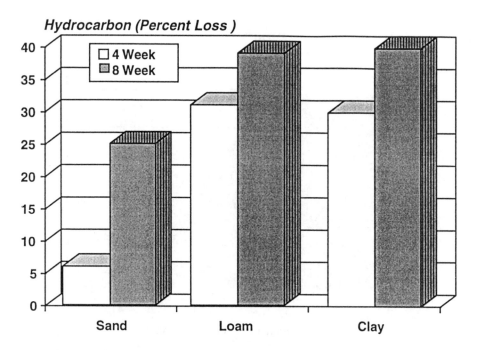

FIGURE 4. Impact of soil texture on biodegradation of crude oil.

lower organic content and lower WHC than loam or clay soils, and this may account for lower initial microbial populations. (The total heterotrophic population in the sand was more than an order of magnitude lower throughout the study.) Lower microbial activity may result in slower biodegradation in sandy soils.

The Offshore Operators Committee (OOC) found that produced waters from oil reservoirs are usually saline, with salinities varying from 15,000 to 203,000 mg/L total dissolved solids (TDS) (OOC 1990). In soils, salinities are usually reported as an EC measurement of a saturated paste of the soil rather than as a TDS value. The EC of normal, native soils will be <4 mmhos/cm (Foth 1990). In this study, the EC of the soil was adjusted from 0.3 to 41.7 mmhos/cm. The results were compared to the cumulative oxygen uptake observed for the native soil and plotted as a percent of the observed native soil value (Figure 5). Oxygen uptake measurements in soil samples having EC values of 12.1 to 31.5 mmhos/cm were roughly 60 to 80% of the value obtained for native soil. At an EC of 41.7 mmhos/cm, soil microbial activity essentially ceased.

DISCUSSION

API gravity appears to be a reliable, simple predictive measure of the relative biodegradability of crude oils. Because API gravity typically is known or

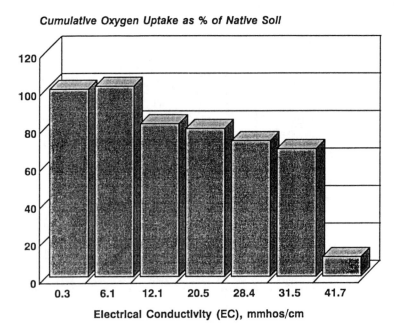

FIGURE 5. Impact of increased soil salinity on microbial activity.

can be ascertained readily, more expensive analytical measurements are not required when determining whether a small crude oil spill will be amenable to bioremediation processes. Crude oils with >30 API gravity can be considered readily biodegradable, while those with <20 API gravity will be slow to biodegrade. Biodegradation rates of crude oils will also vary with soil type and salinity. Soils with high EC values (~12 to 30 mmhos/cm) will have retarded biodegradation losses, with microbial activity being reduced 60 to 80%. If the soil salinity becomes extremely high, microbial activity will cease (in this study at 41.7 mmhos/cm). Sandy soils may exhibit slower biodegradation due to lower microbial activity in such soils.

REFERENCES

Cook, F. D., A. Jobson, R. Phillippe, and D.W.S. Westlake. 1974. "Biodegradability and Crude Oil Composition." *Canadian Journal of Microbiology* 20: 915-928.
Dibble, J., and R. Bartha. 1979. "Effect of Environmental Parameters on the Biodegradation of Oil Sludge." *Applied and Environmental Microbiology* 37: 729-739.
Farrington, J. W., N. M. Frew, P. M. Gschwand, and B. W. Tripp. 1979. "Hydrocarbons in Cores of Northwestern Atlantic Coastal and Continental Margin Sediments." *Estuarine and Coastal Marine Science* 5:793-808.
Foth, H. D. 1990. *Fundamentals of Soil Science*, John Wiley & Sons, New York, NY.
Fyock, O. L., S. B. Nordrum, S. Fogel, and M. Findlay. 1991. "Pilot Scale Composting of Petroleum Production Sludges." In *Proceedings of Treatment and Disposal of Petroleum Sludge*. University of Tulsa, Center for Environmental Research and Technology, Tulsa, OK.
Gray, N. R., S. J. McMillen, A. G. Requejo, J. M. Kerr, G. Denoux, and T. J. McDonald. 1993. "Rapid Characterization of PAHs in Oils and Soils by Total Scanning Fluorescence," SPE 25990, In *Proceedings of the Society for Petroleum Engineers/Environmental Protection Agency Exploration and Production Environmental Conference*, Society of Petroleum Engineers, Richardson, TX.
Haines, J. R., M. Kadkhodayan, D. J. Mocsny, C. A. Jone, M. Islam, and A. D. Venosa. 1994. "Effect of Salinity, Oil Type, and Incubation Temperature on Oil Degradation." In R. E. Hinchee, D. B. Anderson, F. B. Metting, Jr., and G. D. Sayles (Eds.), *Applied Biotechnology for Site Remediation*, pp. 75-83. Lewis Publishers, Boca Raton, FL.
Hunt, J. M. 1979. *Petroleum Geochemistry and Geology*, W. M. Freeman, San Francisco, CA.
Loehr, R. C., J. H. Martin, and E. F. Neuhauser 1992. "Land Treatment of an Aged Oily Sludge-Organic Loss and Change in Soil Characteristics." *Water Research* 26(6): 805-815.
McMillen, S. J., and N. R. Gray. 1994. "Biotreatment of Exploration and Production Wastes." In *Proceedings of the 2nd International Conference on Health, Safety & Environment in Oil & Gas Exploration in Jakarta Indonesia*, Society of Petroleum Engineers, Richardson, TX.
Offshore Operators Committee (OOC). 1990. *Offshore Platform Study.* New Orleans, LA.
Page, A. L., R. H. Miller, and D. R. Keeney (Eds.). 1982. *Methods of Soil Analysis, Parts 1 & 2*, 2nd ed. American Society of Agronomy, Madison WI.
Song, H., X. Wang, and R. Bartha. 1990. "Bioremediation Potential of Terrestrial Fuel Spills." *Applied and Environmental Microbiology* 56: 652-656.
Zimmerman, P. K., and J. D. Robert. 1991. "Oil-Based Drill Cuttings Treated by Landfarming." *Oil & Gas Journal* 89(32): 81-84.

Oil Degradation in Microcosms and Mesocosms

Masami Ishihara, Keiji Sugiura, Masayoshi Asaumi,
Masafumi Goto, Etsuro Sasaki, and Shigeaki Harayama

ABSTRACT

To develop bioremediation technologies for the treatment of spilled oil, a microbial consortium called SM8 was isolated from a sediment in Shizugawa Bay, Japan. This population degraded 50 to 60% of the saturate fraction and 30 to 40% of the aromatic fraction of crude oil in 30 days in batch culture. The *Alcaligenes* sp. strain, 4L, isolated from the SM8 population, degraded 50 to 60% of the saturate fraction and 10 to 20% of the aromatic fraction. Dependency of the growth of SM8 on nitrogen and phosphorus concentrations in batch cultures was estimated by using Monod expressions. These bacteria also degraded crude oil attached to gravel at a similar extent. When the biodegradation of gravel-associated crude oil by gravel-attached bacteria was examined in a flow column system, only 10% of crude oil was degraded. Beach simulator units that simulate the biodegradation in an intertidal shore were constructed. When SM8 and fertilizers were added to oil-contaminated gravel in the simulating unit, gas chromatography (GC)-resolvable crude oil disappeared within 80 days. In contrast, no significant biodegradation of crude oil was observed when SM8 and fertilizers were not added.

INTRODUCTION

After the oil spill at Prince William Sound of the Gulf of Alaska, residual oil on the beaches was removed by microbial activity. To accelerate the catabolic activities of indigenous bacteria, nitrogen- and phosphorus-containing fertilizers were successfully applied. Considering the apparent success of bioremediation in Alaska (Prince 1993), we were interested in developing bioremediation technologies for cleanup of spilled oil, and we initiated biological and bioengineering studies of crude oil degradation by microorganisms. In this paper, we present data on the biodegradation of crude oil in liquid media, microcosms, and mesocosms.

EXPERIMENTAL PROCEDURES AND MATERIALS

Bacteria

An oil-degrading population called SM8 has been isolated from a sediment in Shizugawa Bay, Japan (Venkateswaran 1991). This population has been sub-cultured every 2 weeks on natural seawater-based medium containing crude oil (see below for the definition of the medium and crude oil) at a concentration of 4,000 mg/L. Component bacteria of SM8 were isolated on Marine Agar (Difco, Detroit, USA), and one of the isolates called strain 4L that exhibited the highest oil degradation was used in this study. Conventional taxonomic studies and the determination of the 16S rRNA sequence of this strain tentatively classified it as *Alcaligenes*.

Crude Oil

First, 38 L of Arabian light crude oil was distilled up to 230°C. The unevaporated part (22 L), denoted as crude oil hereafter, was used in this study.

Media

Basal Salt Medium. A basal salt medium (BSM) consists of NH_4NO_3 (1 g/L), NaCl (30 g/L), $MgSO_4 \cdot 7H_2O$ (0.5 g/L), KCl (0.3 g/L), K_2HPO_4 (1.5 g/L), ferric citrate (0.02 g/L), and $CaCl_2$ (0.2 g/L).

Enriched Natural Seawater. Seawater was collected from Kamaishi Bay at a depth of 15 m. To prepare the natural seawater-based medium (enriched NSW), fresh seawater was filtered and added to NH_4NO_3 (1 g/L), K_2HPO_4 (0.2 g/L), and ferric citrate (20 mg/L). The pH of the medium was adjusted to 7.6.

Artificial Seawater. An artificial seawater (ASW) was based on Tropic Marine (Dr. Biener Gmbh, Aquarientechnik, Wartenberg, Germany). The half-strength ASW is called 0.5 × ASW, whereas ASW supplemented with crude oil (5,000 mg/L), 100 mM N-2-hydroxyethylpiperazine-N'-2-ethanesulfonic acid (pH 7.8), ferric citrate (20 mg/L), and varying concentrations of $NaNO_3$ and KH_2PO_4 is called enriched ASW.

Cultivation of Bacteria on Crude Oil

Cultivation in Respirometers. Oxygen uptake of SM8 grown in enriched ASW containing various concentrations of nitrate and phosphate was measured by a respirometer (model OM3001 BOD analyzer, Ohkura, Tokyo, Japan). In the experiments, 50 mL of enriched ASW was inoculated with 1 mL of fully grown SM8 culture, and the oxygen uptake by the SM8 was monitored for 5 to 10 days by the respirometer at 20°C, with continuous agitation by means of a magnetic stirrer.

Batch Cultures. For batch cultures, 500-mL "Sakaguchi" flasks containing 100 mL enriched NSW supplemented with crude oil (5,000 mg/L) were inoculated with 3 mL of a fully grown culture of either SM8 or *Alcaligenes* 4L and then incubated at 20°C for 30 days with shaking at 100 strokes/min.

Rolled Cultures. In a sterile 200-mL rubber-stopped bottle, 25 g of gravel of a diameter between 2.36 and 4.75 mm, 100 mg of crude oil, and 3 mL of a SM8 or *Alcaligenes* sp. 4L culture grown on enriched NSW containing 4,000 mg/L crude oil were added; and the bottle was rolled at 20°C on a roller (Low Profile Roller, Stoval Life Science, USA) operated at 40 rpm. Gas inside the bottle was sampled through a gastight stopper.

Flow Column System. This system consists of a temperature-controlled jar, columns in a temperature-controlled chamber, and peristaltic pumps as shown in Figure 1. Gravel (130 g) of a diameter between 2.36 and 4.75 mm was mixed

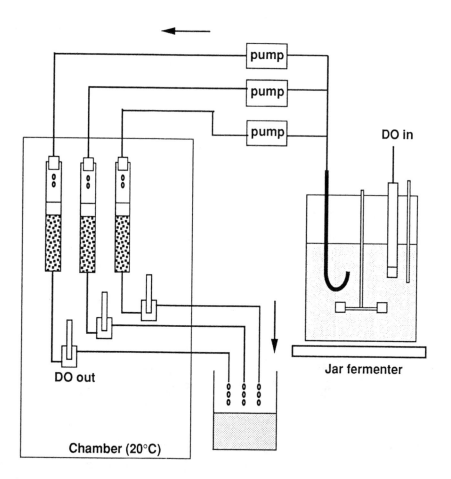

FIGURE 1. Flow column system. DO = dissolved oxygen.

with crude oil and packed into the columns (length, 30 cm; diameter, 3 cm; packed volume 85 mL). Approximately 1 g of crude oil adhered to the gravel; 5 mL of an SM8 or *Alcaligenes* sp. 4L culture grown on enriched NSW supplemented with crude oil (4,000 mg/L) was introduced at the top of each column. BSM at 20°C was subsequently introduced via the inlet tubing through a peristaltic pump into each column at a flowrate of 200 mL/h. The water level of each column was adjusted at 4 cm above the gravel layer. The dissolved oxygen concentrations of the BSM medium at the inlet and outlet were continuously monitored by Clark-type oxygen electrodes.

Beach Simulator System. The system that simulates the intertidal shore consists of a double-walled plastic tank, a reservoir, and a level-controlling device (Figure 2). Seawater in the reservoir was aerated by bubbling and temperature-controlled at 20°C. The internal volume of each tank was 1.5 m³ in which approximately 1 m³ of gravel (2 to 8 mm in diameter) was filled. Seawater filtered through layers of coarse and fine sand was continuously added to the reservoir at a flowrate of 60 L/hr. The level of seawater was adjusted by the level-controlling device. The device moved between 20 and 80 cm high two cycles per day, thus realizing a high and low tide cycle of seawater in these tanks. Excess seawater overflowed from the level-controlling device. At the beginning of the experiments, 1.5 kg of crude oil was poured in each tank. After 4 days, 3 L of the SM8 cultures (approximately 10^8 cells/mL) and two types of fertilizers, namely 300 g of a slow-release solid granular nitrogen fertilizer (Super IB, Mitsubishi Kasei, Tokyo, Japan) and 60 g of a solid granular phosphorus fertilizer (Linstar 30, Mitsubishi Kasei), were added in one tank but not in the other (control) tank.

Preparation of Lyophilized Bacteria

SM8 and *Alcaligenes* sp. 4L were cultivated in Marine Broth (Difco, Detroit, USA) for 5 days. Cultured bacteria were washed twice with centrifugation and

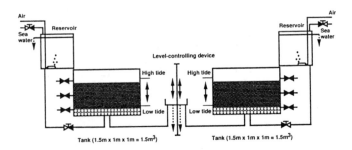

FIGURE 2. Beach simulator units. The main tank of the beach simulator unit is a double-walled plastic tank. The center part is the level-controlling device from which excess seawater overflows.

freeze-dried in vacuo. Powdered bacteria (1 g) was dissolved in 30 mL of 0.5 ×
ASW and used as an inoculum.

Analysis of Respirometry

After examination of more than 30 combinations of nitrogen and phosphorus concentrations, with maximum nitrogen and phosphorus concentrations of 250 mg N/L as nitrate and 45 mg P/L as phosphate, respectively, the observed peak values of oxygen uptake rate (OUR) were used to compute theoretical OUR under any nitrogen and phosphorus concentrations. To do this, a mathematical expression that predicts the OUR based on nitrogen and phosphorus concentrations was identified by applying a nonlinear parameter estimation technique.

Analysis of Crude Oil

Crude oil was extracted from cultural fluids or from gravel with chloroform, and chloroform was evaporated at 50°C. The crude oil in dried samples was analyzed by the following methods.

Gravimetric Measurements. The amount of crude oil was determined by gravimetric measurements in some experiments.

Thin-Layer Chromatography/Flame-Ionization Detection. In most experiments, the amount of crude oil was determined by using the Iatroscan MK-5 system (Iatron Lab., Tokyo, Japan) which applies thin-layer chromatography in combination with flame-ionization detection (TLC/FID, Goto et al. 1994). Briefly, crude oil in cultural fluid was extracted, and the extracts were heated to 60°C to evaporate chloroform. The dried samples were dissolved in chloroform containing stearyl alcohol (1,000 mg/L) being used as a reference for quantification of crude oil. The volume of chloroform was determined according to the initial amount of crude oil added. Chloroform (1 L) was used when 4,000 mg of crude oil was initially added. Chloroform solution (1 μL) was applied to one end of a silica gel rod (Chromarods SIII, Iatron Lab.) and developed in two steps, first with *n*-hexane over 10 cm to displace the saturate fraction of crude oil, and subsequently with *n*-hexane:toluene (20:80 in volume) over 5 cm to separate the aromatic fraction from the resin and asphaltene fractions of crude oil.

The amounts of hydrocarbons thus separated were determined by using FID. The crude oil used in these experiments contained 61% of saturate hydrocarbons, 33% of aromatic hydrocarbons, 4% of resins, and 2% of asphaltenes when analyzed by silica gel column chromatography (Murakami et al. 1985). With the present Iatroscan analysis, only the saturate and aromatic hydrocarbons amounting to 94% of the crude oil were analyzed.

Gas Chromatography/Flame-Ionization Detection. For the analysis of crude oil by gas chromatography (GC), a fused silica capillary column (30 m long, 0.35 mm in diameter) fitted to a gas chromatograph (GC-14A, Shimadzu, Kyoto, Japan) was used. The operational temperature of the FID was 300°C and that of the injector 280°C. The column temperature was increased at 8°C/min from 40 to 105°C and 2.5°C/min from 105 to 280°C. The carrier gas was helium (1 mL/min).

Gas Chromatography/Mass Spectroscopy. GC/MS analysis was performed using QP-2000A (Shimadzu) fitted with a fused silica capillary column (30 m long, 0.35 mm in diameter). Conditions for the GC analysis were as described above. The GC/MS instrument was operated in the selective ion monitoring (SIM) mode. The target analyses were n-alkanes (C_{12} to C_{32}), naphthalene derivatives (C_1 to C_4), phenanthrene derivatives (C_0 to C_2), and dibenzothiophene derivatives (C_0 to C_2).

Determination of the O_2 and CO_2 Concentrations

The CO_2 concentration was determined by GC using a gas chromatograph (GC-14A, Shimadzu) equipped with a molecular sieve 5A SUS column (3 m long, 3 mm in diameter, GL Science, Tokyo, Japan), a Porapak Q SUS column (2 m long, 3 mm in diameter, GL Science), and a thermal conductivity detector (TCD). The column oven temperature was kept at 70°C, and the injection and the detection temperatures were set at 150°C. Helium was used as the carrier gas with a flowrate of 70 mL/min.

Recovery of Viable Bacteria from Grains of Gravel

Five pieces of gravel were placed in a small tube and agitated with 2 mL of 0.5 × ASW for 30 min. The suspension was used for plate counts.

Enumeration of Bacteria

Viable counts were determined on Marine Agar plates by counting colonies on the plates after incubation for 7 days at 20°C.

Scanning Electron Microscopy

Bacterial cells adhered to the surface of grains of gravel were observed by scanning electron microscopy (SEM) after fixing them with 2.5% (w/v) glutaraldehyde. The cells were dehydrated with a 50 to 100% ethanol series, critical point dried, and stained with platinum/palladium. Observations were carried out with a Hitachi S-2500.

Determination of Nitrogen and Phosphorus Concentrations

The amounts of total nitrogen (NH_3-N, NO_2-N, NO_3-N, and organic N), total phosphorus, and phosphate in seawater were determined according to the method of Japanese Industrial Standard (JIS K 0102) and the manual for oceanographic observation (Japan Meteorological Agency 1988).

RESULTS

Effect of Nitrate and Phosphate Concentrations on the Rate of Crude-Oil Biodegradation

It is generally believed that the biodegradation of hydrocarbons in the natural environment is limited by poor growth rate of microorganisms caused by nutrient deficiencies, especially of nitrogen, phosphorus, and iron (Leahy & Colwell 1990). Thus, when bioremediation is attempted, these nutrients usually are applied to the contaminated environment to stimulate biodegradation (Prince 1993). In the present experiment, varying concentrations of nitrate and phosphate were supplemented to a low nutrient medium ASW to make enriched ASW, and the biodegradation of crude oil by the SM8 population was examined.

The data (Figure 3) showed that the addition of both phosphate and nitrate was effective in increasing the rate of biodegradation of the crude oil. This

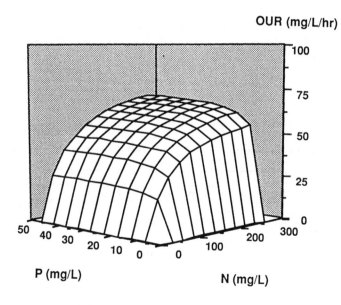

FIGURE 3. Effect of nitrate and phosphate concentration on the rate of crude oil biodegradation. Peak oxygen uptake rates (OURs) were experimentally obtained from more than 30 combinations of nitrogen (as nitrate) and phosphorus (as phosphate) concentration; from these data, theoretical OURs at different nitrogen and phosphorus concentrations were calculated.

study, under the conditions where phosphorus or nitrogen was the sole rate-limiting factor for microbial growth, showed 2 mg P/L of phosphorus was enough to achieve approximately two-thirds of the theoretically achievable maximum growth rate, whereas 78 mg N/L of nitrogen was required to obtain the same effect.

Degradation of Crude Oil by the SM8 Population and *Alcaligenes* sp. 4L in Batch Cultures

When grown in enriched NSW at 20°C for 30 days, the SM8 population degraded approximately 50% of the saturate fraction and more than 30% of the aromatic fraction of crude oil. *Alcaligenes* strain 4L, on the other hand, degraded approximately 60% of the saturate fraction but less than 20% of the aromatic fraction of crude oil (Table 1).

Biodegradation of Crude Oil Associated with Gravel

As shown above, the SM8 population and *Alcaligenes* sp. 4L effectively degraded crude oil in batch cultures. These results, however, do not necessarily indicate that these microbial systems are effective in a contaminated environment, because natural ecosystems are much more complex than batch cultures. For the first approximation of the bioremediation of an oil-contaminated beach, the biodegradation of crude oil by SM8 or *Alcaligenes* 4L was investigated in rolled cultures. As shown in Table 1, SM8 degraded approximately 50% of the saturate fraction and 30% of aromatic fraction after incubation at 20°C for 30 days. The biodegradation was most active in the first week, as judged from the oxygen consumption rate and the CO_2 evolution rate

TABLE 1. Oil degradation in microcosms. [a]

Cultivation Method	Bacteria						Analytical method
	SM8			4L			
	Degradation (%) of crude oil fraction						
	SA	AR	Total	SA	AR	Total	
Sakaguchi flask	51	39		63	18		TLC/FID
Rolled bottle	49	30		46	19		TLC/FID
Rolled bottle (lyophilized cells)	44	35		43	20		TLC/FID
Flow column			12			9.4	Gravimetry

(a) Numbers shown are the average of more than four analytical data except for those of gravimetric measurement, which are the average of two determinations. Fresh cells of the SM8 population or strain 4L were used to inoculate these cultures except one series of rolled bottle cultures where lyophilized cells were used. SA = saturates, AR = aromatics.

(Figure 4). Crude oil was degraded by *Alcaligenes* sp. 4L at a slightly lower efficiency.

Furthermore, the biodegradation of crude oil by lyophilized cells of both SM8 and *Alcaligenes* sp. 4L was examined. The lyophilized cells (0.1 g) were sprayed on oil-contaminated gravel (25 g), and incubated at 20°C. After 30 days of incubation, more than 40% of saturates and 30% of aromatics were degraded by lyophilized SM8, and approximately 40% of saturates and 20% of aromatics were degraded by lyophilized *Alcaligenes* sp. 4L (Table 1).

Biodegradation of Crude Oil in Flow Column System

In an intertidal shore contaminated by crude oil, it is expected that numbers of oil-degrading bacteria in seawater are not high and that the bacteria adhering to the surface of grains of gravel may be the main contributor to the degradation of crude oil. To evaluate the adhesion of SM8 and *Alcaligenes* sp.

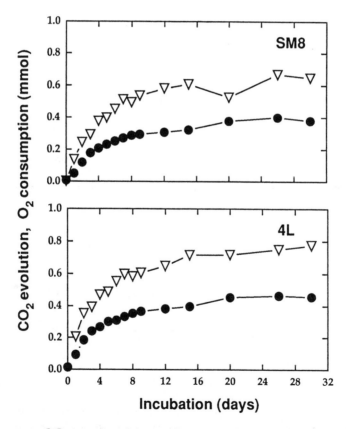

FIGURE 4. O_2 consumption of and CO_2 evolution from rolled cultures.

4L to grains of gravel and the biodegradation of crude oil by the adhered bacteria, flow column setups were made as shown in Figure 1. In this system, both crude oil and the bacteria were fixed, at the beginning of experiments, on the surface of grains of gravel, and BSM was continuously supplied from the top of the columns to wash the gravel. The oxygen concentrations in influent and effluent were then measured by using oxygen electrodes to evaluate the degree of biodegradation in the columns.

The oxygen consumption rate increased rapidly to a maximum within 2 to 3 days after the onset of the experiments, then decreased gradually (Figure 5). These tendencies are common to both SM8 and *Alcaligenes* sp. 4L. After 21 days, approximately 10% of the crude oil in this column was merely degraded (Table 1), and GC-resolvable alkanes were not completely degraded (Figure 6).

The development of bacteria on the surface of grains of gravel was observed by SEM. It was observed that microbial cells were clustered on the gravel surface, but these clusters were separated from each other without covering the whole surface of the gravel.

Evaluation of Crude Oil Degradation in Beach Simulator System

In a pair of mesoscale tanks that simulate an intertidal shore, 1.5 kg of crude oil was poured into each tank, and the SM8 population and two types of

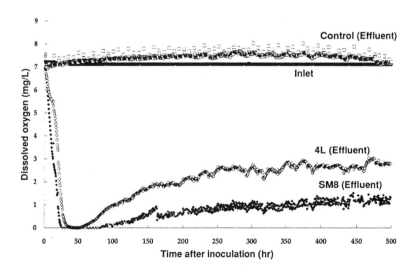

FIGURE 5. Dissolved oxygen concentrations at inlet and outlets of flow columns. The dissolved oxygen concentrations in an inlet and effluents were measured as shown in Figure 1. The difference in the dissolved oxygen concentrations between the inlet and outlets shows the respiration rate of bacteria packed in the columns.

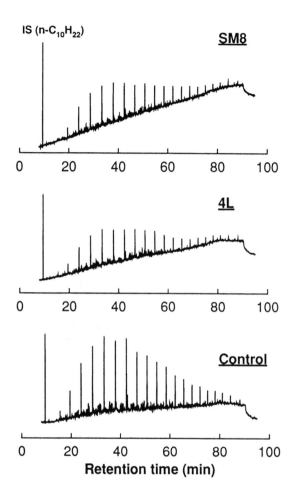

FIGURE 6. Biodegradation of GC-resolvable alkanes in the flow column system. IS is *n*-tetradecane added as an internal standard. Samples were taken after 21 days of incubation.

fertilizers were added in one tank but not added in the other (control) tank. When the dissolved oxygen concentration at 5 cm below the surface of high-tide seawater was measured in the amended tank, it was at 30% of the saturated level in the first 2 days after inoculation; it went up to 70% of the saturated level in 10 days. The dissolved oxygen concentration subsequently decreased again to 20% of the saturated level for the next 70 days. The dissolved oxygen concentration of seawater in the control tank remained unchanged (Figure 7). The numbers of total heterotrophic bacteria were high immediately after the inoculation, then drifted between 10^5 and 10^6 cfu/mL. In the control tank, the viable counts of indigenous bacteria were around 2×10^4 cfu/mL (Figure 7).

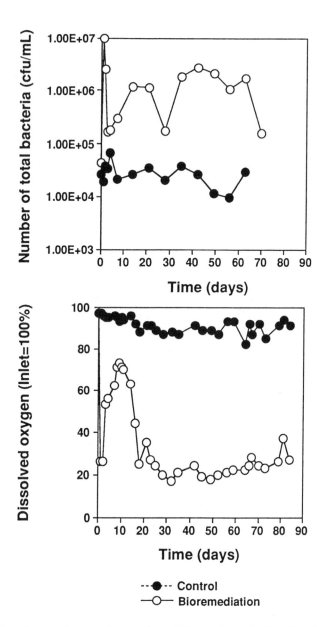

FIGURE 7. Change in numbers of total bacteria and dissolved oxygen concentrations in seawater in beach simulator units.

In the amended tank, GC-resolvable crude oil associated with gravel disappeared within 80 days, but no significant biodegradation was observed in
the control tank (Figure 8). When the biodegradation of some components of
crude oil was analyzed by means of GC/MS, it was revealed that dibenzothiophene derivatives, phenanthrene derivatives, and naphthalene derivatives all

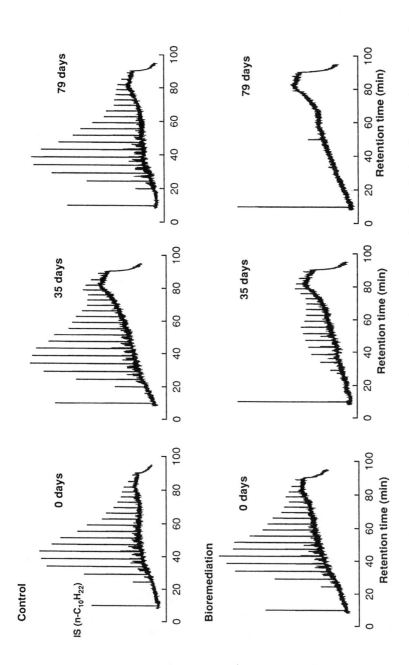

FIGURE 8. Biodegradation of GC-resolvable alkanes on the top layer of gravel in beach simulator units: 400 g of gravel on the top layer was sampled, residual oil attached to the gravel was extracted by chloroform, and the extracts were analyzed by GC.

disappeared within 80 days in seawater of the amended tank (data not shown). This observation indicated that the addition of SM8 and/or fertilizers was effective in removing contaminated crude oil.

The nitrogen concentration in seawater in the amended tank increased immediately after the addition of the fertilizers and gradually decreased to the control level at day 85 (data not shown).

In the nonamended tank, the oil slick remained at the surface for 80 days but disappeared within 15 days from the water surface in the amended tank.

DISCUSSION

In the first part of this work, the concentrations of nitrate and phosphate required for the stimulation of biodegradation were determined. This information is important for the practical application of fertilizers because it is preferable to use their minimum amounts to prevent the secondary pollution that causes the eutrophication of the environment.

The SM8 population used in this study is one of many populations isolated from marine environment. We selected this population because it provided reproducible results in crude oil degradation, whereas the degradative activity of many other populations was unstable. One isolate from the SM8 population, *Alcaligenes* sp. 4L, exhibited biodegradative activity as high as the SM8 population toward the saturate fraction of crude oil. The degradation of the aromatic fraction of crude oil by 4L was not so high and varied between 10 and 20% of the total aromatic fraction. This strain did not express any enzyme activity for degradation of naphthalene and phenanthrene (unpublished data). Therefore, the apparent biodegradation of the aromatic fraction by *Alcaligenes* sp. 4L may be due to degradation of alkyl side chains associated with the aromatic ring.

Crude oil mixed with gravel was degraded by SM8 and *Alcaligenes* sp. 4L. Thus, the adsorption of crude oil on the surface of gravel does not strongly influence the biodegradability of crude oil as long as a minimal amount of enriched NSW was added.

The flow column system made for the study of crude oil biodegradation allowed continuous monitoring of the biodegradative activity by measuring consumption of molecular oxygen. It became clear from experiments using this instrument that the biodegradation of crude oil was highest between the first and third days after inoculation, but the activity remained for 2 weeks. We believed that this system mimics an intertidal shore. The biodegradation of crude oil in the flow column was, however, very low compared to that in rolled cultures. In the rolled cultures, bacteria could grow both in seawater and on gravel, and degraded crude oil either dispersed in seawater or attached to the gravel surface. In the flow column system, bacteria detached from the solid surface were immediately washed out. It seems that bacteria attached to a gravel surface could degrade only the surrounding crude oil, and crude oil in a noncolonized area is not degraded.

Although the microcosms described above provided useful information concerning the biodegradation of crude oil, it was not possible to predict the biodegradation of crude oil in a natural environment. As field experiments are impossible, we undertook a basic study of oil degradation in the beach simulator units in which both physicochemical modification and biodegradation could take place. With the size of this simulating unit, we believe that extrapolation of the results obtained from these simulators into the prediction of the consequence of biotreatment in the natural environment is possible. As a first step toward this goal, the effect of seeding plus nutrients on the crude oil degradation in seawater was studied under a controlled temperature. We found that an oil slick on the surface of seawater disappeared 15 days after the addition of SM8 plus fertilizers, while it remained for more than 80 days in the control tank.

This dramatic visual effect of bioremediation, however, did not correlate with the chemical analysis of residual crude oil. The concentrations of C_{12} to C_{32} alkanes, easily biodegradable compounds, were not significantly reduced in the first 10 days after the start of the bioremediation. This contradictory situation resembles an observation in the bioremediation of spilled oil from *Exxon Valdez*: although fertilizer-treated beaches became visually clean 15 days after the application of fertilizers, it was difficult to prove the effectiveness of the treatment by any chemical methods (Prince 1993). Therefore, it is important to understand the mechanism by which oil slicks dissipated in our simulator. It is, for example, possible that crude oil in the amended tank was emulsified by biosurfactants produced by bacteria and adsorbed to the gravel layer.

In any case, the effectiveness of bioremediation of crude oil was demonstrated in the beach simulator units, both by visual inspection and by chemical and biological methods. Further analysis of the bioremediation process using the simulators would allow the accumulation of useful information concerning in situ and on-site bioremediation.

ACKNOWLEDGMENTS

This work was performed as a part of the Industrial Science and Technology Frontier Program supported by the New Energy and Industrial Technology Development Organization of Japan.

REFERENCES

Goto, M., M. Kato, M. Asaumi, K. Shirai, and K. Venkateswaran. 1994. "TLC/FID method for evaluation of the crude-oil-degrading capability of marine microorganisms." *Journal of Marine Biotechnology* 2: 45-50.

Japan Meteorological Agency. 1988. *Manual for Oceanographic Observation* (in Japanese). Japan Meteorological Agency, Tokyo, Japan.

Leahy, J. G., and R. R. Colwell. 1990. "Microbial Degradation of Hydrocarbons in the Environment." *Microbiological Reviews* 54(3): 305-315.

Murakami, A., K. Suzuki,. A. Yamane, and T. Kusama. 1985. "Degradation of crude oils by *Pseudomonas* sp. in enriched seawater medium." *Journal of Oceanographic Society Japan* 41:337-344.

Prince, R. 1993. "Petroleum spill bioremediation in marine environments." *Critical Reviews in Microbiology* 19(4): 217-242.

Venkateswaran, K., T. Iwabuchi, Y. Matsui, H. Toki, E. Hamada, and H. Tanaka. 1991. "Distribution and biodegradation potential of oil-degrading bacteria in North Eastern Japanese coastal waters." *FEMS Microbiological Ecology* 86: 113-122.

Effectiveness of Bioremediation in Reducing Toxicity in Oiled Intertidal Sediments

Kenneth Lee, Robert Siron, and Gilles H. Tremblay

ABSTRACT

A 123-day field study was conducted with in situ enclosures to compare the effectiveness of bioremediation strategies based on inorganic (ammonium nitrate and triple-superphosphate) and organic (fishbone meal) fertilizer additions to accelerate the biodegradation rates and reduce the toxicity of Venture™ condensate stranded within sand-beach sediments. Comparison of the two fertilizer formulations with identical nitrogen and phosphorus concentrations showed that the organic fertilizer stimulated bacterial productivity within the oiled sediments to the greatest extent. However, detailed chemical analysis indicated that inorganic fertilizer additions were the most effective in enhancing condensate biodegradation rates. The Microtox® Solid-Phase Test (SPT) bioassay was determined to be sensitive to Venture Condensate in laboratory tests. Subsequent application of this procedure to oiled sediment in the field showed a reduction in sediment toxicity over time. However, the Microtox® bioassay procedure did not identify significant reductions in sediment toxicity following bioremediation treatment. An observed increase in toxicity following periodic additions of the organic fertilizer was attributed to rapid biodegradation rates of the fertilizer, which resulted in the production of toxic metabolic products (e.g., ammonia).

INTRODUCTION

Microbiological studies in the laboratory, experimental field trials, and cleanup operations following actual marine oil spill incidents have demonstrated the effectiveness of a bioremediation strategy based on the enhancement of oil biodegradation rates by nutrient addition (Lee and Levy 1991; Bragg et al. 1994). However, some environmentalists have expressed their concern about the net benefit of this bioremediation strategy because of (1) the production of toxic metabolic by-products associated with an accelerated oil biodegradation process and (2) the ineffectiveness of natural metabolic processes to degrade the most toxic components within residual oils.

The objective of this study was to evaluate the effectiveness of bioremediation strategies based on the addition of inorganic and organic fertilizer applications to reduce the concentration of residual oil components and toxicity as measured by the Microtox® bioassay test procedure.

EXPERIMENTAL PROCEDURES AND MATERIALS

Field Site Design and Sampling Procedures

Field studies were conducted in the intertidal zone of a low-energy beach, located on the eastern shore of Nova Scotia, Canada (Lee and Levy On Day 0 (August 6, 1992), Venture™ condensate, a liquid of low viscosity and high volatility, containing a high proportion of low-molecular-weight hydrocarbons, was mixed with sand (7-L samples) collected from the study site, presieved to remove large particles and macrophyte debris, at a concentration of 3% (w/v). Following this experimental treatment and initial subsampling, the sediments were transferred into replicate in situ enclosures, constructed of 240-μm Nitex® mesh fabric. These enclosures were anchored in random order, 2 m apart, 3 cm below the beach surface, midway between the high-water and low-water mark.

On Day 12, inorganic (ammonium nitrate and triple-superphosphate) and organic (fishbone meal) fertilizers were added to the experimental enclosures. During the first 9 of 10 subsequent sampling periods, nutrients at the single application dosage were added directly to the sediments within the enclosures designated to receive periodic nutrient treatment (Table 1). These nutrient additions were carried out immediately after the collection of samples for microbiological, toxicological, and chemical analyses.

Microbiological and Chemical Analyses

Bacterial productivity in the beach sediments were monitored by determining the uptake rate of the radiolabeled methyl-[^{3}H]thymidine (Lee and Levy 1987). Changes in the concentration and composition of remaining Venture™ condensate in the sediments were determined by capillary gas chromatography (Lee et al. 1993). Potential nutrient release rates of nitrate, nitrite, phosphate, and ammonia from the sediments were determined by recovery of a supernatant sample for analysis, 6 h after the addition of a 5.0-g wet sediment sample to a 50-mL beaker containing 7.0 mL of standard reference seawater (Hansen and Grasshoff 1983).

Toxicity Tests by Microtox® Bioassay

Immediately after collection, sediment samples collected for toxicity analysis were frozen in 50-mL glass vials topped with Teflon™ caps for storage until

TABLE 1. Fertilizer additions to oiled sediments within the in situ sediment enclosures.

Enclosure Treatments	Fertilizer Additions (mg/L wet sediment)		
	N	P	FB
V[a]	—	—	—
V - NP[b]	316	373	—
V - NP[c]	3,160	3,730	—
V - FB[d]	—	—	1,731
V - FB[c]	—	—	17,310

(a) V = Venture Condensate 3%(w/v).
(b) N = Ammonium nitrate 33-0-0 (N:P:K); P = Triple superphosphate 0-46-0 (N:P:K).
(c) Fertilizers (single application dosage) reapplied after each sampling event.
(d) FB = Fishbone meal 6-10-1 (N:P:K).

analysis. The Microtox® SPT was used to measure sediment toxicity. Tests were performed using the Microtox® Model 500 Analyzer (Microbics Corporation) to measure the bioluminescence of a bacterial reagent following exposure to a series of suspended sediment concentrations for 20 min in a thermoregulated bath at 15°C (Microbics Corporation 1992). Both the basic SPT protocol and the Large Sample Procedure (LSP) from Microbics Corporation were evaluated. In the basic SPT procedure, 0.3 g of wet sediment was added to 3 mL of diluent and, for each sample, 12 serial (1:2) dilutions were tested for their toxicity. In the LSP protocol (Microtox update manual, 1992), 7 g of wet sediment was resuspended into 35 mL of diluent and stirred in a 50-mL glass bottle for about 30 s. An aliquot fraction (1.5 mL) of the homogeneous sediment suspension was pipetted, then 12 serial (1:2) dilutions were made. Data collection and reduction from dose response curves were made, using version 7.0 of the Microtox® software, to calculate the effective concentration (as ppm of wet sediment) that induces a 50% reduction (EC_{50}) of bioluminescence of the test marine bacterium *Photobacterium phosphoreum* relative to a control population.

RESULTS

Periodic additions of the inorganic fertilizers (NP*) stimulated bacterial productivity rates above those found in the unfertilized enclosures, whereas a single application had little effect (Figure 1a). Sustained elevated bacterial production rates were observed in the oiled enclosures following the addition of the organic fertilizer with the periodic addition of the fishbone meal, further

120

Microbial Processes for Bioremediation

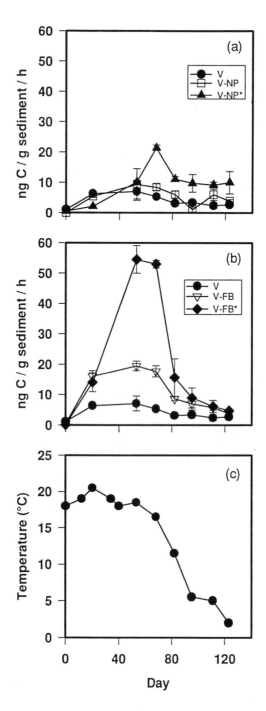

FIGURE 1. Change in total bacterial productivity in response to Venture™ condensate and/or nutrient additions (a,b). Seasonal surface-seawater temperature (c).

increasing bacterial productivity rates by a factor of 3 over a single application (Figure 1b). However, despite the stimulatory effect attributed to nutrient additions, bacterial productivity rates declined with the seasonal decrease in seawater temperatures (Figure 1c).

Changes in the composition and concentration of the n-alkane fraction of the residual oil (n-C_{12} to n-C_{35}) were used to illustrate the effect of nutrient treatment on the biodegradation of the residual condensate within the sediments (Figure 2). The results on Day 123 clearly showed that periodic additions of the fishbone meal fertilizer retarded the biodegradation rates of the n-alkane fraction (n-C_{12} to n-C_{21}). A single addition of the fishbone meal had no effect during the experimental period. In contrast, the application of the fertilizer mixture significantly enhanced the biodegradation rates of aliphatic components up to n-C_{26}. Periodic applications of the inorganic fertilizer failed to offer substantial improvement, suggesting that optimal nutrient concentrations were achieved and maintained during the study period by a single application of the inorganic fertilizer. This effectiveness of the inorganic nutrient relative to the organic nutrient as a bioremediation agent was observed also in the results of the chemical analysis of the residual aromatic fractions on Day 123 (Figure 3).

Microtox® toxicity test bioassays on sediments contaminated with known concentrations of Venture™ condensate showed a direct increase in sediment

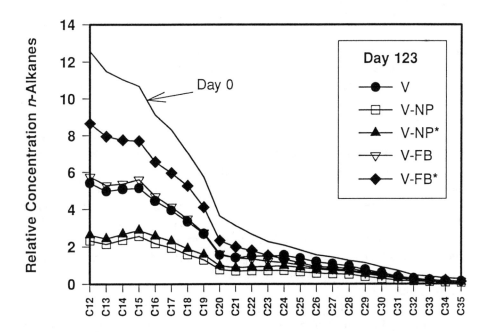

FIGURE 2. Change in the relative concentration of n-alkanes (numbered by chain length) observed on Day 123 in response to nutrient treatment (* = fertilizers in single application dosage reapplied after each sampling event).

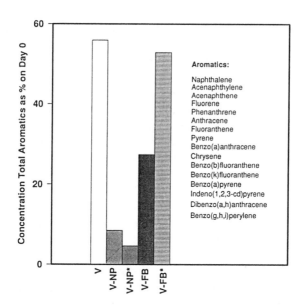

FIGURE 3. Concentration of total aromatic hydrocarbons on Day 123 (as a percent of Day 0 values) (* = fertilizers in single application dosage reapplied after each sampling event).

toxicity (i.e., an EC_{50} decrease) with concentration. No significant difference in toxicity results was obtained when using either of the standard Microtox® SPT or LSP protocols (Figure 4). Toxicity measurements on unoiled sediments confirmed that the Microtox® response was due entirely to Venture™ condensate additions.

The residual toxicity of contaminated sediments from the field experiment was measured by the standard SPT procedure. The Microtox® data (EC_{50}) showed a reduction of toxicity in the oiled sediments over time that was due presumably to natural environmental processes, including physical removal of the oil from the sediments, degradation, and evaporation (Figure 5). In terms of bioremediation by nutrient enrichment, fertilizer applications under field conditions failed to reduce sediment toxicity significantly ($P > 0.10$ on comparison of regression coefficients) under field conditions. Periodic additions of the fish meal fertilizer increased toxicity significantly ($P < 0.10$) relative to the oiled control plots.

Release of ammonia mediated by microbial degradation of the fishbone meal and rapid accumulation was detected within the enclosures treated with periodic additions of the fishbone meal (Figure 6). This chemical evidence was confirmed by direct field observations—by Day 34, sand treated with periodic additions of the organic fertilizer was beginning to turn black and emitted a faint "anoxic" odor.

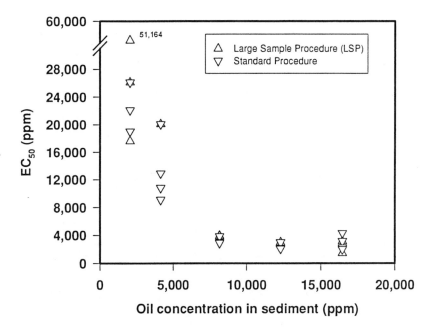

FIGURE 4. Sediment toxicity (EC_{50}—ppm wet sediment) as a function of Venture condensate concentration. Comparison of results obtained with the basic Microtox® Solid-Phase Test (SPT) and the Microtox® Large Sample Procedure (LSP).

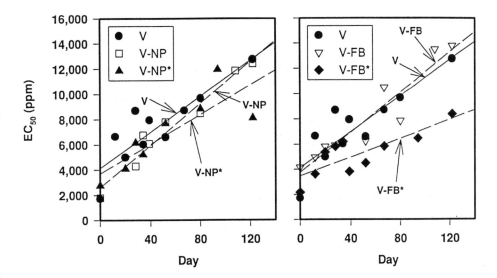

FIGURE 5. Change in sediment toxicity (EC_{50}—ppm wet sediment) in response to fertilizer applications (* = fertilizers in single application dosage reapplied after each sampling event).

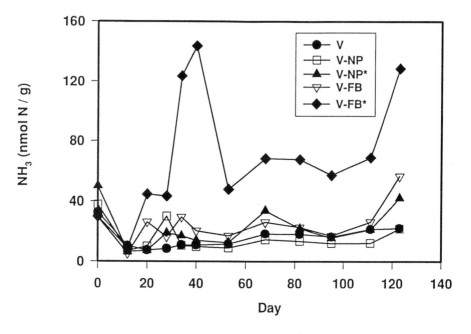

FIGURE 6. Potential nutrient (NH_3) release in response to tidal exposure (* = fertilizers in single application dosage reapplied after each sampling event).

DISCUSSION

Microbiological measurements derived from the uptake of methyl-[^{3}H]thymidine provide only an index of bacterial productivity and not a direct measure of oil degradation. Therefore, detailed chemical analysis was used to confirm the success of the bioremediation strategies. Observed alterations in the composition of the aliphatic hydrocarbons suggested that periodic additions of the fishbone meal suppressed the rates of residual condensate degradation.

These results contradict those of previous bioremediation studies on shorelines that have demonstrated the advantage of the oleophilic nutrient Inipol EAP 22, a microemulsion containing organic nitrogen (urea) and phosphorus [tri(Laureth-4)-phosphate] encapsulated within oleic acid, to induce rapid microbial growth (Sveum and Ladousse 1989; Ladousse and Tramier 1991). At similar nitrogen and phosphorus concentrations, the fishbone meal was expected to be a more effective bioremediation agent than the inorganic fertilizers as it would provide limiting macronutrients, trace elements (e.g., iron), and organic growth factors required by bacteria (Dibble and Bartha 1976; Prince 1993).

The failure of the fishbone meal fertilizer to enhance oil degradation rates, when applied at nitrogen and phosphorus concentrations similar to that of the inorganic fertilizer, may be attributed to the enrichment of specific microbial communities, diauxic responses by the indigenous microorganisms, and toxicity.

In most aquatic environments, enrichments of oil-degrading microbial communities occur soon after oil contamination of the shorelines. In this experiment, the increase of species within the microbial population adapted to the use of the organic carbon substrates in the fishbone meal may have been so dramatic (as indicated by bacterial productivity measurements and anoxia within the enclosures fertilized periodically) that other microorganisms, including the oil-degraders, did not have a chance to get established. Unfortunately, this hypothesis was not validated by taxonomic studies of the bacterial populations within the nutrient-treated enclosures.

The preferential use of components within the organic fertilizer by indigenous bacteria over the oil within the experimental sediment enclosures may also explain the observations of higher microbial productivity rates without significant effects on residual oil concentrations and composition during the experimental period. For example, previous studies with Inipol EAP 22 have shown that hexadecane biodegradation did not to occur until fatty acids in the nutrient formulation were metabolized (Rivet et al. 1993).

Bioremediation strategies based on nutrient enrichment have received a generally positive response from the public because an implied goal of bioremediation treatment is to reduce toxic effects. Nevertheless, there are still concerns that the metabolic by-products from oil degradation, or components within the fertilizer formulations, could be toxic (U.S. Congress 1991; Hoff 1993; Wrenn et al. 1994). Although adverse effects from bioremediation by nutrient enrichment have not been observed in actual field operations to date (Prince 1993), the possibility of a future incident does exist.

Oil biodegradation was also suppressed in the present study following periodic additions of fishbone meal due to oxygen limitation and the development of potential toxic conditions associated with the rapid degradation of the organic substrate. Microbial oil degradation rates under anoxic conditions are very slow (Aeckersberg et al. 1991; Lee and Levy 1991). A single application of the fishbone meal fertilizer (at 6.7% by weight of the condensate) had no effect on oil degradation rates. These results differed from those obtained in a previous study, which showed that natural organic nutrient additions (fish and animal bonemeal at 10.0% by weight of the oil) accelerated the growth of bacteria and hydrocarbon degradation rates (Basseres et al. 1993). In addition to differences in the bioavailability and toxicity of the oil and nutrient formulations, the variations in results between the studies may be attributed to site-specific differences in tidal exchange rates, sediment grain size, temperature, oxygen availability, and microflora composition.

Although the application of the inorganic fertilizer (ammonium nitrate, triple-superphosphate) did not sustain elevated bacterial production rates, based on the results of the gas chromatographic analyses, it appeared to sustain the highest oil degradation rates in this study. Because periodic reapplication of the inorganic fertilizers had little or no effect, we hypothesize that the microbial oil-degrading consortium became limited by availability of degradable components within residual condensate stranded in the sediments and low temperatures after its initial stimulation by inorganic fertilizer application.

The observed Microtox® response to oiled marine sediments in the laboratory confirmed the sensitivity of this microbioassay to toxic components within Venture™ condensate and supported its subsequent use as tool to monitor bioremediation treatment efficacy. However, application of the standard Microtox® SPT procedure failed to show the reduction of toxicity following bioremediation treatment of the condensate-treated sediments. This may be attributed to the fact that the most toxic components of the residual condensate may be refractory to natural biodegradation processes. Thus, toxicity reduction in the contaminated intertidal sediments was achieved primarily by physical/chemical processes. Venture™ Condensate contains numerous toxic components, and detailed studies by Strain (1986) have demonstrated that even its very light components (e.g., C_7 and C_8 acyclic and cyclic saturated hydrocarbons) may persist in the intertidal zone of a sandy beach for more than 6 months. Observed increases in toxicity in response to periodic additions of the organic fertilizer most likely were attributed to microbial production and accumulation of ammonia and/or sulfide which is toxic to the test organism at elevated concentrations.

Future operational guidelines for the application of bioremediation technologies will require the development of reliable ecotoxicological monitoring programs to document their efficacy for toxicity reduction.

In summary, this study and the results of our previous experiments (Lee and Levy 1989, 1991; Lee et al. 1993), have demonstrated that each contaminated environment (including the same site at different times) is likely to respond to bioremediation treatment differently. Successful bioremediation programs will require application methodologies (form and type of fertilizer, type and frequency of application) specifically tailored to the environmental parameters (including oil characteristics) at each contaminated site. Although bioremediation is the most promising of the new oil spill countermeasures under development, it should not be used without adequate knowledge of the environment in which it is to be employed and its potential toxic effects.

ACKNOWLEDGMENTS

We are grateful for financial support from the Interdepartmental Panel on Energy Research and Development (PERD) Canada and Canada's Green Plan. We would like to express our gratitude to D. Guay, R. Dumas, and J. Lavoie for their technical assistance.

REFERENCES

Aeckersberg, F., F. Bak, and F. Widdel. 1991. "Anaerobic Oxidation of Saturated Hydrocarbons to CO_2 by a New Type of Sulfate-Reducing Bacterium." *Archives Microbiology* 156: 5-14.

Basseres, A., P. Eyraud, A. Ladousse, and B. Tramier. 1993. "Enhancement of Spilled Oil Biodegradation by Nutrients of Natural Origin." *Proceedings of the 1993 Oil Spill Conference,* American Petroleum Institute, Washington, DC, pp. 495-501.

Bragg, J. R., R. C. Prince, E. J. Harner, and R. M. Atlas. 1994. "Effectiveness of Bioremediation for the *Exxon Valdez* Oil Spill." *Nature 368:* 413-418.

Dibble, J. T. and R. Bartha. 1976. "Effect of Iron on the Biodegradation of Petroleum in Sea Water." *Applied Environmental Microbiology 31:* 544-550.

Hansen, H. P. and K. Grasshoff. 1983. "Automated Chemical Analysis." In K. Grasshoff, M. Ehrhardt, and K. Kremling (Eds.), *Methods of Seawater Analysis,* pp. 347-379. Verlag Chemie, Weinheim.

Hoff, R. 1993. "Bioremediation: An Overview of its Development and Use for Oil Spill Cleanup." *Marine Pollution Bulletin 26*(9): 476-481.

Ladousse, A. and B. Tramier. 1991. "Results of 12 Years of Research in Spilled Oil Remediation Inipol EAP 22." *Proceedings of the 1991 Oil Spill Conference,* American Petroleum Institute, Washington, DC, pp. 577-581.

Lee, K. and E. M. Levy. 1987. "Enhanced Biodegradation of a Light Crude Oil in Sandy Beaches." *Proceedings of the 1987 Oil Spill Conference,* American Petroleum Institute, Washington, DC, pp. 411-416.

Lee, K. and E. M. Levy. 1989. "Enhancement of the Natural Biodegradation of Condensate and Crude Oils on Beaches of Atlantic Canada." *Proceedings of the 1989 Oil Spill Conference,* American Petroleum Institute, Washington, DC, pp. 479-486.

Lee, K. and E. M. Levy. 1991. "Bioremediation: Waxy Crude Oils Stranded on Low-Energy Shorelines." *Proceedings of the 1991 Oil Spill Conference,* American Petroleum Institute, Washington, DC, pp. 541-547.

Lee, K., G. H. Tremblay, and E. M. Levy. 1993. "Bioremediation: Application of Slow-Release Fertilizers on Low-Energy Shorelines." *Proceedings of the 1993 Oil Spill Conference,* American Petroleum Institute, Washington, DC, pp. 449-454.

Microbics Corporation. 1992. *Microtox Manual: Volume II -Detailed Protocols,* pp. 153-178. Microbics Corporation, Carlsbad, CA.

Prince, R. C. 1993. "Petroleum Spill Bioremediation in Marine Environments." *Critical Reviews in Microbiology 19*(4): 217-242.

Rivet, L., G. Mille, A. Basseres, A. Ladousse, C. Gerin, M. Acquaviva, and J.-C. Bertrand. 1993. "*n*-Alkane Biodegradation by a Marine Bacterium in the Presence of an Oleophilic Nutriment." *Biotechnology Letters 15:* 637-640.

Strain, P. M. 1986. "The Persistence and Mobility of a Light Crude Oil in a Sandy Beach." *Marine Environmental Research 19:* 49-76.

Sveum, P. and A. Ladousse. 1989. "Biodegradation of Oil in the Arctic: Enhancement by Oil-Soluble Fertilizer Application." *Proceedings of the 1989 International Oil Spill Conference,* American Petroleum Institute, Washington, DC, pp. 439-446.

U.S. Congress, Office of Technology Assessment. 1991. *Bioremediation for Marine Oil Spills—Background Paper.* U.S. Congress, Office of Technology Assess-ment Report, OTA-BP-O-70, U.S. Government Printing Office, Washington, DC.

Wrenn, B. A., J. R. Haines, A. D. Venosa, M. Kadkodayan and M. T. Suidan. 1994. "Effects of Nitrogen Source on Crude Oil Biodegradation." *Journal of Industrial Microbiology 13:* 279-286.

Petroleum Hydrocarbon Biodegradation Under Mixed Denitrifying/Microaerophilic Conditions

Dennis E. Miller and Stephen R. Hutchins

ABSTRACT

Data are presented for aqueous-flow, soil-column microcosms in which removal of benzene, toluene, ethylbenzene, and xylenes (BTEX) is observed for two operating conditions: (1) nitrate, 25 to 26 mg(N)/L, as the single electron acceptor and (2) nitrate, 27 to 28 mg(N)/L combined with low levels of oxygen, 0.8 to 1.2 mg O_2/L. Soils used in this study include aquifer material from Traverse City, Michigan; Park City, Kansas; and Eglin Air Force Base (AFB), Florida. BTEX compounds are introduced at concentrations ranging from 2.5 to 5 mg/L, with total BTEX loading from 20 to 22 mg/L. Complete removal of toluene and partial removal of ethylbenzene, m-xylene, and o-xylene were observed for all soils during trials in which nitrate was the only electron acceptor. Combining low levels of oxygen with nitrate produced varying effects on BTEX removal, nitrate utilization, and nitrite production. Benzene proved recalcitrant throughout all operating trials.

INTRODUCTION

Bioremediation of fuel-contaminated aquifers is often limited by the availability of compounds that can accept electrons (electron acceptors) from the microbial oxidation of the organic contaminants.

Oxygen is usually the electron acceptor of choice, but has limited solubility and is unacceptable for use in aquifers that are already anaerobic. It has long been recognized that nitrate and nitrite can act as electron acceptors for the degradation of organic materials (Jeris et al. 1974). Degradation of alkylbenzenes has been shown using nitrate instead of oxygen (Battermann & Werner 1984; Kuhn et al. 1985), but reports of benzene removal have conflicted (Anid et al. 1993). More information is required as to the prevalence and extent of this process. Disappearance of alkylbenzenes upon introduction of nitrate has been observed under field conditions (Hutchins et al. 1991; Gersberg et al. 1993) and in various batch studies (Hutchins 1991a; Hutchins 1991b). However, it is unknown whether mixed oxygen/nitrate systems can facilitate enhanced BTEX removals. Comparative information from column studies that better mimic flow in aquifer systems has proven inconclusive (Hutchins et al. 1992). This research will refine column techniques and contribute to the information database required to design and implement cost-effective remedial strategies. Column microcosm studies will be carried out to (1) evaluate the feasibility of using nitrate-based bioremediation at different sites, (2) examine the combined use of nitrate and oxygen to optimize bioremediation, and (3) evaluate procedures and methods to optimize design of treatability studies.

COLUMN DESIGN AND OPERATION

Glass columns, 3.8 cm ID and 30.5 cm in length, were assembled and operated within an anaerobic glovebox. The columns were configured to operate in an upflow mode, and all associated inlet and effluent lines were constructed of stainless steel tubing. A column configuration schematic is provided in Figure 1. Aquifer material studied included soils from Traverse City, Michigan

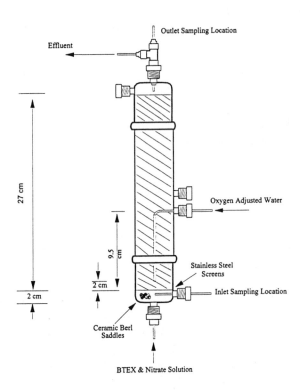

FIGURE 1. Schematic of soil-column setup.

(Hutchins et al. 1991), Park City, Kansas (Hutchins & Wilson 1994), and Eglin AFB, Florida (Hinchee et al. 1989). Selected soil was packed into each column (Hutchins et al. 1992) to a height of 25.4 cm. Aqueous flowrates were used that corresponded to a residence time within the soil matrix of approximately 24 h. Initially, BTEX compounds were introduced into the soil column with no accompanying electron acceptor (no nitrate or oxygen) to allow sorption/desorption processes to stabilize. In addition to monitoring the inlet and outlet BTEX concentrations, column streams were monitored for dissolved oxygen to ensure that anaerobic conditions were maintained. Nitrate addition was initiated after complete breakthrough was observed for the compounds. Changes resulting from the nitrate addition were monitored until a stable concentration of BTEX compounds was observed in the outlet stream. A low concentration of oxygen was then incorporated into the inlet stream, and monitoring of the BTEX concentrations was continued.

RESULTS

Denitrification activity observed within the Park City soil column was consistent with general denitrifying processes in which toluene is most readily removed, *o*-xylene is least removed, and benzene is completely recalcitrant to denitrifying removal processes. Figure 2 illustrates the toluene removal

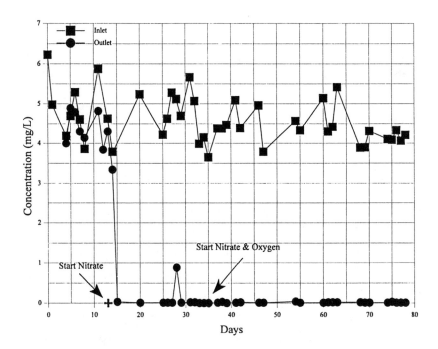

FIGURE 2. Inlet and outlet toluene concentrations. Park City, KS, soil column.

capacity of the soil column in response to introduction of 24.9 mg nitrate(N)/L to the inlet stream and, later, the addition of combined nitrate and oxygen to the inlet stream. The toluene removal response was most rapid following introduction of the nitrate source. Complete removal of toluene was achieved within 2 days of addition of nitrate to the inlet stream. Removal of *m*-xylene also was initiated immediately, but required approximately 7 days to approach complete removal. Partial removal of the ethylbenzene was observed upon introduction of the nitrate, but significant reduction of the outlet ethylbenzene concentration was delayed until depletion of the *m*-xylene component was complete. The inclusion of 1.17 mg O_2/L with the inlet-stream nitrate did not produce any statistical changes in the BTEX removal within the soil column. Table 1 summarizes the removal observed for each component during each operating period. Nitrate utilization was constant through both operating periods, with approximately the same amount of nitrite production in each case. Most of the inlet oxygen was utilized within the column, but a trace amount of oxygen was always present at the column outlet.

The Traverse City soil column also exhibited hydrocarbon removal corresponding to the introduction of 26.5 mg nitrate (N)/L to the inlet stream. The toluene and *m*-xylene components were simultaneously removed from the wastestream, and each component required a 7-day time lag before maximum removal efficiency was achieved. The soil column was able to utilize a portion of the ethylbenzene, but did not approach the efficiency displayed by the Park City soil column, even after the majority of the *m*-xylene was depleted from the wastestream. Removal of *o*-xylene was less efficient than removal of ethylbenzene, and benzene removal was not observed. Figure 3 presents removals of the toluene component throughout the monitoring periods. Combining 0.8 mg O_2/L with 28.4 mg nitrate (N)/L had a very detrimental effect on BTEX removals. As illustrated in Figure 3 for toluene, BTEX removal efficiencies decreased from 18 to 100%. At the same time, the amount of nitrate utilized was not statistically different between the two trials. Nitrite production was a little greater in the presence of oxygen, but this increase was not statistically different at a 90% confidence interval.

TABLE 1. Summary of BTEX removals for Park City, KS, soil column.

	Dentitrification			Denitrification/Microaerophilic		
	Inlet mg/L	Outlet mg/L	Δmg/L	Inlet mg/L	Outlet mg/L	Δmg/L
Nitrate (N)	24.9 ± 2.5	10.0 ± 1.7	15.9 ± 4.0	26.9 ± 0.5	11.3 ± 1.8	15.6 ± 1.9
Nitrite (N)	0.03 ± 0.01	3.85 ± 1.43	−3.52 ± 1.68	0.04 ± 0.01	3.52 ± 0.44	−3.48 ± 0.44
Oxygen	—	—	—	1.17 ± 0.30	0.07 ± 0.03	1.09 ± 0.3
Benzene	4.32 ± 0.31	4.20 ± 0.42	0	4.14 ± 0.15	4.12 ± 0.25	0
Toluene	4.70 ± 0.35	0.01 ± 0.0	4.65 ± 0.37	4.50 ± 0.27	0.02 ± 0.01	4.49 ± 0.27
Ethylbenzene	2.33 ± 0.92	0.10 ± 0.28	2.23 ± 1.20	2.94 ± 0.12	0.13 ± 0.08	2.81 ± 0.18
m-Xylene	4.52 ± 0.37	0.09 ± 0.02	4.39 ± 0.39	4.61 ± 0.20	0.10 ± 0.07	4.51 ± 0.23
o-Xylene	4.49 ± 0.37	3.36 ± 0.30	1.08 ± 0.34	4.50 ± 0.19	3.51 ± 0.24	0.99 ± 0.24
BTEX	20.9 ± 1.6	8.67 ± 0.99	12.1 ± 1.3	20.7 ± 0.8	7.87 ± 0.52	12.8 ± 0.9

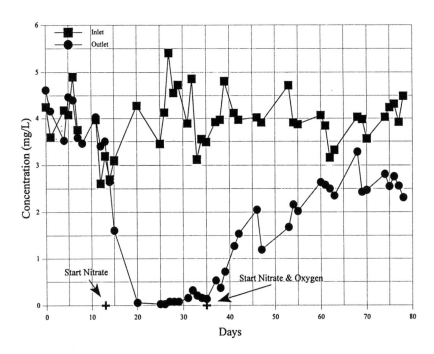

FIGURE 3. Inlet and outlet toluene concentrations. Traverse City, MI, soil column.

Table 2 provides a summary of the inlet concentrations and the utilization observed for the two trials. One possible explanation is that the addition of oxygen resulted in partial oxidation of organic material within the soil. This partially oxidized material was then solubilized within the aqueous stream and made available to microbial utilization. The microbial population preferentially made use of this carbon source at the expense of BTEX removal. The soil column is continuing to be monitored to observe whether this effect is transitory and if BTEX removal will increase over time.

Application of 26.1 mg nitrate(N)/L to the Eglin AFB soil column resulted in removal of BTEX components similar to removals previously observed in the Park City and the Traverse City soil columns. As summarized in Table 3, toluene was the most readily utilized, followed by m-xylene and ethylbenzene, with a slight removal of o-xylene and benzene. The slight removal of benzene was observed prior to addition of nitrate and was not induced by the nitrate. Utilization of ethylbenzene paralleled m-xylene utilization rather than being hindered by the utilization of m-xylene. The soil column required 7 days to reach a maximum toluene removal, as shown in Figure 4. Addition of 0.8 mg O_2/L to the inlet nitrate stream did not change the removal observed for combined BTEX. Individually, a slight statistical decrease in removal was observed for toluene along with a greater decline in the o-xylene removal. Ethylbenzene removal statistically increased and, although the mean removal of m-xylene

TABLE 2. Summary of BTEX removals for Traverse City, MI, soil column.

	Denitrification			Denitrification/Microaerophilic		
	Inlet mg/L	Outlet mg/L	Δmg/L	Inlet mg/L	Outlet mg/L	Δmg/L
Nitrate (N)	26.5 ± 2.5	11.6 ± 3.2	14.9 ± 2.6	28.4 ± 1.3	15.9 ± 1.6	12.4 ± 1.8
Nitrite (N)	0.10 ± 0.01	3.85 ± 1.10	-3.74 ± 1.09	0.10 ± 0.01	4.90 ± 0.48	-4.80 ± 0.48
Oxygen	—	—	—	0.80 ± 0.13	0.04 ± 0.01	0.81 ± 0.12
Benzene	4.71 ± 0.43	4.54 ± 0.42	0	4.55 ± 0.20	4.53 ± 0.19	0
Toluene	4.13 ± 0.39	0.12 ± 0.05	4.01 ± 0.39	4.12 ± 0.18	1.47 ± 0.40	2.65 ± 0.46
Ethylbenzene	3.37 ± 0.33	2.21 ± 0.31	1.16 ± 0.50	3.36 ± 0.17	2.87 ± 0.09	0.49 ± 0.16
m-Xylene	5.05 ± 0.47	0.25 ± 0.07	4.80 ± 0.44	5.05 ± 0.26	1.10 ± 0.36	3.94 ± 0.44
o-Xylene	4.85 ± 0.45	3.87 ± 0.48	0.98 ± 0.60	4.79 ± 0.23	4.74 ± 0.15	0
BTEX	22.1 ± 2.0	11.0 ± 1.1	11.1 ± 2.2	21.9 ± 1.0	14.7 ± 0.9	7.14 ± 1.35

TABLE 3. Summary of BTEX removals for Eglin AFB, FL, soil column.

	Denitrification			Denitrification/Microaerophilic		
	Inlet mg/L	Outlet mg/L	Δmg/L	Inlet mg/L	Outlet mg/L	Δmg/L
Nitrate (N)	27.6 ± 1.6	7.41 ± 1.38	20.2 ± 1.6	27.3 ± 0.5	10.8 ± 0.7	16.5 ± 0.6
Nitrite (N)	0.12 ± 0.02	1.71 ± 0.42	-1.60 ± 0.41	0.23 ± 0.03	0.48 ± 0.07	-0.25 ± 0.08
Oxygen	—	—	—	0.80 ± 0.10	0.02 ± 0.01	0.78 ± 0.10
Benzene	4.06 ± 0.14	3.74 ± 0.13	0.30 ± 0.14	3.30 ± 0.23	3.06 ± 0.17	0.22 ± 0.18
Toluene	4.84 ± 0.14	0.03 ± 0.0	4.80 ± 0.15	4.64 ± 0.15	0.06 ± 0.15	4.55 ± 0.16
Ethylbenzene	2.75 ± 0.08	0.07 ± 0.02	2.68 ± 0.08	2.92 ± 0.09	0.05 ± 0.02	2.86 ± 0.10
m-Xylene	4.70 ± 0.13	0.26 ± 0.07	4.42 ± 0.16	4.58 ± 0.18	0.08 ± 0.03	4.47 ± 0.19
o-Xylene	4.79 ± 0.11	3.13 ± 0.16	1.64 ± 0.19	4.67 ± 0.16	3.49 ± 0.16	1.18 ± 0.15
BTEX	21.2 ± 0.6	7.23 ± 0.34	13.8 ± 0.6	20.1 ± 0.7	6.74 ± 0.19	13.3 ± 0.7

remained the same, the outlet concentration of *m*-xylene was lower during operation with trace amounts of oxygen. Nitrate utilization and nitrite production were both reduced during microaerophilic operation.

Traces of oxygen were present in the outlet stream, although often no oxygen could be detected. Component comparisons are summarized in Table 3 for the two operating conditions. Inclusion of oxygen into the inlet stream produced small changes in the removal of individual compounds but did not affect the overall BTEX removal. One possible explanation for the unchanged BTEX removal is that the oxygen might have been supplementing the denitrification processes by initiating mineralization of the BTEX rather than by being used for complete mineralization. Based on stoichiometry of complete mineralization under both electron acceptor conditions, the amount of oxygen added was insufficient to compensate for the decreased denitrification activity.

CONCLUSIONS

For the soils examined so far, toluene was the most readily removed compound when nitrate was the only electron acceptor added to the inlet streams.

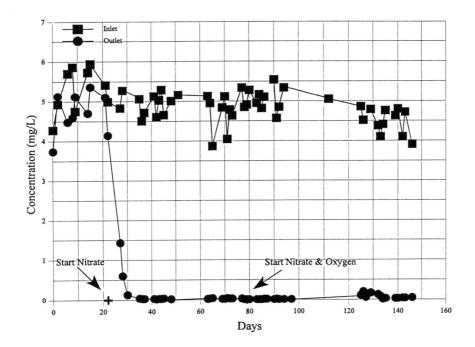

FIGURE 4. Inlet and outlet toluene concentrations. Eglin AFB, FL, soil column.

Ethylbenzene, m-xylene, and o-xylene were removed with varying success by the different soils and, in the case of the Traverse City soil, m-xylene was preferentially removed over ethylbenzene. Addition of low levels of oxygen (< 2.0 mg O_2/L) produced varying effects in the different soils. The BTEX removal, nitrate utilization, and nitrite production were unchanged in the Park City soil. The combination of oxygen and nitrate negatively affected the BTEX removal in the Traverse City soil column but did not influence nitrate utilization or nitrite production. The addition of oxygen into the nitrite stream did not change the BTEX removal but decreased the amount of nitrate utilized and nitrite produced in the Eglin AFB soil column. Benzene removal was not observed in any soil tested.

DISCLAIMER

The research described in this paper has been funded wholly or in part by the U.S. Environmental Protection Agency (U.S. EPA) and the U.S. Air Force (MIPR N92-65, AL/EQ-OL, Environmental Quality Directorate, Armstrong Laboratory, Tyndall Air Force Base). However, it has not been subjected to review by either agency and therefore does not necessarily reflect the views of those agencies; no official endorsement should be inferred.

REFERENCES

Anid, P.J., P.J.J. Alvarez, and T.M. Vogel. 1993. "Biodegradation of Monoaromatic Hydro-carbons in Aquifer Columns Amended With Hydrogen Peroxide and Nitrate." *Water Research* 27(4): 685-691.

Battermann, G., and P. Werner. 1984. "Beseitigung einer Untergrundkontamination mit Kohlenwasserstoffen Durch Mikrobiellen Abbau." *Grundwasserforschung-Wasser/Abwasser* 125:366- 373.

Gersberg, R.M., W.J. Dawsey, and M.D. Bradley. 1993. "Nitrate Enhancement of Mono-aromatic Compounds in Groundwater." *Remediation Spring*: 233-245.

Hinchee, R.E., D.C. Downey, J.K. Slaughter, D.A. Selby, M.S. Westray, and G.M. Long. 1989. *Enhanced Bioreclamation of Jet Fuels: A Full-Scale Test at Eglin AFB Florida.* Air Force Engineering & Services Center, Tyndall Air Force Base, Florida.

Hutchins, S.R. 1991a. "Biodegradation of Monoaromatic Hydrocarbons by Aquifer Micro-organisms Using Oxygen, Nitrate, or Nitrous Oxide as the Terminal Electron Acceptor." *Appl. Environ. Microbiol. 57*: 2403-2407.

Hutchins, S.R. 1991b. "Optimizing BTEX Biodegradation Under Denitrifying Conditions." *Environ. Toxicol. Chem. 10*:1437-1448.

Hutchins, S.R., W.C. Downs, J.T. Wilson, G.B. Smith, D.A. Kovacs, D.D. Fine, R.H. Douglass, and D.J. Hendrix. 1991. "Effect of Nitrate Addition on Biorestoration of Fuel-Contami-nated Aquifer: Field Demonstration." *Ground Water 29*: 571-580.

Hutchins, S.R., S.W. Moolenaar, and D.E. Rhodes. 1992. "Column Studies on BTEX Bio-degradation Under Microaerophilic and Denitrifying Conditions." *Journal of Hazardous Materials 32*: 195-214.

Hutchins, S.R., and J.T. Wilson. 1994. "Nitrate-based Bioremediation of Petroleum Con-taminated Aquifer at Park City, Kansas: Site Characterization and Treatability Study." In R.E. Hinchee, B.C. Alleman, R.E. Hoeppel, and R.N. Miller (Eds.), *Hydrocarbon Bio-remediation*, pp. 80-92. Lewis Publishers, Ann Arbor, MI.

Jeris, J.S., C. Beer, and J.A. Mueller. 1974. "High Rate Biological Denitrification Using a Granular Fluidized Bed." *Journal WPCF 46*(9): 2118-2128.

Kuhn, E.P., P.J. Colberg, J.L. Schnoor, O. Wanner, A.J.B. Zehnder, and R.P. Schwarzenbach. 1985. "Microbial Transformations of Substituted Benzenes During Infiltration of River Water to Groundwater: Laboratory Columns Studies." *Environ. Sci. Technol. 19*(10): 961-968.

Aromatic Hydrocarbon Biotransformation Under Mixed Oxygen/Nitrate Electron Acceptor Conditions

Liza P. Wilson, Neal D. Durant, and Edward J. Bouwer

ABSTRACT

Research is being conducted to investigate the effect of mixed oxygen/ nitrate electron acceptor conditions on the biodegradation of monocyclic and polycyclic aromatic hydrocarbons. Batch studies have determined that microorganisms, sampled from aquifer sediments and enriched under denitrification conditions, were able to degrade benzene, toluene, ethylbenzene, and naphthalene (each present at 2 mg/L) under conditions of 2 mg/L oxygen and high levels of nitrate (≥ 150 mg/L NO_3^-). Complete mineralization of benzene to carbon dioxide was observed in some samples. Biodegradation was inhibited at high (8.6 mg/L) and low (<2 mg/L) oxygen levels. Only toluene was degraded under anaerobic denitrification conditions. Further studies will assess the reaction kinetics and stoichiometry of biodegradation of aromatic hydrocarbons under mixed oxygen/nitrate electron acceptor conditions.

INTRODUCTION

Benzene, toluene, ethylbenzene, and xylenes (BTEX) and polycyclic aromatic hydrocarbons (PAHs) have been detected in the groundwater and subsurface sediments at the site of a former manufactured gas plant in Baltimore, Maryland (Durant et al. 1995). In situ addition of nutrients, electron acceptors, electron donors and/or surfactants may stimulate biodegradation of subsurface contaminants. Studies have demonstrated that the biodegradation of petroleum in groundwater and natural sediments often is limited by the supply of oxygen as an electron acceptor. Addition of oxygen to the subsurface is difficult due to its low solubility in water and its rapid consumption in reduced groundwater environments (Morgan & Watkinson 1992). The presence of both oxygen at microaerophilic levels (≤ 2 mg/L O_2) and nitrate as electron acceptors may result in enhanced ability of subsurface microorganisms to attack the aromatic ring (using oxygenases) of many organic compounds and then to complete degradation using nitrate as an electron acceptor. Although nitrate is a favorable alternative to oxygen as an electron acceptor, some studies have observed benzene and naphthalene to resist biodegradation under strict anaerobic denitrification conditions (Hutchins et al. 1991; Kuhn et al. 1988; Flyvbjerg et al. 1993). In situ bioremediation under mixed oxygen/nitrate electron acceptor conditions may prove more feasible than remediation under strict aerobic or denitrification conditions, particularly in light of the engineering difficulties with maintaining an adequate supply of oxygen in the subsurface.

A laboratory investigation is under way to provide a better understanding of the effect of mixed oxygen/nitrate electron acceptor conditions on the biodegradation of monocyclic and polycyclic aromatic hydrocarbons. The specific objectives of the research are to (1) quantify the stoichiometry and kinetics of biodegradation of aromatic hydrocarbons under microaerophilic (defined as ≤ 2 mg/L O_2) conditions and (2) assess the relative efficacy of bioremediation under microaerophilic conditions compared with strict aerobic or denitrification conditions in the laboratory using batch microcosms and aquifer sediment columns. This paper describes the initial results of batch microcosm studies on BTEX and PAH biodegradation under varying combinations of oxygen and nitrate.

EXPERIMENTAL METHODS

Microbial Inocula

Laboratory batch microcosm studies were conducted to assess the effect of varying levels of oxygen and nitrate on biodegradation of a mixture of aromatic hydrocarbons representative of the groundwater contaminants at the Baltimore site. Initial batch sediment/groundwater microcosm studies determined that aquifer bacteria from the Baltimore site were able to degrade significant concentrations of ^{14}C-naphthalene under aerobic (90% ±4) and anaerobic denitrification (15% ±5) conditions. These data suggested that the population consisted of facultative denitrification bacteria capable of using both oxygen and nitrate as electron acceptors. A denitrification enrichment of these aquifer bacteria maintained in an anaerobic glove box was used as inocula for the studies of biodegradation under mixed oxygen/nitrate electron acceptor conditions. Aseptic field collection of aquifer sediments and the preparation of microbial enrichments, and sediment/groundwater microcosms are described elsewhere (Durant et al. 1995; Wilson et al. 1993).

Microcosm Preparation

All glassware and laboratory equipment used to prepare the microcosms were sterilized by autoclave. Microcosms were prepared without headspace in 160-mL glass serum bottles capped with Teflon™-faced, butyl rubber septa and aluminum crimp seals. Each bottle was filled with an anaerobic mixture of aromatic compounds (approximately 2 mg/L benzene, 2 mg/L toluene, 2 mg/L ethylbenzene, 2 mg/L *m*-xylene, 2 mg/L naphthalene, 0.3 mg/L phenanthrene, and 3 mCi/L ^{14}C-labeled benzene), nutrient media (17 mg/L KH_2PO_4, 43.6 mg/L K_2HPO_4, 66.8 mg/L $Na_2HPO_4*7H_2O$, 0.5 mg/L $FeCl_2*4H_2O$, 0.0625 mg/L $MnCl_2*$ $4H_2O$, 0.125 mg/L $Na_2MoO_4*2H_2O$, 0.0025 mg/L $CuCl_2*2H_2O$, 0.0025 mg/L Na_2SeO_3, 0.00625 mg/L H_3BO_3, 0.00625 mg/L $ZnCl_2$, 0.00625 mg/L $AlCl_3$, 0.00625 mg/L $CoCl_2*6H_2O$, 0.004 mg/L $Ni(NO_3)2*6H_2O$, and 3.125 mg/L $FeCl_3*$ $6H_2O$), and 10 mL of the denitrification enrichment of indigenous subsurface bacteria. Three separate sets of microcosms were prepared with different levels of oxygen: 0 mg/L, 2 mg/L and 8.6 mg/L. For a given oxygen concentration, three to four different levels of nitrate were studied: 0 mg/L, 50 mg/L, 150 mg/L, and/or 400 mg/L. For each combination, replicate (n = 10) microcosms were prepared along with replicate (n = 10) killed controls (4.6 g/L $HgCl_2$). All bottles except those containing 8.6 mg/L oxygen (aerobic) were incubated in an anaerobic glove box (Coy). Duplicate batch bottles and associated controls were sacrificed over time.

Microcosm Analysis

Disappearance of the organic compounds was quantified by gas chromatography using a Hewlett Packard 5890A gas chromatograph (with flame ionization detector); dissolved oxygen concentration was determined using a Microelectrodes, Inc. dissolved oxygen microprobe and meter; phosphate, nitrate, and nitrite concentrations in the microcosms were established using a Dionex 2010i ion chromatography system; and ^{14}C-benzene and ^{14}C-benzene degradation products were quantified using a Beckman LS 3801 liquid scintillation counter.

RESULTS AND DISCUSSION

Biodegradation of the test compounds favored microaerophilic levels of oxygen (2 mg/L) and moderate to high levels of nitrate (150 mg/L or 400 mg/L) (Tables 1 and 2). Of the microcosms incubated under anaerobic conditions, only those incubated with 400 mg/L nitrate demonstrated biodegradation activity (50% loss of toluene) over 78 days of incubation with concurrent consumption of 14 mg/L nitrate. These data suggest that there is some threshold level of oxygen and nitrate or optimal combination that enhances use of both oxygen and nitrate as electron acceptors in the biodegradation of aromatic hydrocarbons.

All the test compounds resisted biodegradation during 68 days of incubation under aerobic conditions (8.6 mg/L O_2), regardless of nitrate levels. Aerobic conditions appeared to inhibit consumption of oxygen by these bacteria regardless of the level of nitrate (Figure 1), however, some oxygen consumption was observed in aerobic microcosms, likely due to endogenous decay. These data suggest that high levels of oxygen (8.6 mg/L) inhibited biodegradation of the aromatic mixture, and that varying the level of nitrate failed to counteract this inhibition. Hernandez and Rowe (1987) also observed that increasing nitrate concentrations could not overcome inhibition by oxygen. Oxygen likely inhibits

TABLE 1. Aromatic hydrocarbons biodegraded under varying levels of nitrate and oxygen.

	0 mg/L Oxygen	2.0 mg/L Oxygen	8.6 mg/L Oxygen
0 mg/L Nitrate	None	—	None
50 mg/L Nitrate	None	None	None
150 mg/L Nitrate	None	B, T, E, N	None
400 mg/L Nitrate	T	B, T, E, N	None

B = benzene, T = toluene, E = ethylbenzene, N = naphthalene, None = no evidence of biodegradation of target compounds, — = combination not studied.

denitrifying bacteria by repressing their ability to synthesize denitrifying enzymes and their capacity to efficiently utilize oxygen as an alternative electron acceptor (Kuhn et al. 1988).

The fluctuation of carbon-14 in active microcosms relative to killed controls containing 2 mg/L oxygen and 400 mg/L nitrate is illustrated in Figure 2. No production of $^{14}CO_2$ or ^{14}C-benzene intermediates was observed in killed controls. The conversion of ^{14}C-benzene to nonvolatile intermediates or cells as a result of ^{14}C-benzene biodegradation in microcosms containing 2 mg/L oxygen is illustrated in Figure 3. A greater percent of ^{14}C-benzene was converted to intermediates or cells in microcosms containing 150 mg/L nitrate than those containing 400 mg/L nitrate. However, a significant concentration of ^{14}C-benzene was mineralized to $^{14}CO_2$ in microcosms containing 400 mg/L nitrate, whereas no benzene mineralization was observed in microcosms with less nitrate (Figure 4). Although

TABLE 2. Percent biodegradation of aromatic hydrocarbons under varying levels of nitrate and 2 mg/L oxygen.

Aromatic Compound	50 mg/L Nitrate	150 mg/L Nitrate	400 mg/L Nitrate
Benzene	< 1%	16%	51%
Toluene	< 1%	65	70
Ethylbenzene	< 1%	98	62
m-Xylene	< 1%	< 1%	< 1%
Naphthalene	< 1%	13	12
Phenanthrene	< 1%	< 1%	< 1%

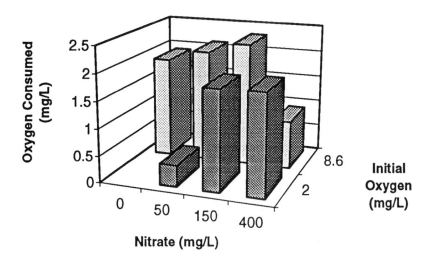

FIGURE 1. Percent oxygen consumed by denitrification bacteria in microcosms with various combinations of oxygen and nitrate.

biodegradation of ^{14}C-benzene occurred with 2 mg/L oxygen and 150 mg/L nitrate, additional levels of nitrate were required to yield complete mineralization of ^{14}C-benzene in the presence of this mixture of aromatic hydrocarbons. These data suggest that high levels of nitrate in conjunction with 2 mg/L oxygen result in mineralization of benzene by this enrichment of denitrifying bacteria.

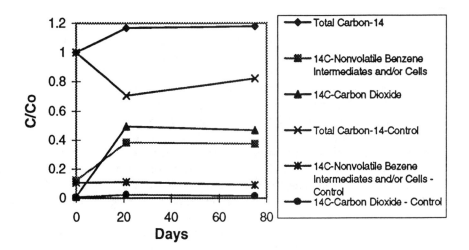

FIGURE 2. Fractionation of ^{14}C-benzene in active and killed control microcosms with 2 mg/L oxygen and 400 mg/L nitrate.

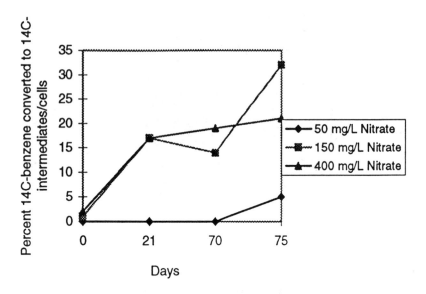

FIGURE 3. Percent ^{14}C-benzene converted to nonvolatile intermediate and/or cells in microcosms with 2 mg/L oxygen.

CONCLUSIONS

Biodegradation of a mixture of BTEX, naphthalene, and phenanthrene by an enrichment of denitrifying aquifer bacteria was inhibited by high levels of oxygen (8.6 mg/L). Under strict anaerobic denitrification conditions, biodegradation of toluene was observed after 78 days of incubation in microcosms containing 400 mg/L nitrate. Lower levels of nitrate inhibited biodegradation under denitrification conditions. Biodegradation of BTEX and PAHs was optimal in microcosms incubated with 2 mg/L oxygen and high levels (150 and 400 mg/L) of nitrate. Benzene mineralization was maximized with 2 mg/L oxygen and 400 mg/L nitrate. The conversion of benzene to a nonvolatile intermediate or cells was observed in microcosms incubated with 2 mg/L oxygen and 150 mg/L nitrate. No evidence of *m*-xylene or phenanthrene biodegradation was observed in this study.

Initial results of this study indicate that there are optimal combinations of nitrate and oxygen, which enable this enrichment of denitrifying bacteria to degrade benzene, ethylbenzene, toluene, and naphthalene. Data suggest that BTEX and PAH in situ bioremediation schemes relying solely on oxygen addition might be less effective than those using a combination of oxygen and nitrate. Further studies are required to confirm these results and to assess the reaction kinetics and stoichiometry of biodegradation of aromatic hydrocarbons under mixed oxygen/nitrate electron acceptor conditions. Data from these batch studies will be used in the design of laboratory sediment columns, which will be used to simulate field bioremediation under mixed oxygen/nitrate electron acceptor conditions.

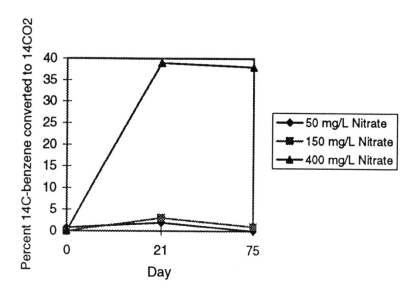

FIGURE 4. Percent ^{14}C-benzene mineralized (converted to [14]CO_2) in microcosms with 2 mg/L oxygen.

ACKNOWLEDGMENTS

This research was made possible in part by the generous support of the U.S. EPA Robert S. Kerr Environmental Laboratory (Project CR-821907) and in part by the Baltimore Gas and Electric Company. Although the research described in this paper has been funded wholly or in part by the U.S. EPA, it has not been subjected to Agency review and therefore does not necessarily reflect the views of the Agency, and no official endorsement should be inferred.

REFERENCES

Durant, N. D., L. P. Wilson, and E. J. Bouwer. 1995. "Microcosm studies of subsurface PAH degrading bacteria from a former manufactured gas plant." *Journal of Contaminant Hydrology* 17:213-237.

Flyvbjerg, J., E. Arvin, B. K. Jensen, and S. K. Olsen. 1993. "Microbial degradation of phenols and aromatic hydrocarbons in creosote-contaminated groundwater under nitrate-reducing conditions." *Journal of Contaminant Hydrology* 12:133-150.

Hernandez, D., and J. J. Rowe. 1987. "Oxygen regulation of nitrate uptake in denitrifying *Pseudomonas aeruginosa.*" *Applied and Environmental Microbiology* 53:745-750.

Hutchins, S. R., G. W. Sewell, D. A. Kovacs, and G. A. Smith. 1991. "Biodegradation of aromatic hydrocarbons by aquifer microorganisms under denitrification conditions." *Environmental Science and Technology* 25:68-76.

Kuhn, E. P., J. Zeyer, P. Eicher, and R. P. Schwarzenbach. 1988. "Anaerobic degradation of alkylated benzenes in denitrification laboratory aquifer columns." *Applied and Environmental Microbiology* 54:490-496.

Morgan, P., and R. J. Watkinson. 1992. "Factors limiting the supply and efficiency of nutrient and oxygen supplements for the in situ biotreatment of contaminated soil and groundwater." *Water Research* 26:73-78.

Wilson, L. P., N. D. Durant, and E. J. Bouwer. 1993. "Relative effects of nitrogen, phosphorus, trace metals and temperature on biodegradation of coal tar constituents." *Proceedings of the Sixth IGT Symposium on Gas, Oil and Environmental Biotechnology.* Institute of Gas Technology, Chicago, IL.

Biodegradation of Polycyclic Aromatic Hydrocarbons in the Presence of Nonionic Surfactants

Frank Volkering, Rike van de Wiel, Anton M. Breure,
Johan G. van Andel, and Wim H. Rulkens

ABSTRACT

The effect of the nonionic surfactants Triton X-100 and Tergitol NPX on the biodegradation of crystalline and sorbed polycyclic aromatic hydrocarbons (PAHs) was investigated. The presence of surfactants in high concentrations increased the apparent solubility and the maximum dissolution rate of crystalline naphthalene and phenanthrene. Biodegradation of these PAHs in the dissolution-limited growth phase was also stimulated by surfactant addition. In desorption experiments it was found that more naphthalene was desorbed from an inert matrix in the presence of surfactants than in the absence of surfactants, but it could not be shown that the presence of surfactant resulted in increased initial desorption rates. Biodegradation of sorbed naphthalene was stimulated by the addition of surfactant. In contrast to the bio-degradation of crystalline PAHs, this stimulation also occurred at low surfactant concentrations. It was concluded that in this system the bioavailability of the PAHs was enhanced by the use of surfactants.

INTRODUCTION

The slow desorption of hydrophobic organic compounds (HOCs) is one of the most important reasons for the slow biodegradation of these compounds in the biological cleanup of soil (Mihelcic et al. 1993). One possible way to solve this problem is the use of surfactants to increase the mobility of such pollutants. Surfactant molecules are amphiphilic and tend to concentrate at surfaces and interfaces, lowering the surface tension and the interfacial tensions. At surfactant concentrations above the critical micelle concentration or CMC, aggregates of 20 to 200 surfactant molecules are formed, which are named micelles. Micelles can be regarded as little vesicles with a hydrophobic interior and the presence of micelles leads to an increase in the apparent solubility of hydrophobic compounds. It is assumed that micellar solubilization and mobilization due to lowered interfacial tensions (facilitated transport) will enhance the bioavailability of sorbed HOCs (Volkering et al. 1995).

Due to complexity of the interactions among soil, bacteria, substrate, and surfactant, different results have been found in studies on the effect of surfactants on the biodegradation of HOCs sorbed onto soil. Some authors have found that surfactants had no effect on desorption, but stimulated mineralization (Aronstein et al. 1991), whereas others found that desorption was stimulated, but that biodegradation was not affected (Dohse & Lion 1994) or even inhibited (Laha & Luthy 1992). In this study the PAHs are used in crystalline form or sorbed onto XAD-4, an inert matrix.

MATERIALS AND METHODS

Bacterial Cultures and Culture Conditions

The isolation of strain 8909N, growing on naphthalene, and strain 8803F, growing on phenanthrene, has been described previously (Volkering et al. 1992, 1995). They have been identified as gram negative *Pseudomonas* species. Organisms were grown at 30°C in mineral medium (Volkering et al. 1993) that was buffered at pH 7.0 with 50 mM sodium phosphate.

Dissolution Experiments
and Desorption Experiments

These experiments were performed in 250-mL serum flasks on a rotary shaker (200 rpm, 30°C). The experiments were started by adding 0.15 or 0.25 g of PAH crystals, or 0.05 or 0.1 g of naphthalene-loaded Amberlite XAD-4 (Supelco, Bellefonte, USA) with an initial loading 145 mg·g^{-1}, to the flasks containing 150 mL of sterile buffered mineral medium and the appropriate amount of surfactant. At regular intervals, samples of 0.75 mL were taken to determine the PAH concentration in the liquid phase. For the desorption experiments, the residual naphthalene concentrations of both the liquid phases and solid phases were determined after approximately 1 week.

Biodegradation Experiments

Biodegradation experiments were performed in 250-mL serum flasks on a rotary shaker (150 or 200 rpm, 30°C). The flasks contained 100 mL of mineral medium with the appropriate amount of surfactant and were supplied with the appropriate amount of loaded matrix or crystals. To eliminate the possibility of oxygen limitation, the headspace of the flasks was filled with oxygen. The experiments were started by inoculation with 1 mL of active batch-grown cells. The biodegradation was followed by measuring the percentage of CO_2 in the headspace of the bottle. At the end of the experiment, when the CO_2 production was stopped, 1 mL of 12 M HCl was added to the flasks to remove the dissolved CO_2 and the CO_2 concentration was measured. For experiments with sorbed naphthalene, the residual naphthalene concentration of the liquid and the solid phase was determined at the end of the experiment.

Analytical Procedures

Aqueous PAH concentrations were determined by injecting samples (filtered with a 0.2 μm filter and 1:1 diluted with acetonitrile on a high-performance liquid chromatograph with a C_{18} reversed-phase column. The eluent was an 85/15 mixture of acetonitrile/water. Peaks were detected with an ultraviolet detector by measuring the absorbance at 274 nm for naphthalene and at 254 nm for phenanthrene. Extraction samples were measured likewise, but were diluted to a concentration lower than 50 mg·L^{-1}.

Extraction of naphthalene was performed by adding 50 or 100 mL of acetonitrile to the XAD-4, incubating for 1 week, and then measuring the naphthalene concentration in the acetonitrile. When the extraction was performed with a highly loaded matrix (determination of starting concentrations), the acetonitrile was replaced by the same volume of clean acetonitrile and the procedure was repeated. The extraction efficiency of this method was usually better than 95% for XAD-4.

CO_2 in the headspace of the serum flasks was determined using a gas chromatograph fitted with a thermal conductivity detector and a Hayesep Q column (Chrompack, Middelburg, The Netherlands). Helium was used as the carrier gas with a flow rate of 30 mL·$min.^{-1}$. The injector temperature was 150°C, the oven temperature 80°C, and the detector temperature 200°C.

RESULTS AND DISCUSSION

Crystalline PAHs

Dissolution. It is well known that the presence of micelles can increase the (apparent) aqueous solubility of HOCs. Under bioavailability-limiting conditions, however, the biodegradation rate of HOCs is not dependent on the solubility, but on the dissolution rate. For crystalline substrates, such as naphthalene and phenanthrene, the maximal dissolution rate (J_{max}) can be described as $K_l \cdot A \cdot C_{max}/V$, in which K_l is the mass transfer coefficient, A is the surface area of the crystals, C_{max} is the maximum solubility (but not the maximal *apparent* solubility), and V is the aqueous volume (Volkering et al. 1992). The factor $K_l \cdot A$ is constant when the particle diameter is considered constant and can be determined in dissolution experiments by following the increase in the aqueous concentration (C_t). The results can be fitted using equation 1 to obtain the value for $K_l \cdot A$.

$$C_t = C_{max} \cdot (1 - e^{\frac{K_l \cdot A \cdot t}{V}}) \tag{1}$$

Graphically, the maximum dissolution rate is depicted by the slope of the dissolution curve at time t = 0. If increasing the apparent solubility is the only effect of the micelles, the maximum dissolution rate will not be affected by the presence of surfactants. To test this, dissolution experiments with naphthalene and phenanthrene in solutions with surfactants in different concentrations have been performed. As an example, the results of the experiment with naphthalene and Triton X-100 (CMC = ± 0.15 g.L^{-1}) are shown in Figure 1. The straight line represents J$_{max}$, as calculated from the experiment without surfactant. It is clear that the dissolution in the presence of high surfactants concentrations proceeds faster than the calculated maximal dissolution. Triton X-100 present in concentrations below the CMC had no effect on the dissolution of naphthalene. Similar results have been found in dissolution experiments with phenanthrene and with Tergitol NPX as the surfactant.

Biodegradation. The effect of surfactant on the biodegradation of PAH was tested by addition of the two surfactants in the dissolution-limited growth phase. In this phase, the aqueous PAH concentration is low (Volkering et al. 1992) and the solubilizing effect of the surfactants will be small. The effect of adding different concentrations of Triton X-100 on the CO_2 production rate of strain 8803F on phenanthrene is shown in Figure 2. The addition of the surfactant clearly stimulates bacterial growth as can be seen from the increase in the CO_2 production rate directly after surfactant addition. This was not caused by surfactant

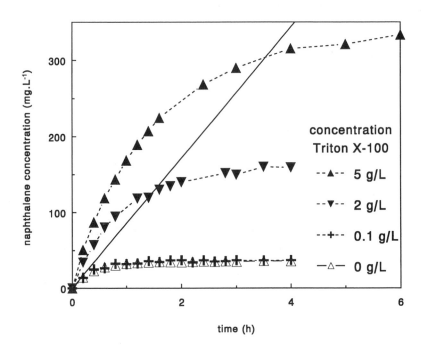

FIGURE 1. Effect of Triton X-100 on the dissolution of crystalline naphthalene.

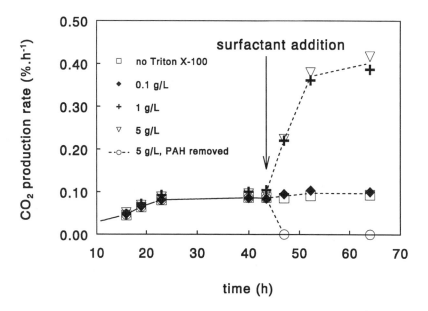

FIGURE 2. CO_2 production rates of strain 8803F on crystalline phenanthrene under addition of different amounts of Triton X-100.

biodegradation as was shown in separate experiments in which the phenanthrene crystals were removed before adding Triton X-100. Experiments with naphthalene and with Tergitol NPX showed similar results. This confirms that the presence of surfactant in high concentrations affects the dissolution process.

Sorbed PAHs

Desorption. Batch desorption experiments may give information on both equilibrium concentrations and desorption kinetics. The effect of Triton X-100 on the desorption of naphthalene from XAD-4 is shown in Figure 3. The presence of surfactant changes the partitioning of the naphthalene over the solid phase and the liquid phase. The effect of the presence of surfactant on the initial desorption rate that was observed may also have been caused by micellar solubilization, and therefore no conclusions about facilitated transport of naphthalene can be drawn on the basis of the experiments presented here.

Biodegradation. The biodegradation of sorbed naphthalene was monitored by measuring the CO_2 concentration in the headspace of batch cultures. Typically, growth on sorbed naphthalene starts with an exponential growth phase, followed by a phase in which the bacterial growth is limited by the desorption of naphthalene from the matrix. Experiments were performed in which the surfactant was added to the medium in the desorption-limited growth phase. As an example the results of the CO_2 measurements of an experiment with Triton X-100 are

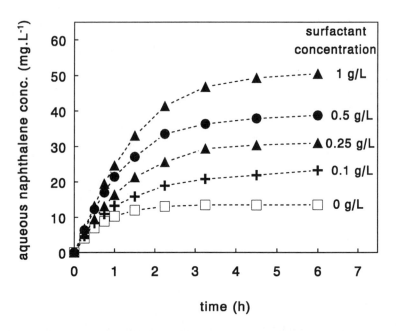

FIGURE 3. Desorption of naphthalene from XAD-4 in the presence of different concentrations of Triton X-100.

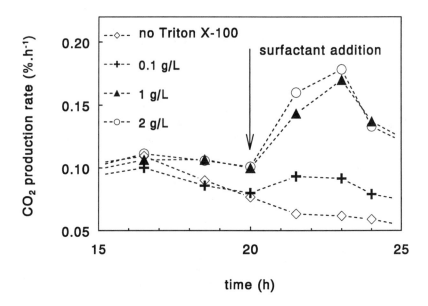

FIGURE 4. Effect of Triton X-100 addition on CO_2 production rate of strain 8909N during biodegradation of sorbed naphthalene.

shown in Figure 4; similar results have been found with Tergitol NPX. In none of the batches naphthalene could be measured after extraction of the XAD-4 at the end of the experiment. The final CO_2 concentrations were equal in batches with and without surfactant, showing that the observed increase in the CO_2 production rate was not caused by growth on the surfactant. Moreover, no CO_2 production was found in blanks that contained surfactant but no naphthalene.

CONCLUSIONS

In this study it was shown that the dissolution and desorption of PAHs was stimulated by the presence of synthetic nonionic surfactants. The enhanced mass transfer enabled faster biodegradation, as shown in biodegradation experiments. An important difference was found between the biodegradation of crystalline and sorbed PAHs. In the first case, stimulation was found only at surfactant concentrations higher than the CMC, whereas the biodegradation of sorbed naphthalene was also stimulated at a concentration below the CMC. As the costs of surfactants may play an important role in the applicability of the technique, it is interesting that surfactants at low concentrations may enhance the bioavailability of sorbed HOCs.

ACKNOWLEDGMENT

This work was supported by the Netherlands Integrated Soil Research Program (grant no. 8977).

REFERENCES

Aronstein, B. N., and M. Alexander. 1993. "Effect of a Nonionic Surfactant Added to the Soil Surface on the Biodegradation of Aromatic Hydrocarbons within the Soil." *Appl. Microbiol. Biotechnol.* 39:386-390.

Dohse, D. M., and L. W. Lion. 1994. "Effect of Microbial Polymers on the Sorption and Transport of Phenanthrene in a Low-Carbon Sand." *Environ. Sci. Technol.* 28: 541-548.

Laha S., and R. G. Luthy. 1992. "Effects of Nonionic Surfactants on the Mineralization of Phenanthrene in Soil-Water Systems." *Biotechnol. Bioeng.* 40:1367-1380.

Mihelcic, J. R., D. R. Lueking, R. J. Mitzell, and J. M. Stapleton. 1993. "Bioavailability of Sorbed- and Separate-Phase Chemicals." *Biodegradation* 4:141-153.

Volkering, F., A. M. Breure, A. Sterkenburg, and J. G. van Andel. 1992. "Microbial Degradation of Polycyclic Aromatic Hydrocarbons: Effect of Substrate Availability on Bacterial Growth Kinetics." *Appl. Microbiol. Biotechnol.* 36:548-552.

Volkering, F., A. M. Breure, and J. G. van Andel. 1993. "Effect of Microorganisms on the Bioavailability and Biodegradation of Crystalline Naphthalene." *Appl. Microbiol. Biotechnol.* 40:535-540.

Volkering, F., A. M. Breure, J. G. van Andel, and W. H. Rulkens. 1995. "Influence of Nonionic Surfactants on the Bioavailability and Biodegradation of Polycyclic Aromatic Hydrocarbons." *Appl. Environ. Microbiol.* 61(5) (in press).

Diversity of Metabolic Capacities Among Strains Degrading Polycyclic Aromatic Hydrocarbons

Murielle Bouchez, Denis Blanchet,
Bernard Besnaïnou, and Jean-Paul Vandecasteele

ABSTRACT

Strains of *Pseudomonas* and *Rhodococcus* genera were isolated for their capacity to use, as a sole carbon and energy source, one of the following polycyclic aromatic hydrocarbons (PAHs): naphthalene (NAP), fluorene (FLU), phenanthrene (PHE), anthracene (ANT), fluoranthene (FLT), and pyrene (PYR). The range of PAHs supporting growth of these pure strains was usually restricted, but several other hydrocarbons were used by *Rhodococcus* sp. All strains could grow on simple organic acids. Maximal specific growth rates (μ_{max}) of all strains on their PAH growth substrates were determined by respirometry. No clear relationships between μ_{max} values and the molecular weight or water solubility of PAHs were apparent, but *Pseudomonas* sp. exhibited the highest μ_{max} values. Carbon balances for PAH biodegradation were established. Differences between strains were observed, but high mineralization rates (56% to 77%) and low production of soluble metabolites (7% to 23%) were obtained for all PAHs. Bacterial biomass represented 16% to 35% of the carbon consumed. Strain diversity was also apparent in the interactions observed in the degradation of a mixture of two PAHs by individual strains, which often involved inhibition of PAH substrate degradation, with or without cometabolization of the second PAH.

INTRODUCTION

Progress in PAH biodegradation in recent years has been largely conditioned by the isolation of new bacterial strains capable of using, as single carbon and energy sources, PAHs containing up to 4 rings. In the present study, various aspects of bacterial diversity with respect to the overall performance of the degradation of 2-, 3-, and 4-ring PAHs were examined.

MATERIALS AND METHODS

Microbiology

Isolation and Maintenance of Microorganisms. Pure strains were isolated, by selective enrichment, from soils polluted or not by coal tar for their ability to use various individual PAHs as sole sources of carbon and energy. Strain identifications were conducted by Institut Pasteur, Laboratoire des Identifications (Paris, France). Pure strains, periodically checked for purity, were maintained with their isolation PAH as a carbon source in a vitamin-supplemented mineral salt medium (MSM) as previously described (Bouchez et al. 1995).

Growth Capacities. Growth on different carbon substrates was investigated in Erlenmeyer flasks in MSM. Substrates were added at 500 ppm, except for methanol and monoaromatic hydrocarbons, which were used at 200 ppm to avoid toxicity. Growth was assessed by visually observing biomass increase and also, for PAHs, by quantitative determination of their utilization. Time courses of oxygen consumed (Oc) during growth on PAHs were conducted using sensitive respirometric equipment (D-12-S Sapromat, Voith, Heidenheim, Germany). In the Sapromat culture flasks, electrolytically produced oxygen was maintained at a constant concentration, and its consumption was continuously recorded. The μ_{max} values were determined by plotting Oc versus time in a semilogarithmic scale.

PAH Interactions. As previously described (Bouchez et al. 1995), interactions between 2 PAHs, with one used as a growth substrate, were studied in 30-mL, tightly closed tubes containing 1.5 mL of MSM. The inoculum (5% v/v) was grown on the PAH to be employed as the main growth substrate in the interaction experiments. Each PAH was added at 500 ppm. Reference growth experiments on individual PAHs were run simultaneously. Assays were run in triplicate. After incubation, PAHs were quantified.

Carbon Balances. Experiments to determine the carbon balance of the biodegradation of individual PAHs (500 ppm) were conducted in 60-mL penicillin flasks containing 3 mL of MSM and 150 µL of an inoculum grown on the PAH studied. Flasks were tightly closed with Teflon™ caps equipped with "mininert" valves (Supelco Inc., Bellefonte, Pennsylvania), which allowed injection through a septum without gas exchange. Determinations were made when metabolic activity ceased, as indicated in simultaneous growth experiments conducted in the Sapromat equipment. Three flasks were used to measure CO_2 produced and, afterward, for quantification of residual PAH. Three other flasks, run in parallel, were used to determine the metabolites and biomass produced.

Analytical Procedures

In the tightly closed penicillin flasks, CO_2 was trapped in the aqueous phase as Na_2CO_3 by adding 100 µL of 4N NaOH. Then, after centrifugation, the dissolved carbon was evaluated in the supernatants with a DC-80 carbon analyzer (Xertex, Santa Clara, California) before and after acidification and bubbling. The difference between the two values represented the carbon present as CO_2. Metabolites were evaluated with the DC-80 carbon analyzer on nonalkalinized supernatants of cultures after acidification and bubbling. Biomass was evaluated by the Lowry method (Herbert et al. 1971), using bovine serum albumin as standard. The ratio of protein/carbon in the biomass was taken to be 1 (w/w). As previously detailed (Bouchez et al. 1995), PAHs were quantified by high-performance liquid chromatography, using UV detection at 254 nm, or by gas-phase chromatography, using a flame ionization detector.

RESULTS AND DISCUSSION

Growth Capacities

The PAHs used for strain isolation and the growth capacities of some of the isolated strains are presented in Table 1. All PAH-contaminated soils tested yielded strains growing on 2-, 3-, and 4-ring PAHs. Strains growing on fluorene were the least frequently isolated (data not shown). Soils that were not contaminated by PAHs yielded strains able to grow only on naphthalene. The bacteria belonged mainly to the *Pseudomonas* and *Rhodococcus* genera. *Pseudomonas* strains were obtained only on the 2- and 3-ring PAHs naphthalene, and phenanthrene. Growth capacities of all strains were wide for organic acids and more restricted for saturated or monoaromatic hydrocarbons. It is interesting to note that the *Rhodococcus* species had the capacity to degrade hexadecane. Several of these observations are in line with those of Kästner et al. (1994). The range of PAHs used as growth substrates by individual strains was usually quite narrow. Strains with a wider specificity have been described by some authors (Kästner et al. 1994; Kiyohara et al. 1992).

In our experiments, the 500 ppm of PAH supplied was completely utilized by the appropriate strains within a period extending from 3 days for naphthalene up to 40 days for pyrene. Such a concentration, which allowed satisfactory growth and carbon balance determinations, largely exceeded the solubilities of PAHs which ranged from 30 ppm for naphthalene, 2 ppm for fluorene, down to 0.05 ppm for anthracene. Except, as detailed later, for the most soluble compound naphthalene, no toxic effects were observed at the high PAH concentrations employed, most probably because solid phases of PAHs were not inhibitory to the bacteria. All strains grew on the PAH they were isolated on regardless of whether the nitrogen source was ammonium nitrate (as in MSM), sodium nitrate, or ammonium sulfate (data not shown).

The μ_{max} values determined in the early growth phase on individual PAHs are presented in Table 2. Exponential growth could be observed only in the early phase, because thereafter degradation was limited by substrate availability. The μ_{max} values could not be related to the solubility of PAHs, as is often suggested, and no simple relationship with the number of aromatic rings was

TABLE 1. Growth capacities of various isolated strains.

Substrate	Growth capacity of strains:							
	S Nap Ru 1 Rh. sp.[a]	S Nap Ka 1 Ps. stutzeri	S Flu Au 1	S Flu Na 1 Rh. sp.[b]	S Phe Na 1 Ps. sp.	S Ant Mu 3 Bacillus[c]	S Flt Na 1 Rh. sp.[b]	S Pyr Na 1 Rh. sp.[b]
Naphthalene	I,++	I,++	-	-	-	-	-	-
Fluorene	-	-	I,++	I,+	-	-	-	-
Phenanthrene	-	-	-	-	I,++	-	+/-	+/-
Anthracene	-	-	-	-	-	I,+	-	-
Fluoranthene	-	-	-	-	-	+	I,+	+
Pyrene	-	-	-	-	-	+	+	I,+
Benzene	+	-			-	-	-	-
Toluene	+	-			-	-	-	-
m-Xylene	-	-			-	-	-	-
n-Octane	-	-			-	-	-	+
n-Hexadecane	+	-			-	+	+	+
Acetate	++	++			++	++	++	++
Butyrate	+	+			+	+	+	+
Propionate	++	++			++	++	++	++
Lactate	++	++			++	-	+	+
Succinate	++	++			++	-	-	-
Benzoate	++	++			-	++	++	++
Glucose	++	++			-	+	++	++
Methanol	-	-			-	-	-	-
Ethanol	++	++			++	++	++	++

(a) *Rhodococcus* sp. similar to *Rh. rhodochrous.*
(b) *Rhodococcus* sp. similar to *Rh. equi.*
(c) *Bacillus* similar to *Aureobacterium.*

++ Growth visually detectable in less than 10 days.
+ Growth visually detectable between 10 and 30 days.
+/- Growth decreasing with the number of transfers.
- No growth observed after 2 months.
I Isolation PAH.

TABLE 2. Maximal specific growth rates of pure strains on various PAHs.

Strain	Growth PAH	μ_{max} (h^{-1})
Ps. S Nap Ka 1	NAP	0.23
Rh. S Nap Ru 1	NAP	0.13
Rh. S Flu Na 1	FLU	0.050
Ps. S Phe Na 1	PHE	0.36
S Ant Mu 3	ANT	0.056
S Ant Mu 3	PYR	0.026
Rh. S Flt Na 1	PYR	0.042
Rh. S Flt Na 1	FLT	0.025

apparent. But, for a given substrate, the μ_{max} values were related to the strains; the latter finding has been previously shown in literature. In our studies, *Pseudomonas* strains presented higher μ_{max} values than the *Rhodococcus* strains.

Carbon Balances

The fate of carbon during PAH degradation is crucial. Accordingly, carbon balances (Figure 1) were established for all strains. Results were reproducible

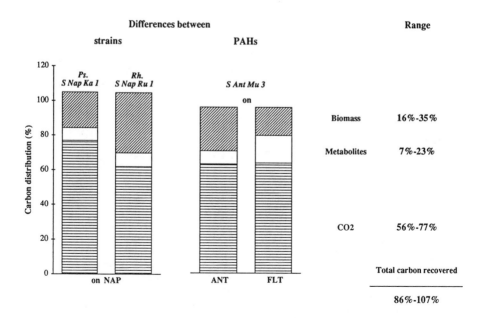

FIGURE 1. Carbon balances of the biodegradation of individual PAHs by pure strains.

158   *Microbial Processes for Bioremediation*

(individual values of the different terms within 5%) and with good mass balances. An important point is that all strains showed high rates of mineralization and low production of soluble metabolites. However, carbon balances were dependent on PAHs: as shown for strain *S Ant Mu 3*, less biomass was formed during degradation of 4-ring PAHs than of 2- or 3-ring PAHs. Carbon balances were also dependent on the strains, as illustrated in the case of naphthalene degradation by a *Pseudomonas* sp. and a *Rhodococcus* sp.

PAH Interactions

Interactions between PAHs are shown in Table 3. All strains were capable of PAH cometabolic degradation, an interesting point in view of their restricted PAH carbon source range. The range of PAHs cometabolized varied among strains. Nevertheless, for most of the strains, fluorene and phenanthrene were cometabolized more easily than the higher molecular weight PAHs. An interesting situation was observed with strain *S Pyr Na 1* grown on phenanthrene. The addition of fluorene as a cosubstrate increased the utilization of phenanthrene (synergy). The pattern most usually observed, however, was a lower degradation of the substrate PAH when a second PAH was added. This phenomenon (inhibition) occurred whether the second PAH was cometabolized (as often observed with fluorene) or not (as for example in the case of anthracene). With naphthalene as cosubstrate, the inhibition was so strong that it prevented, in all cases,

TABLE 3. Interactions during degradation of PAH pairs.

Strain	Main growth substrate	Utilization of PAHs as cosubstrates					
		NAP	FLU	PHE	ANT	FLT	PYR
S Nap Ru 1	NAP		–	+	+	–	–
S Nap Ka 1	NAP		–	+	–	+	–
S Flu Na 1	FLU	–,T		+,I	–,I	–,I	–,I
S Phe Na 1	PHE	–,T	+,I		+	+	–
S Ant Mu 3	ANT	–,T	+	+,I		S	S
S Flt Na 1	FLT	–,T	+,I	A	–,I		
S Flt Na 1	PYR	–,T	–,I	A	–,I		
S Pyr Na 1	FLT	–,T	+,I	A	–,I		
S Pyr Na 1	PHE	–,T	+,E		–,I	A	

+   Cometabolism   –   No cometabolism
+,I Cometabolism with inhibition   –,I No cometabolism and inhibition
+,E Cometabolism with synergy   –,T No cometabolism and toxicity

A   Antagonism of the two substrates.
S   Simultaneous degradation of the two substrates.

growth and degradation of the substrate PAH. This behavior (toxicity) of naphthalene is quite likely to be related to the relatively high solubility of this compound. Strain diversity concerning this point is apparent in the literature (Walter et al. 1991; Weissenfels et al. 1991). The situation, in our case, may be related to the origin of the strains, which were isolated from soils where little naphthalene was detected. When the second PAH was also a growth substrate, the inhibition of the utilization of both PAHs was called, when observed, substrate antagonism.

It is apparent from these results that diverse interactions were taking place during PAH degradation. From this point of view, each strain presented specific characteristics with, in all cases, a limited capacity for PAH degradation. This situation indicates that degradation of PAH mixtures in the environment implies a wide variety of strains and complex metabolic interactions.

REFERENCES

Bouchez, M., D. Blanchet, and J. P. Vandecasteele. 1995. "Degradation of Polycyclic Aromatic Hydrocarbons by Pure Strains and by Defined Strains Associations: Inhibition Phenomena and Cometabolism." *Applied Microbiology and Biotechnology*. Accepted.

Herbert, D., P. J. Phipps, and R. E. Strange. 1971. "Chemical Analysis of Microbial Cells." In: J. R. Norris and D. W. Ribbons (Eds.), *Methods in Microbiology*, Vol. 5B. Academic Press, London. pp. 209-344.

Kästner, M., M. Breuer-Jammali, and B. Mahro. 1994. "Enumeration and Characterization of the Soil Microflora from Hydrocarbon-Contaminated Soil Sites Able to Mineralize Polycyclic Aromatic Hydrocarbons (PAH)." *Applied Microbiology and Biotechnology* 41: 267-273.

Kiyohara, H., N. Takizawa, and T. Nagao. 1992. "Natural Distribution of Bacteria Metabolizing Many Kinds of Polycyclic Aromatic Hydrocarbons." *Journal of Fermentation and Bioengineering* 74: 49-51.

Walter, U., M. Beyer, J. Klein, and H. J. Rehm. 1991. "Degradation of Pyrene by *Rhodococcus* sp. UW 1." *Applied Microbiology and Biotechnology* 34: 671-676.

Weissenfels, W. D., M. Beyer, J. Klein, and H. J. Rehm. 1991. "Microbial Metabolism of Fluoranthene: Isolation and Identification of Ring Fission Products." *Applied Microbiology and Biotechnology* 34: 528-535.

Enhancement of BTX Biodegradation by Benzoate

Kenneth H. Rotert, Leslie A. Cronkhite, and Pedro J. J. Alvarez

ABSTRACT

Aquifer microcosms were used to investigate the effect of adding environmentally benign aromatic substrates on the phenotypic composition of indigenous microbial communities. Addition of aromatic compounds (i.e., benzoate or phenylalanine) exerted preferential selective pressure for benzene, toluene, and xylene (BTX) degraders. Addition of a non-aromatic substrate (i.e., acetate), however, did not stimulate a significant increase in the fraction of total heterotrophs capable of degrading BTX. A selective proliferation of BTX degraders would enhance biodegradation kinetics, which should decrease the duration (and cost) of BTX bioremediation. Proof of concept was obtained with laboratory aquifer columns that were continuously fed benzene, toluene, and o-xylene (at about 200 µg/L each). Benzoate addition to the column's influent (1 mg/L) enhanced aerobic BTX degradation and attenuated BTX breakthrough relative to acetate-amended (2 mg/L) or unamended control columns.

INTRODUCTION

BTX bioremediation projects often focus on overcoming limitations to natural degradative processes associated with the insufficient supply of inorganic nutrients and electron acceptors. Although subsurface addition of oxygen or nitrate has proven successful in numerous occasions, it has been only marginally effective at some sites (e.g., Aelion & Bradley 1991; Stapps 1990; Wilson et al. 1990). Sometimes, the concentration of a target BTX compound fails to decrease below a threshold level even after years of continuous addition of inorganic nutrients and electron acceptors (e.g., Barbaro et al. 1990). Trace residual concentrations of carcinogenic compounds, such as benzene, could exceed applicable cleanup standards and remain a threat to public health.

The lack of BTX degradation below a threshold concentration might be due, in part, to (1) the requirement for a minimum concentration of suitable substrates to sustain a steady population of BTX degraders (Bouwer & McCarty 1984), and/or

(2) the existence of a threshold substrate concentration below which induction of the necessary catabolic enzymes does not occur (Linkfield et al. 1989). Therefore, limitations associated with the presence and expression of appropriate microbial catabolic capacities may hinder the effectiveness of bioremediation. In some cases, the marginal effectiveness of bioremediation may be due to (1) negligible concentration of microorganisms with the required catabolic capacity, and/or (2) failure of the microbial consortium to induce appropriate degradative enzymes, suggesting that the addition of environmentally acceptable structural analogues, such as benzoate, might enhance BTX bioremediation. Such stimulatory substrates could enhance the induction of appropriate catabolic enzymes, act as primary substrates in the cometabolism of BTX, and/or serve as growth substrate for a selective proliferation of BTX-degraders which, in turn, would improve biodegradation kinetics.

This paper addresses the feasibility of using benzoate as biostimulatory substrate to overcome limitations associated with the presence and expression of the appropriate catabolic capacities.

EXPERIMENTAL METHODOLOGY

Effect of Substrate Addition on Phenotypic Composition

The ability of potential stimulatory substrates to exert selective pressure for BTX degraders was evaluated in aquifer microcosms. Three microcosm sets, seeded with pristine soil, were prepared in duplicate as described by Alvarez and Vogel (1991). The first two sets were amended with aromatic substrates (benzoate or phenylalanine at 1 mg/L each). The third set was amended with acetate (1 mg/L) as a control substrate to determine whether selection for BTX degraders was due to the aromatic structure of the added substrates. All microcosms were incubated for 5 days at 20°C. This time was sufficient for the aerobic degradation of the added substrates. Viable plate counts were subsequently done to measure changes in the relative concentration of total heterotrophs and BTX degraders. *Pseudomonas* spp. also were enumerated because they tend to have a broad catabolic specificity and are generally capable of degrading BTX (Ridgeway et al. 1990). Serial dilutions were plated on trypticase soy agar to grow and count total heterotrophic colonies. *Pseudomonas* isolation agar (DIFCO) was used to enumerate *Pseudomonas* spp., and carbon-free agar (BBL purified agar) amended with BTX vapors was used to grow and count colonies of BTX degraders.

Effect of Substrate Addition on BTX Degradation

Four flow-through columns were used to mimic in situ conditions of microbial exposure to BTX and biostimulatory substrates (Figure 1). All columns were packed with pristine, sandy aquifer material, and were fed a BTX-enriched mineral medium (Anid et al. 1993). One (treatment) column received this mixture along with benzoate (1 mg/L). The second column was fed acetate instead of benzoate

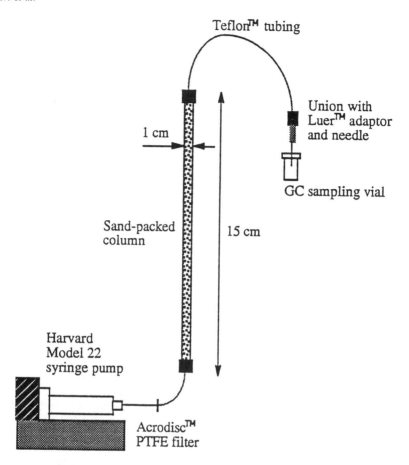

FIGURE 1. Laboratory apparatus for flowthrough aquifer columns.

at an equivalent electron donor concentration (2 mg/L as acetate). The third column was a no-treatment control, and received no supplemental substrates. The fourth column was poisoned with mercuric chloride (200 mg/L) to discern biodegradation from potential volatilization losses. Supplemental substrates were added at low concentrations (about 2 mg/L as chemical oxygen demand, [COD]) to avoid exacerbating the oxygen demand and to preclude potential diauxic effects which are common at high substrate concentrations (Egli et al. 1993). All columns were operated aerobically, which was verified by measuring the effluent dissolved oxygen concentration. The hydraulic characteristics of the columns, which were determined from bromide tracer data, and the influent BTX concentrations are shown in Table 1.

Analytical Procedures

BTX compounds were analyzed with a Hewlett Packard 5890 gas chromato-graph equipped with a flame ionization detector as described by Alvarez and

TABLE 1. Hydraulic characteristics and operating conditions of aquifer columns.

Parameter	Value
Porosity (fraction of total volume)	0.37
Pore velocity (cm/h)	6.8
Dispersion coefficient (cm²/h)	0.29
Retardation factors (relative to Br⁻)	
Benzene	2.1
Toluene	4.2
o-Xylene	8.5
Influent substrate concentrations [a]	
Benzene (µg/L)	193 ± 16
Toluene (µg/L)	183 ± 18
o-Xylene (µg/L)	192 ± 15
Acetate (mg/L)	2 ± 0.1
Benzoate (mg/L)	1 ± 0.1

(a) All three BTX compounds were fed concurrently. One column was fed acetate, and another one was fed benzoate. Values depict the mean ± one standard deviation.

Vogel (1991). Benzoate and phenylalanine were analyzed by high-performance liquid chromatography (Gilson) equipped with an ultraviolet-visible detector (Spectra-Physics). Bromide was analyzed by ion chromatography using a Dionex 4500i ion chromatograph. A biological oxygen monitor (YSI 530), equipped with a microchamber and an oxygen microprobe was used to measure dissolved oxygen concentrations. The detection limits were about 1 µg/L for each BTX compound, 0.1 mg/L for benzoate and phenylalanine, and 0.1 mg/L for dissolved oxygen.

RESULTS AND DISCUSSION

BTX degradation rates are directly proportional to the "active" microbial concentration (Alvarez et al. 1991, 1994; Chen et al. 1992). Therefore, efforts leading to an increase in the concentration of desirable phenotypes may reduce the duration (and cost) of bioremediation. Bioaugmentation, in which specialized exogenous microbes are inoculated into aquifers, often fails to significantly enhance bioremediation because the concentration of the target contaminants may be too low to support their growth; or because the added microbes may be susceptible to toxins or predators in the environment, may use other organic compounds in preference of the pollutant, or may be unable to move through soil sites containing the contaminants (Goldstein et al. 1985). An alternative technique to speed up the microbial degradation process is the proposed selective proliferation of appropriate indigenous microorganisms. This could be achieved by the addition of environmentally benign aromatic substrates, provided that their addition did not have the adverse ecological effect of selecting a microbial population that was ineffective in degrading the target BTX. To address this concern, we studied the effect of potential stimulatory substrates on the phenotypic composition of indigenous microbial communities.

Addition of benzoate or phenylalanine (at 1 mg/L) exerted preferential selective pressure for BTX degraders in batch studies (Figure 2). The concentration of BTX degraders increased by two orders of magnitude in these microcosms. Acetate addition, however, did not stimulate a statistically significant increase in the concentration of BTX degraders. Apparently the aromatic nature of benzoate and phenylalanine provided a competitive advantage for the proliferation of BTX degraders.

BTX removal in viable, but not sterile, columns provided evidence of biodegradation. Benzoate addition (1 mg/L) enhanced aerobic BTX degradation and attenuated the breakthrough of benzene (Figure 3a), toluene (Figure 3b), and *ortho*-xylene (Figure 3c) relative to acetate-amended (2 mg/L) or unamended control columns. Both the maximum BTX concentrations that eluted and the

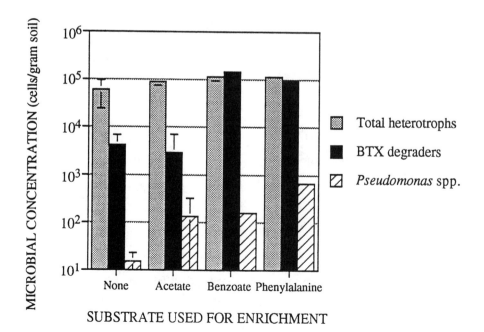

FIGURE 2. Effect of substrate addition on phenotypic selection in microcosms.

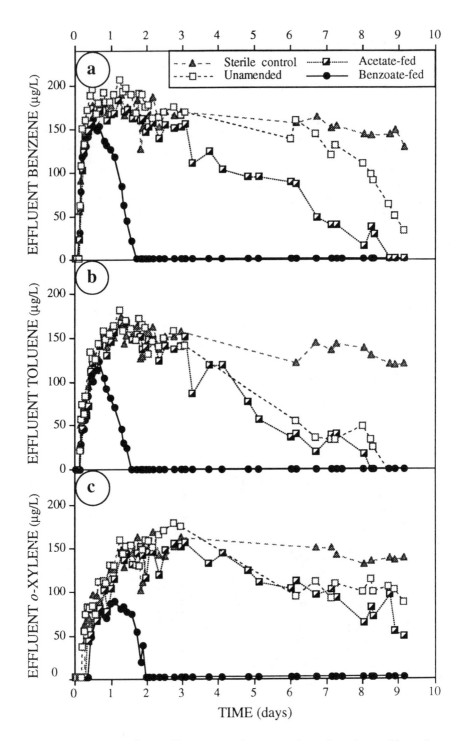

FIGURE 3. Concentrations of benzene, toluene, and *o*-xylene in aquifer column effluent.

TABLE 2. Attenuation of BTX breakthrough by benzoate addition to aquifer columns.

	Treatment[a]			
	Benzoate-fed	Acetate-fed	Unamended	Sterile Control
Benzene	85% (1.0 days)	91% (>10 days)	92% (8.0 days)	98% (NA)
Toluene	64% (1.4 days)	89% (8.0 days)	95% (9.0 days)	98% (NA)
o-Xylene	45% (2.0 days)	83% (>10 days)	94% (9.2 days)	100% (NA)

(a) Values depict the maximum BTX concentration that broke through as a percentage of the influent concentration. Numbers in parentheses represent the time required for the effluent BTX concentration to drop below the detection limit (i.e., 1 µg/L).

time required for the effluent BTX concentration to drop below detection levels were significantly lower in the benzoate-amended column. BTX compounds were not detected after 2 days in the effluent of the benzoate-fed column, but it took longer than 8 days for the acetate-fed or unamended control to remove BTX below detection levels (Table 2). No benzoate or acetate was detected in the column effluents.

It is interesting that benzoate addition has been reported to have a potential adverse effect: *Pseudomonas putida* mt-2 cells that contain the TOL plasmid may lose it when grown on benzoate, which could cause them to lose their ability to degrade toluene, *m*-xylene, and *p*-xylene. The TOL plasmid was cured when this strain was grown on benzoate because of a greater growth rate associated with the benzoate-induced *ortho*-fission pathway (Williams & Murray 1974). Nevertheless, little is known about the effect that benzoate has on BTX degradation by other genera, or the overall effect of benzoate addition when mixed indigenous cultures are involved. These results indicate that addition of relatively small amounts of benzoate can have an overall positive effect on BTX bioremediation.

In conclusion, bioremediation shows great promise as an approach to hazardous waste management. The full potential of BTX bioremediation, however, can be limited by the presence and expression of appropriate catabolic capacities. Such limitations could be overcome by the addition of environmentally benign aromatic substrates (e.g., benzoate), which could stimulate metabolic and population shifts that increase the efficiency of bioremediation.

ACKNOWLEDGMENT

This work was funded by the Iowa State Water Resources Research Institute under Grant #14-08-0001-G2019-02.

REFERENCES

Aelion, C. M., and P. M. Bradley. 1991. "Aerobic Biodegradation Potential of Subsurface Micro-organisms from a Fuel-Contaminated Aquifer." *Appl. Environ. Microbiol.* 57: 57-63.

Alvarez, P.J.J., and T. M. Vogel. 1991. "Substrate Interactions of Benzene, Toluene and para-Xylene During Microbial Degradation by Pure Cultures and Mixed Culture Aquifer Slurries." *Appl. Environ. Microbiol.* 57: 2981-2985.

Alvarez, P.J.J., P. J. Anid, and T. M. Vogel. 1991. "Kinetics of Aerobic Biodegradation of Benzene and Toluene in Sandy Aquifer Material." *Biodegradation* 2: 43-51.

Alvarez, P.J.J., P. J. Anid, and T. M. Vogel. 1994. "Kinetics of Toluene Degradation by Denitrifying Aquifer Microorganisms." *ASCE J. Environ. Engrg.* 120: 1327-1336.

Anid, P. J., P.J.J. Alvarez, and T. M. Vogel. 1993. "Biodegradation of Monoaromatic Hydrocarbons in Aquifer Columns Amended with Hydrogen Peroxide and Nitrate." *Wat. Res.* 27: 685-691.

Barbaro, J. R., J. F. Barker, L. A. Lemon, R. W. Gillham, and C. I. Mayfield. 1990. *In Situ Cleanup of Benzene in Groundwater by Employing Denitrifying Bacteria.* PACE Report No. 91. Canadian Petroleum Products Institute, Ottawa, Canada.

Bouwer, E. J., and P. L. McCarty. 1984. "Modeling of Trace Organics Biotransformation in the Subsurface." *Groundwater* 22: 433-440.

Chen, Y. M., L. M. Abriola, P.J.J. Alvarez, P. J. Anid, and T. M. Vogel. 1992. "Biodegradation and Transport of Benzene and Toluene in Sandy Aquifer Material: Model-Experiment Comparisons." *Water Resources Research* 28: 1833-1847.

Egli, T., U. Lendenmann, and M. Snozzi. 1993. "Kinetics of Microbial Growth with Mixtures of Carbon Sources." *Antonie van Leeuwenhoek* 63: 289-298.

Goldstein, R. M., L. M. Mallory, and M. Alexander. 1985. "Reasons for Possible Failure of Inoculation to Enhance Biodegradation." *Appl. Environ. Microbiol.* 50: 977-983.

Linkfield, T. G., J. M. Suflita, and J. M. Tiedje. 1989. "Characterization of the Acclimation Period Before Anaerobic Dehalogenation of Chlorobenzoates." *Appl. Environ. Microbiol.* 55: 2773-2778.

Ridgeway, H. F., J. Safarik, D. Phipps. P. Carl, and D. Clark. 1990. "Identification and Catabolic Activity of Well-Derived Gasoline-Degrading Bacteria from a Contaminated Aquifer." *Appl. Environ. Microbiol.* 56: 3565-3575.

Stapps, S. J. 1990. *International Evaluation of In Situ Biorestoration of Contaminated Soils and Groundwater.* EPA/540/2-90/012. U.S. Environmental Protection Agency. September.

Williams, P. A., and K. Murray. 1974. "Metabolism of Benzoate and the Methylbenzoates by *Pseudomonas putida (arvilla)* mt-2: Evidence for the Existence of a TOL Plasmid." *J. Bacteriol.* 120: 416-423.

Wilson, J., L. Leach, J. Michalowski, S. Vandergrift, and R. Callaway. 1990. "In Situ Reclamation of Spills from Underground Storage Tanks: New Approaches for Site Characterization, Project Design, and Evaluation of Performance." In: I. P. Muraka and S. Cordle (Eds.), *Proc. Environmental Research Conference on Groundwater Quality and Waste Disposal.* Washington DC May 2-4, 1989. pp. 30-1–30-16.

Substrate Range of the (Chloro)biphenyl Degradation Pathway of *Alcaligenes* sp. JB1

John R. Parsons, Heleen Goorissen, Arjan R. Weiland,
Jolanda A. de Bruijne, Dirk Springael,
Daniel van der Lelie, and Max Mergeay

ABSTRACT

The genes encoding the biphenyl degradation pathway of *Alcaligenes* sp. strain JB1 were cloned in an R-prime plasmid in *Alcaligenes eutrophus* strain AE53. Subsequently, recombinant *Alcaligenes eutrophus* strains containing the *bphABC* genes of JB1 were selected by the accumulation of a yellow ring-cleaved product from biphenyl and 2,3-dihydroxy-biphenyl. Gluconate-grown cultures of one of these recombinants were tested for their ability to degrade 2,2′,3,3′-tetrachlorobiphenyl, 2-chloro-dibenzo-*p*-dioxin, 1-chloronaphthalene, 1,1,1-trichloro-2,2-*bis*(4-chloro-phenyl)ethane (*p,p*′-DDT), and 2,4′-dichlorodiphenyl ether. Depending on their solubility, the initial concentrations of these compounds were between 10 and 100 µg/L. In all cases more than 65% of these compounds was removed within a 24-h incubation period. These results show that the enzymes responsible for degrading (chloro)biphenyls in strain JB1 are able to attack a wide range of aromatic compounds. This strain, or those containing similar enzymes, may be applicable to the bioremediation of mixtures of contaminants.

INTRODUCTION

The biodegradation of polychlorinated biphenyls (PCBs) has been studied extensively in many bacterial strains (Abramowicz 1990). In almost all cases, degradation takes place via an initial dioxygenation in the *ortho* and *meta* positions of the least chlorinated ring to yield dihydroxy derivatives, followed by a second dioxygenation, resulting in ring cleavage and the formation of chlorinated benzoates. Many strains able to carry out these reactions have been isolated and have been studied for their ability to degrade different PCB congeners. In contrast, relatively little attention has been paid to the ability of such strains to degrade other aromatic compounds. However, there is some evidence to suggest that the catabolic enzymes used to degrade chlorinated biphenyls are closely related to those involved in the biodegradation of other chlorinated and polycyclic aromatic compounds.

Recently, it was reported that the biphenyl-degrading bacterium *Pseudomonas paucimobilis* Q1 was also able to grow on naphthalene and that the same ring cleaving dioxygenase was active on both substrates (Kuhm et al. 1991). Other authors reported that *Pseudomonas cepacia* Et4 degraded diphenyl ether by a pathway similar to that used by other bacteria to degrade biphenyl (Pfeifer et al. 1993) and that the chlorobiphenyl-degrading strain *Alcaligenes eutrophus* A5 degrades 1,1,1-trichloro-2,2-*bis*(4-chlorophenyl)ethane (*p,p'*-DDT) to 4-chloro-benzoate, again apparently by the same pathway (Nadeau et al. 1994).

Beijerinckia sp. strain B1 is capable of growth on biphenyl, naphthalene, anthracene, and phenanthrene (Gibson et al. 1973) and of transforming dibenzo-*p*-dioxin and some of its chlorinated derivatives (Klečka & Gibson 1980). A common set of enzymes may be involved in all these reactions (Zylstra et al. 1994). Furthermore, recently isolated dibenzo-*p*-dioxin- and dibenzofuran-utilizing strains degrade these compounds by pathways resembling the biphenyl degradation pathway mentioned above (e.g., Fortnagel et al. 1990, Strubel et al. 1991, Wittich et al. 1992).

Alcaligenes sp. strain JB1 is able to grow on biphenyl and naphthalene and to cometabolize chlorinated biphenyls (Parsons et al. 1988), dibenzo-*p*-dioxins (Parsons & Storms 1989), and dibenzofurans (Parsons et al. 1990). We report here the first results of an investigation of whether the same set of enzymes is involved in the degradation of these and some related compounds.

MATERIALS AND METHODS

Alcaligenes sp. strain JB1 (originally tentatively identified as a *Pseudomonas* strain) was isolated from soil with biphenyl as the growth substrate (Parsons et al. 1988). The chromosomal genes encoding the biphenyl degradation pathway of strain JB1 were cloned by RP4::Mu3A-mediated transfer to *Alcaligenes eutrophus* strain CH34 (Springael et al. 1994). The wide host range cosmid vector pLAFR3 was then used to prepare *E. coli* recombinants containing RP4::Mu3A::*bph*⁺ DNA fragments. These recombinants were tested for the presence of the *bphC* gene. Transconjugation with *A. eutrophus* CH34 yielded recombinants which were selected for the presence of the *bphA,B,C* genes and the absence of the *bphD* gene. These strains were detected by their accumulation of yellow 2-hydroxy-6-oxo-6-phenylhexa-2,4-dienoic acid from biphenyl and 2,3-dihydroxybiphenyl and their lack of growth on biphenyl. One of these recombinant strains, AE1669, was used in the experiments described here. This procedure will be described more fully elsewhere (Springael et al. in preparation).

Cultures of AE1669 and CH34 were grown on gluconate (2 g/L) in the following media: $Na_2SO_4 \cdot 10H_2O$ (1.0 mM), $CaCl_2 \cdot 2H_2O$ (0.01 mM), $MgCl_2 \cdot 6H_2O$ (0.625 mM), NH_4Cl (50 mM), KH_2PO_4 (30 mM), $N(CH_2COOH)_3$ (1 mM), ZnO (12.7 μM), $FeCl_3 \cdot 6H_2O$ (50 μM), $MnCl_2 \cdot 4H_2O$ (25.3 μM), $CuCl_2 \cdot 2H_2O$ (2.54 μM),

CoCl$_2$·6H$_2$O (5 μM), HBO$_3$ (2.58 μM), and Na$_2$MoO$_4$·2H$_2$O (0.046 μM) at a pH of 7.0 ± 0.1. Tetracycline (10 mg/L) was added to the medium to select for the pLAFR3::*bphABC* plasmid in cultures of AE1669.

Cometabolism experiments were carried out by adding 25 mL of an overnight culture of AE1669 to 75 mL fresh medium. The compounds to be tested were then added to an initial concentration of 10 to 100 μg/L, and the cultures were incubated at 30°C. Control experiments were carried out with cultures of CH34, which is not able to degrade biphenyl, in order to determine possible losses of substrate caused by other enzymes in CH34 or abiotic processes.

Duplicate 10-mL samples were taken after 0, 1, 2, 4, and 24 h and were extracted twice with 10 mL pentane after addition of 10 ng 2,2',4,5'-tetra-chlorobiphenyl as a recovery standard. The extracts were eluted through columns containing 100 to 120 mesh silica + 44% w/w H$_2$SO$_4$ above 100 to 120 mesh silica + 33% w/w 1 M NaOH to remove lipids and polar compounds. Analysis was by gas chromatography/electron-capture detection, using a 30 m × 0.32 mm DB5 column and pentachlorobenzene as internal standard.

RESULTS AND DISCUSSION

The results described here show that gluconate-grown cultures of *Alcaligenes eutrophus* strain AE1669, containing the *bphABC* genes of *Alcaligenes* sp. strain JB1, are able to transform 2,2',3,3'-tetrachlorobiphenyl (2,2',3,3'-TCB, Figure 1), 1,1,1-trichloro-2,2-*bis*(4-chlorophenyl)ethane (*p,p'*-DDT, Figure 2), 2,4'-dichloro-diphenyl ether (2,4'-DCDE, Figure 3) and 1-chloronaphthalene (1-CN, Figure 4). Transformation of these compounds was not detected in control cultures of *Alcaligenes eutrophus* CH34. Unfortunately, a simultaneous control experiment with 1-CN was not carried out due to poor growth of the culture of CH34. However, in another experiment CH34 showed no transformation of 1-CN (an initial concentration of 1.72 nmol/L was unchanged after 24 h). Furthermore, in an incubation of an overnight culture of AE1669 with 2-chlorodibenzo-*p*-dioxin (2-CDD), an initial concentration of 13.9 μmol/L 2-CDD had declined to a level under the detection level (ca. 0.5 nmol/L) after 6 h. In an experiment using a sterilized culture of AE1669, an initial concentration of 1.42 nmol/L 2-CDD had decreased only to 1.37 nmol/L after 24 h.

These results indicate that the enzymes involved in the degradation of chlorobiphenyls in strain JB1 are able to attack a wide range of aromatic compounds, including polycyclic aromatic compounds such as 1-CN and 2-CDD. To our knowledge, the transformation of a similarly wide range of substrates by one or more of the enzymes of one pathway has only been reported for *Beijerinckia* sp. strain B1 (Zylstra et al. 1994), although it is not clear to what extent this strain is able to transform chlorinated substrates. Chlorinated biphenyl-degrading strains have been divided into several classes based either on their specificity towards different PCB congeners (Bedard & Haberl 1990) or on the properties of their *bph* operon (Hayase et al. 1989). At the moment, it is not clear to which class of PCB-degrading bacteria *Alcaligenes* sp. strain JB1 belongs and

2,2',3,3'-TCB

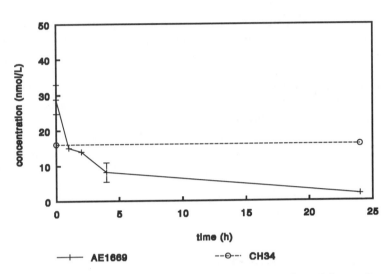

FIGURE 1. Transformation of 2,2',3,3'-tetrachlorobiphenyl by a gluconate-grown culture of *Alcaligenes eutrophus* strain AE1669, with *Alcaligenes eutrophus* strain CH34 as control.

p,p'-DDT

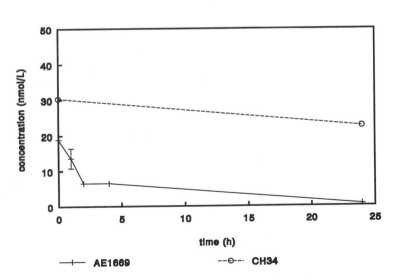

FIGURE 2. Transformation of 1,1,1-trichloro-2,2-*bis*(4-chlorophenyl)ethane by a gluconate-grown culture of *Alcaligenes eutrophus* strain AE1669, with *Alcaligenes eutrophus* strain CH34 as control.

2,4'-DCDE

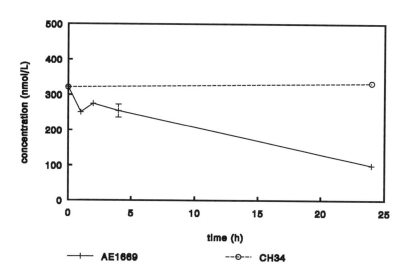

FIGURE 3. Transformation of 2,4'-dichlorodiphenyl ether by a gluconate-grown culture of *Alcaligenes eutrophus* strain AE1669, with *Alcaligenes eutrophus* strain CH34 as control.

1-CN

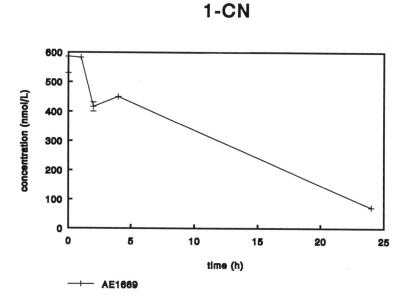

FIGURE 4. Transformation of 1-chloronaphthalene by a gluconate-grown culture of *Alcaligenes eutrophus* strain AE1669.

whether other PCB-degrading strains are also able to degrade other polycyclic and chlorinated aromatic compounds. Further studies will be directed towards a more complete characterization of the substrate range of the chlorobiphenyl pathway of strain JB1.

Because soil, sediment, and groundwater are often contaminated by mixtures of different aromatic compounds, *Alcaligenes* sp. strain JB1, or bacteria containing similar pathways, may be applicable to the bioremediation of such mixtures of contaminants.

REFERENCES

Abramowicz, D. A. 1990. "Aerobic and anaerobic biodegradation of PCBs: A review." *Crit. Rev. Biotechnol.* 10: 241-251.

Bedard, D. L., and M. L. Haberl. 1990. "Influence of chlorine substitution pattern on the degradation of polychlorinated biphenyls by eight bacterial strains." *Microb. Ecol.* 20: 87-102.

Fortnagel, P., H. Harms, R.-M. Wittich, S. Krohn, H. Meyer, V. Sinnwell, H. Wilkes, and W. Francke. 1990. "Metabolism of dibenzofuran by *Pseudomonas* sp. strain HH69 and the mixed culture HH27." *Appl. Environ. Microbiol.* 56: 1148-1156.

Gibson, D. T., R. L. Roberts, M. C. Wells, and V. M. Kobal. 1973. "Oxidation of biphenyl by a *Beijerinckia* species." *Biochem. Biophys. Res. Commun.* 50: 211-219.

Hayase, N., K. Taira, and K. Furukawa. 1989. "*Pseudomonas putida* KF715 *bphABCD* operon encoding biphenyl and polychlorinated biphenyl degradation: cloning, analysis and expression in soil bacteria." *J. Bacteriol.* 172: 1160-1164.

Klečka, G. M., and D. T. Gibson. 1980. "Metabolism of dibenzo-*p*-dioxin by a *Beijerinckia* species. *Appl. Environ. Microbiol.* 39: 288-296.

Kuhm, A. E., A. Stolz and H.-J. Knackmuss. 1991. "Metabolism of naphthalene by the biphenyl-degrading bacterium *Pseudomonas paucimobilis* Q1." *Biodegradation* 2: 115-120.

Nadeau, L. J., F.-M. Menn, A. Breen, and G. S. Sayler. 1994. "Aerobic degradation of 1,1,1-trichloro-2,2-bis(4-chlorophenyl)ethane (DDT) by *Alcaligenes eutrophus* A5." *Appl. Environ. Microbiol.* 60: 51-55.

Parsons, J. R., D. T. H. M. Sijm, A. van Laar, and O. Hutzinger. 1988. "Biodegradation of chlorinated biphenyls and benzoic acids by a *Pseudomonas* strain." *Appl. Microbiol. Biotechnol.* 29: 81-84.

Parsons, J. R., and M. C. M. Storms. 1989. "Biodegradation of chlorinated dibenzo-*p*-dioxins in batch and continuous cultures of strain JB1." *Chemosphere,* 19: 1297-1308.

Parsons, J. R., C. Ratsak, and C. Siekerman. 1990. "Biodegradation of chlorinated dibenzofurans by an *Alcaligenes* strain." In O. Hutzinger and H. Fiedler (Eds.), *Organohalogen Compounds,* Vol. 1, Proc. Dioxin '90 - EPRI Seminar, Bayreuth, Germany, Sept. 10-14, 1990, pp. 377-380. Ecoinforma Press, Bayreuth, Germany.

Pfeifer, F., H. G. Trüper, J. Klein, and S. Schacht. 1993. "Degradation of diphenylether by *Pseudomonas cepacia* Et4: enzymatic release of phenol from 2,3-dihydroxydiphenylether." *Arch. Microbiol.* 159: 323-329.

Springael, D., L. Diels, and M. Mergeay. 1994. "Transfer and expression of PCB-degradative genes into heavy metal resistant *Alcaligenes eutrophus* strains." *Biodegradation,* 5: 343-357.

Springael, D., J. van Thor, A. Ryngaert, H. Goorissen, L.C.M. Commandeur, J. R. Parsons, and M. Mergeay. "Transfer of the chlorobiphenyl catabolic pathway of *Alcaligenes denitrificans* JB1 into soil bacteria by means of RP4::Mu3A mediated prime plasmid formation." In preparation.

Strubel, V., K.-H. Engesser, P. Fischer, and H.-J. Knackmuss. 1991. "3-(2-Hydroxyphenyl)-catechol as substrate for proximal *meta* ring cleavage in dibenzofuran degradation *Brevibacterium* sp. strain DPO 1361." *J. Bacteriol.* 173: 1932-1937.

Wittich, R.-M., H. Wilkes, V. Sinnwell, W. Francke, and P. Fortnagel. 1992. "Metabolism of dibenzo-*p*-dioxin by *Sphingomonas* sp. strain RW1." *Appl. Environ. Microbiol.* 58: 1005-1010.

Zylstra, G. J., X. P. Wang., E. Kim, and V.A. Didolkar. 1994. "Cloning and analysis of the genes for polycyclic aromatic hydrocarbon degradation." In R. K. Bajpai and A. Prokop (Eds.), *Recombinant DNA Technology.* Annals of the New York Academy of Sciences. Vol. 721. pp. 386-398. New York Academy of Sciences. New York, NY.

Controlled Degradation of Toxic Compounds in a Nutristat, a Continuous Culture

Michiel Rutgers, Anton M. Breure, and Johan G. van Andel

ABSTRACT

A continuous culture with control of the substrate concentration, a "nutristat," was constructed to study the inhibition kinetics of pentachlorophenol (PCP). A PCP-degrading isolate (strain P5) was grown in this system on PCP as the only source of carbon and energy at several set-point PCP concentrations ranging from 18 to 1,060 µM. The controlled value deviated from the set-point concentration by 1% maximally. In the PCP-nutristat, the steady-state dilution rate (and hence specific growth rate of the bacteria) had a maximum of $0.142 \pm 0.004 \text{ h}^{-1}$ at PCP concentrations between 30 and 200 µM. At PCP concentrations over 200 µM, the steady-state specific growth rate decreased due to inhibition. It was concluded that the nutristat is very useful in establishing steady-state situations at growth-inhibiting PCP concentrations, conditions for which the chemostat is not suitable.

INTRODUCTION

Many manmade and hazardous organic chemicals are potentially subjected to microbial degradation in natural environments. However, recalcitrance of otherwise biodegradable compounds often is observed for a number of possible reasons (Alexander 1985). One of these reasons might be the high toxicity of the chemical at the prevailing concentration, which inhibits the microbial population.

To assess the potential of microbial populations to remediate heavily contaminated sites, it is necessary to determine kinetic parameters such as the affinity of the degrader population for the pollutant and inhibition constants. In addition, this type of knowledge may help to address the origin of the recalcitrance of chemicals in natural environments. For instance, recalcitrance is expected when concentrations of chemicals importantly exceed the inhibition constants determined in the laboratory.

Batch cultures can be used to determine kinetic parameters from the curves of disappearance of a chemical (Simkins & Alexander 1984). A serious drawback of the batch culture technique for determining kinetic constants is that micro-organisms do not have much time to adapt to the desired conditions and subsequently have to cope with a continuously changing environment. This is the result of a microbial process in a closed environment where substrates are consumed and biomass and/or products are produced. The kinetics in these batch cultures might thus be affected by the physiological state of the inoculum (Owens & Legan 1987).

Chemostat cultures are generally considered superior to batch cultures for determining kinetic constants of microbial growth, especially the affinity constant (K_s). Replenishment of exhausted nutrients takes place continuously in these type of systems, ultimately resulting in a steady-state situation in which organisms grow at a constant rate in a constant environment (Tempest 1970). However, with toxic substrates (such as PCP), chemostats are inherently unstable (Figure 1; Edwards 1970). The answer to this problem is the nutristat (Rutgers et al. 1993), which is a continuous culture with control of the substrate concentration (Figure 2). The important differences between the chemostat and the nutristat are best characterized by the controlled parameters and the variables (Table 1). In the chemostat, the investigator controls the growth rate of the microorganisms by setting the dilution rate of the system at a desired value. In contrast, the controlled parameter in the nutristat is the concentration of the (limiting or toxic) substrate. Consequently, the growth rate of the microorganisms (and hence the dilution rate) in the nutristat system is a variable. With the nutristat system

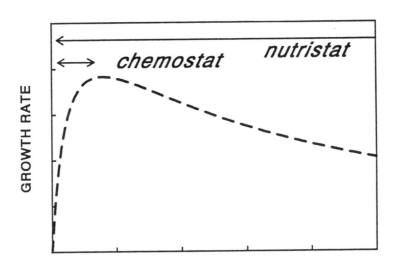

CONCENTRATION OF TOXIC SUBSTRATE

FIGURE 1. Hypothetical relationship between growth rate and the concentration of a toxic substrate according to the Haldane equation ($\mu = \mu_{max} \cdot s/(s + K_s + s^2/K_i)$). Domains are indicated for which the chemostat or nutristat applies.

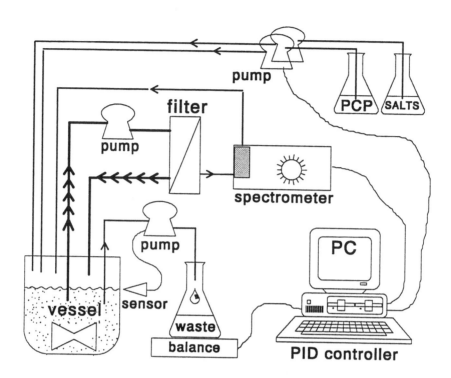

FIGURE 2. Scheme of the nutristat. The culture suspension is continuously circulated through a tangential (crossflow) filter by means of a high-rate peristaltic pump. The clear cell-free filtrate is analyzed by a flowthrough spectrophotometer at 313 nm. The A_{313} signal (V) is converted to digital units and processed by a personal computer using a PID closed-loop control algorithm. The output is fed to the peristaltic medium pump for the control of the feeding rate of PCP and salts solution. Also drawn is the system for overflow of culture suspension. The effluent generated by the overflow device is continuously weighed in order to produce a signal for the on-line calculation of the dilution rate.

one can thus control the environment completely and look for the response (growth rate) of the microorganisms. Clearly, this type of apparatus accommodates well in the discipline of microbial physiology, because the organism is the focal point in this system. When the preset conditions are harsh, the organisms stop growing but still remain in the system. Organisms in the chemostat wash out from the system when conditions are unfavorable; i.e., the process is the focal point.

In this paper we present a nutristat with control of the PCP concentration. With this apparatus we determined the effect of the concentration of PCP on the growth of *Flavobacterium* P5 with PCP as only source of carbon and energy.

TABLE 1. Some essential characteristics of nutristat and chemostat systems (see text).

Characteristic	Chemostat	Nutristat
Dilution rate	Fixed	Variable
[Substrate]	Variable	Fixed
Domain	Limited growth ($< \mu_{max}$)	All growth rates (often μ_{max})
Requirement	—	Fast and sensitive substrate sensor

EXPERIMENTAL PROCEDURES AND MATERIALS

Medium, Strain, and Equipment

A low-chloride mineral salts medium was used containing $(NH_4)_2HPO_4$ (5 mM), KH_2PO_4 (2 mM), nitrilotriacetic acid (1 mM), $MgSO_4$ (1 mM), and a trace-element solution (1 mL L^{-1}) was used for the nutristat experiments. The trace-element solution contained $FeSO_4$ (20 mM), $MnSO_4$ (10 mM), $CoSO_4$ (10 mM), $CaSO_4$ (10 mM), $CuSO_4$ (1 mM), and $ZnSO_4$ (1 mM) dissolved in 1 M HNO_3. A solution of 10 mM pentachlorophenol (Fluka, Buchs, Switzerland) in 60 mM KOH was supplied separately to the continuous cultures as the carbon and energy source. Mineral medium and the PCP solution were fed to the nutristat culture with a multichannel peristaltic pump (Minipuls 3; Gilson, Villiers le Bel, France) at approximately equal rates.

The PCP-degrading strain, *Flavobacterium* sp. strain P5, was recently isolated and described by Van Gestel et al. (1993). The nutristat experiments were carried out in continuous culture equipment as described before (Rutgers et al. 1993).

On-Line Monitoring System and
Control of the PCP Concentration

On-line monitoring of the PCP concentration in the culture was achieved by continuous recirculation of culture fluid (approx. 300 mL min^{-1}) through a cross-flow filter (Mini Ultrasette, pore size 0.16 μm; Filtron Technology, Northborough, Massachusetts) providing a clear filtrate (approx. 1 mL min^{-1}). This filtrate was monitored continuously at 313 nm using a flowthrough spectrophotometer (UV-MII, Pharmacia, Uppsala, Sweden) and recirculated to the culture.

As a precaution to the accumulation of active biomass and particles, the filter flow was reversed (daily) or the tangential filter was removed and replaced by a rinsed and sterilized filter (each 2 days). Replacement of the filter did not affect the PCP concentration or the dilution rate of the nutristat, indicating negligible accumulation of PCP and PCP-degrading activity (bacteria) in the filter during the nutristat runs.

A personal computer (Intel 486DX, 33 MHz) equipped with an AD/DA converter (PCL-812PG, Advantech Corp., Sunnyvale, California) and data acquisition and control software (Labtech Notebook 7.2, Labtech Corp., Wilmington, Massachusetts) was used as monitoring and controlling interface. Closed-loop control with a proportional integral derivative (PID) algorithm was used: the input signal was generated by the PCP-monitoring system, the output signal was used after conversion to analog units to control the peristaltic pump (Minipuls 3, Gilson, France) for PCP and mineral medium.

Determinations

PCP concentrations in supernatants of the nutristat cultures were routinely quantified by measuring the absorbance at 318 nm as described by Rutgers et al. (1993). In addition, a high-performance liquid chromatography (HPLC) method was used to determine PCP concentrations in supernatants of the cultures and to validate the determinations of PCP from measurement of the A_{318}. Samples (1 mL) were mixed with acetonitrile (0.5 mL; HPLC grade, acidified with 0.01% v/v *ortho*-phosphoric acid). The HPLC (1050 series; Hewlett Packard Co., Avondale, Pennsylvania) was equipped with a variable-wavelength detector, a C18 (100 × 3 mm) column (Chromspher 5; Chrompack, Middelburg, The Netherlands), and an autosampler. The mobile phase was an isocratic mixture of 75% acetonitrile and 25% KH_2PO_4 (10 mM) with a flow of 0.4 mL min^{-1}. The injection volume was 10 µL. The detection wavelength was 215 nm. Integration and a subsequent data analysis were performed using the Hewlett Packard software (HPLC^{3D} Chemstation). Carbon dioxide in the headspace of the nutristat was determined with a gas chromatograph equipped with molecular sieve and a thermal conductivity detector as described by Volkering et al. (1995). Measurement of the absorbance of samples from the nutristat culture at 540 nm was used as a biomass indicator. Protein, dry weight, chloride concentrations, and gas flowrates were determined as described previously (Rutgers et al. 1993).

RESULTS AND DISCUSSION

Strain P5 was grown aerobically in the nutristat with PCP as the sole carbon and energy source. Figure 3 shows a typical example of the initial phase of a nutristat run. At time zero, the system was inoculated with cells pregrown on PCP in mineral medium in a batch culture. In the beginning, the system was operated as a chemostat with low dilution rates (0.001 to 0.01 h^{-1}). When the optical density at 540 nm (OD_{540}) had reached 0.08 after about 125 h, the nutristat system was started by forcing the set-point value for the A_{313} of the spectrophotometer (which is correlated to the PCP concentration) to 2.25 volt (the units for the control system) with support of the on-line measurement and control system. A personal computer was incorporated for on-line data acquisition and closed-loop control. A PID algorithm processed by a personal computer generated the output signal to control the peristaltic medium pump. In the nutristat

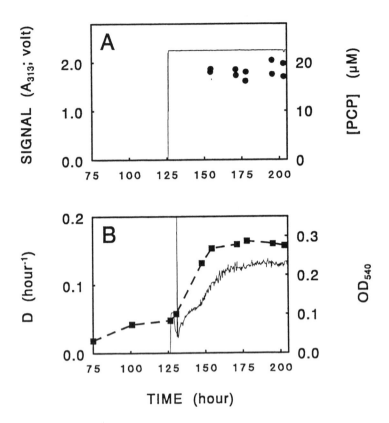

FIGURE 3. Example of the startup of a nutristat run. (A) Picture of the controlled signal, representing the A_{313} reading (solid line) of the spectrophotometer (in volt) and the off-line measured PCP concentration (●) as
a function of time. The nutristat experiment was started at 125 hours using
a set point of 2.25 V. (B) Course of the dilution rate (D; solid line) and
OD_{540} (-■-) during the nutristat run.

run shown in Figure 3, the set point value was 2.25 volt and the controlled value
deviated from this value by 0.01 volt (1%) maximally. It was concluded that
tight control of the A_{313} was accomplished.

The PCP concentration in the culture during the nutristat run was determined
off line by HPLC analysis and by A_{318} measurements (Figure 3A). The 2.25 volt
set-point corresponded to 18 ± 1 μM from 150 to 200 h. The error was bigger
than 1% for a number of reasons: (1) additional error introduced by the off-line
determination of the PCP concentration, (2) the background value for the A_{313}
was not zero and not completely constant, and (3) crossflow filter performance
affected to a slight extent the final A_{313} reading. However, it was assumed that
the final control of the PCP concentration was sufficiently tight for establishing
steady-state conditions.

Comparison of the present nutristat system with the prototype nutristat (Rutgers et al. 1993) revealed a considerable improvement with respect to the control of the PCP concentration. The prototype used an on/off control mechanism, which resulted in oscillating PCP concentrations in the culture between 45 and 77 µM in the offered case. The establishment of steady-state situations in the nutristat, and hence the demonstration of constant flow rates, may allow for the satisfactory use of the PID algorithm.

The effluent of the nutristat system was continuously weighed to allow on-line quantifying of the dilution rate (D). After starting up the nutristat system at 125 h, the D was 0.01 h^{-1} but increased in about 40 h to about 0.13 h^{-1} (Figure 3B). From 165 to 200 hour, the average D and the average OD_{540} remained constant at 0.131 ± 0.005 h^{-1} (n = 160) and 0.281 (n = 4), respectively. When the D and the OD_{540} did not show a decreasing or increasing tendency, a steady state was assumed after at least 5 vessel volume changes of the medium. Thus at about 200 h the nutristat was in a steady state with a D of 0.131 h^{-1} at 18 µM PCP (Figure 3). Mass balances of carbon and chlorine showed recoveries of 91 to 104%, and 91 to 98%, respectively. These results indicate that PCP was completely converted to biomass, carbon dioxide, and aqueous chloride without significant accumulation of intermediates or sorption of PCP onto biomass.

Nutristat runs with set-point PCP concentrations ranging from 18 µM to 1,060 µM PCP were performed. Figure 4 shows the relationship of the steady-state D (and hence specific growth rate) and the PCP concentration derived from these nutristat runs. The maximum D was 0.142 ± 0.004 h^{-1} at PCP concentrations between 30 and 200 µM. At concentrations higher than 200 µM, PCP caused inhibition of growth as is demonstrated by a decreasing D with increasing PCP concentration.

Relations between the concentration of a (toxic) substrate (such as PCP) and growth rate are often modeled with the Haldane equation (Figure 1; Edwards 1970), which was derived from enzyme kinetics. We attempted to fit the data of Figure 4 with the Haldane equation, but the best fit seriously overpredicted the growth rate at high PCP concentrations. Considering the microbial cell as a single enzyme for which the inactive substrate-enzyme complex accounts for inhibition is probably an oversimplification of reality.

Radehaus and Schmidt (1992) analyzed the influence of the concentration of PCP on the growth of *Pseudomonas* RA2 in batch cultures. They did not use the Haldane equation, but fitted the descending part of the relationship of the PCP concentration versus growth rate with a straight line. They defined then an inhibition constant (K_i) as the concentration at which the growth rate was half the maximum observed growth rate at optimum PCP concentrations. For *Pseudomonas* RA2 the K_i was 580 µM. To compare our isolate, strain P5, with *Pseudomonas* RA2 we followed the same procedure. The K_i for strain P5 in the steady-state nutristat culture was 830 µM. Edgehill and Finn (1982) used batch cultures with *Arthrobacter* sp. ATCC33790, which were fed manually to maintain the PCP concentration within a desired range and determined the growth rate from the exponential increase in the biomass concentration. According to the above definition, the K_i of *Arthrobacter* sp. was about 1,300 µM. Consequently,

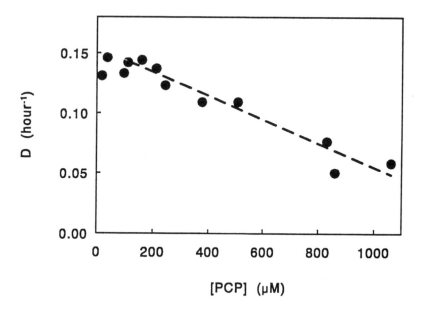

FIGURE 4. Relationship between the steady-state PCP concentration and the dilution rate in nutristat cultures with strain P5 growing on PCP as the only carbon and energy source.

the K_i values of these three strains are in the same order of magnitude. Nevertheless, this method using linear regression may not be generally applicable and does not make use of knowledge of enzyme inhibition kinetics.

The nutristat system is very useful to establish steady-state conditions while maintaining the substrate concentrations at toxic levels. Furthermore, the nutristat can be used to study microbial physiology of inhibition in more detail than is possible with batch cultures, because the amount of biomass in the steady-state culture is much higher then in batch cultures. In addition, the nutristat can be used to establish a stable environment for the formation of mutants with an improved resistance against toxic compounds. These mutants do have a growth advantage and eventually will outgrow the more sensitive cells.

REFERENCES

Alexander, M. 1985. "Biodegradation of Organic Chemicals." *Environ. Sci. Technol. 18*: 106-111.

Edgehill, R. U., and R. K. Finn. 1982. "Isolation, Characterization and Growth Kinetics of Bacteria Metabolizing Pentachlorophenol." *Eur. J. Microbiol. Biotechnol. 16*: 179-184.

Edwards, V. H. 1970. "The Influence of High Substrate Concentrations on Microbial Kinetics." *Biotechnol. Bioeng. 12*: 679-712.

Owens, J. D., and J. D. Legan. 1987. "Determination of the Monod Substrate Saturation Constant for Microbial Growth." *FEMS Microbiol. Rev. 46*: 419-432.

Radehaus, P. M., and S. K. Schmidt. 1992. "Characterization of a Novel *Pseudomonas* sp. that Mineralizes High Concentrations of Pentachlorophenol." *Appl. Environ. Microbiol. 58*: 2879-2885.

Rutgers, M., J. J. Bogte, A. M. Breure, and J. G. Van Andel. 1993. "Growth and Enrichment of Pentachlorophenol-degrading Microorganisms in the Nutristat, a Substrate Concentration-controlled Continuous Culture." *Appl. Environ. Microbiol. 59*: 3373-3377.

Simkins, S., and M. Alexander. 1984. "Models for the Mineralization Kinetics with the Variables of Substrate Concentration and Population Density." *Appl. Environ. Microbiol. 47*: 1299-1306.

Tempest, D. W. 1970. "The Continuous Cultivation of Microorganisms. I. Theory of the Chemostat." In: J. R. Norris and D. W. Ribbons (Eds.), *Methods in Microbiology* Vol. 2. Academic Press, London and New York, NY. pp. 259-276.

Van Gestel, Y.P.C.M., A. M. Breure, J. G. van Andel, and M. Rutgers. 1994. "Isolation and Characterization of a Pentachlorophenol-Degrading Bacterium." In: T. Stephenson (Ed.), *Proc. 2nd. Int. Symp. Environ. Biotechnol.* Institution of Chemical Engineers, Warwickshire, UK. pp. 119-121.

Volkering, F., A. M. Breure, J. G. van Andel, and W. H. Rulkens. 1995. "Influence of Nonionic Surfactants on Bioavailability and Biodegradation of Polycyclic Aromatic Hydrocarbons." *Appl. Environ. Microbiol. 61*(5) (in press).

Interactions of Peroxidases with Dyes and Plastics

Andrzej Paszczynski, Stefan Goszczynski,
Ronald L. Crawford, and Don L. Crawford

ABSTRACT

Azo dyes and polystyrenes that were recalcitrant to degradation were chemically modified, and subsequent changes in their susceptibility to degradation were investigated. Methoxylation and acetoxylation of polystyrene increased its degradability by *Phanerochaete chrysosporium* as compared to unmodified polystyrene of a similar molecular weight. Acetoxy-substituted polystyrene under ligninolytic conditions (low nitrogen medium) was degraded most quickly. Degradation of 4-hydroxy and 2-hydroxy dimethyl derivatives of the azo dye azobenzene-4'-sulfonic acid by whole cultures, lignin peroxidase, and manganese peroxidase was investigated. Decolorization rates and specific activities of *P. chrysosporium* peroxidases were compared. Dyes with a hydroxyl in the 2 position of the aromatic ring were inferior substrates for peroxidases compared to dyes with a hydroxyl in the 4 position. A two-methylgroup substitution pattern increased the oxidation rates of azo dye isomers. A new affinity chromatography matrix was prepared by attachment of 3,5-dimethyl-4-hydroxyazobenzene-4'-carboxylic acid to EAH Sepharose 4B. This matrix was used to separate peroxidases from the culture filtrate of *P. chrysosporium*.

INTRODUCTION

Heme peroxidases, which have been purified from a number of organisms, use H_2O_2 as the terminal acceptor of electrons removed from their substrates. The plant peroxidase superfamily has been divided into three classes on the basis of amino acid sequences. Class 1 includes yeast cytochrome *c* peroxidase, chloroplast ascorbate peroxidase, and gene-duplicated bacterial peroxidase; class 2 includes extracellular fungal lignin peroxidase (LP) and Mn peroxidase (MP); and class 3 is classic plant secretory peroxidase (Welinder 1992). White-rot fungi such as the hymenomycete *Phanerochaete chrysosporium* Burds produce numerous extracellular lignin peroxidases and/or manganese peroxidases that are able to initiate depolymerization and degradation of lignin and its oligomers and monomers (Hammel et al. 1993; Warishi et al., 1991).

Because LP and MP oxidize a variety of persistent environmental pollutants with redox potentials beyond the reach of other plant and bacterial peroxidases, and because they are not very substrate specific, the lignin-degrading white-rot fungi and their enzymes have potential for applications such as biopulping, bio-bleaching, and the cleanup of toxic organic chemicals in soils and waters. Previously, we found that by using substitution patterns resembling those in the lignin polymer, we could enhance a chemical's biodegradability by microbial lignolytic enzymes (Paszczynski et al. 1991, 1992; Pasti et al. 1992). We investigated mechanisms of azo dye oxidation by lignin peroxidase as a model system (Paszczynski and Crawford 1991) and recently proposed a detailed pathway for degradation of 3,5-dimethyl-4-hydroxyazobenzene-4'-sulfonic acid (designated azo dye 1) by ligninolytic peroxidases of *P. chrysosporium* (Goszczynski et al. 1994).

To further understand enzyme-substrate interactions, we synthesized several isomers of both azo dye 1 (Figure 1) and modified polystyrenes (Figure 2), and measured their rates of oxidation by whole cultures and purified enzymatic preparations. Through these efforts, we have been able to identify both structures very susceptible to peroxidatic degradation and structures resistant to degradation.

EXPERIMENTAL PROCEDURES AND MATERIALS

The six isomers of azo dyes were synthesized according to our previous procedure (Pasti et al. 1992). The purity of each preparation was determined using an HPLC-C18 microbore column with a mobile phase consisting of 50% acetonitrile and 50% 50-mM phosphate buffer (pH 7.5) containing 10 mM tetrabutylammonium hydrogen sulfate. For affinity chromatography, 3,5-dimethyl-4-hydroxyazobenzene-4'-carboxylic acid (m.p., 229 to 230°C) was prepared from diazotized 4-aminobenzoic acid by coupling under alkaline conditions with 2,6-dimethylphenol. After workup and recrystallization, the product was shown to be pure by high-performance liquid chromatography (S. Goszczynski, personal communication). The product was coupled to EAH Sepharose 4B (Pharmacia LKB, Uppsala, Sweden) using the carbodiimide method (Pharmacia 1991). The resulting affinity substrate was used for purification of peroxidases from crude culture filtrates. Dye decolorization rates were measured at their isosbestic points (Pasti et al. 1992); these points and pH-dependent wavelength shifts of the azo dyes used are compared in Table 1.

Polystyrene of various molecular weights was obtained from Aldrich (Milwaukee, Wisconsin), and 4-hydroxystyrene polymer of average molecular weight (MW) 5,800 was obtained from Hoechst and Celanese (Bishop, Texas). The 4-methoxypolystyrene was obtained by methylation of 4-hydroxypolystyrene with dimethyl sulfate under alkaline conditions, as follows. First, 10 g of a commercial sample of 4-hydroxystyrene was dissolved in 72 mL of 10% sodium hydroxide. The solutions were heated to boiling under reflux, and 10 mL of

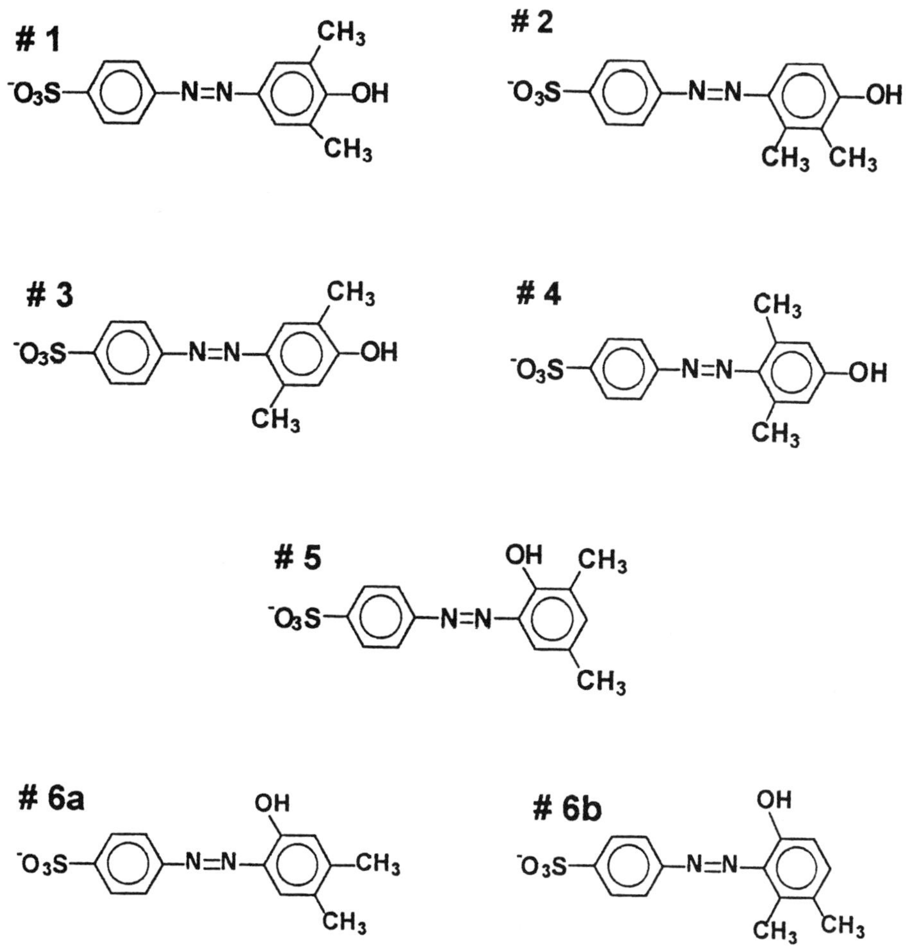

FIGURE 1. Azo dye structures.

dimethyl sulfate was added dropwise within 30 min, with vigorous stirring. Further boiling under reflux was continued for 4 h. The hot suspension was filtered off and washed with hot water until the pH became neutral. The yield was 93.8%. Calculated: C, 80.56%; H, 7.51%. Found: C, 80.69%; H, 7.68%.

Acetylation of 4-hydroxystyrene polymer was done under acidic conditions using acetic anhydride. First, 10 g of 4-hydroxystyrene was dissolved in 20 mL of glacial acetic acid. The solution was heated to boiling under reflux and 15 mL of acetic anhydride was introduced within 30 min. The refluxing was continued for 6 h; the solution was then cooled to room temperature and poured onto crushed ice. The solid material was filtered off, washed with water to neutral pH, and dried. A yield of 12.9 g of colorless material was obtained, 95.8% of the theoretical yield. Calculated: C, 74.05; H, 6.22. Found: C, 74.3; H, 6.42.

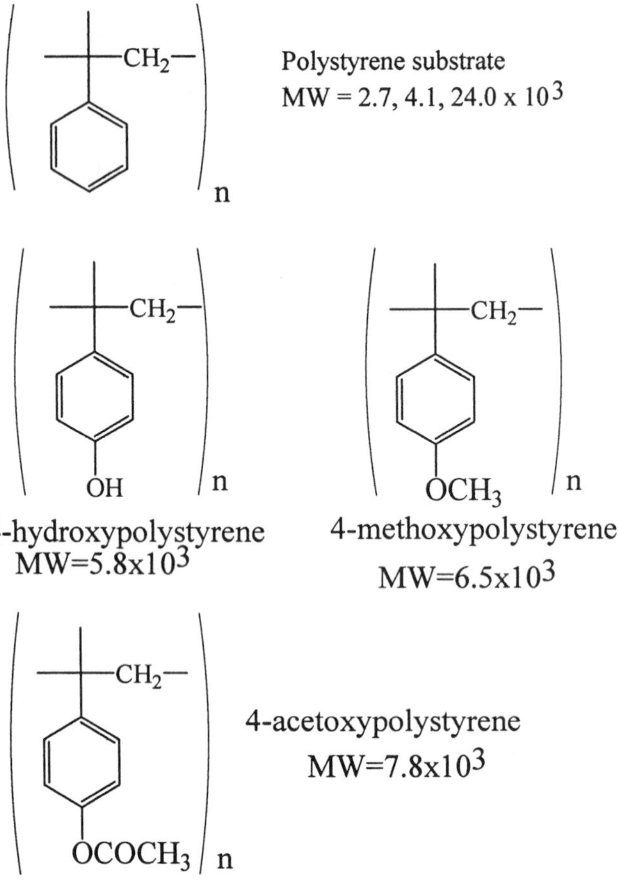

Polystyrene substrate
MW = 2.7, 4.1, 24.0 x 10^3

4-hydroxypolystyrene
MW=5.8x10^3

4-methoxypolystyrene
MW=6.5x10^3

4-acetoxypolystyrene
MW=7.8x10^3

FIGURE 2. Structures of polystyrenes.

RESULTS

Our investigations of the degradation of polystyrene, 4-hydroxypolystyrene, 4-methoxypolystyrene, and 4-acetoxypolystyrene in ligninolytic and nonligninolytic cultures of *P. chrysosporium* BKM-F-1767 (ATCC 24725) indicated that both culture conditions and substituents in benzene rings influenced degradation (Figure 3). Polystyrenes (MW 2,700, 4,100, 24,000) and polyhydroxystyrene (MW 5,800) were not degraded under any conditions investigated. Acetoxy-substituted polystyrene under ligninolytic conditions (low nitrogen medium) was degraded more completely than methoxy-substituted polystyrene.

The rates of decolorization of azo dye isomers by whole cultures and enzyme preparations are compared in Figure 4 and Table 2. Dye 1 was clearly the best substrate for enzymatic oxidations. Azo dye 1 was oxidized by manganese peroxidase approximately six to fifteen times faster than any of the other dyes.

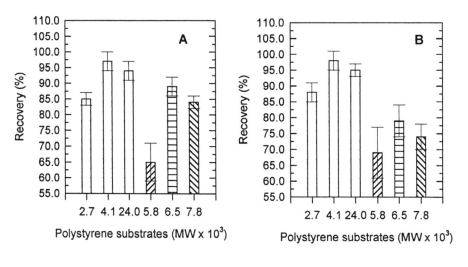

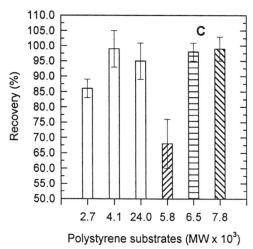

FIGURE 3. Degradation of polystyrene and modified polystyrene by *P. chryso-sporium* cultures. (A) Culture with high nitrogen concentration (16 mM). (B) Culture with low nitrogen concentration (2 mM). (C) Extraction efficiency for investigated polymers. Tetrahydrofuran (THF) extracts of 3-week, 60°C dried cultures were analyzed using HPLC-GPC (gel permeation chromatography); 25 μL of each sample was injected onto a column. The mobile phase consisted of THF at 1 mL/min. A wavelength of 254 nm was used for the peak detection. Extraction efficiency was measured by dividing the peak area of substrate directly solubilized in THF after GPC separation by the peak area of extracted uninoculated cultures shaken for 3 weeks. Inoculated samples were homogenized, dried, solubilized in THF, and analyzed. MW × 10^3 = 2.7, 4.1, 24.0, unmodified polystyrenes; MW × 10^3 = 5.8, polyhydroxystyrene; MW × 10^3 = 6.5, polymethoxystyrene; MW × 10^3 = 7.8, polyacetoxypolystyrene.

TABLE 1. pH-dependent wavelength shift of azo dyes tested.

Azo dye	0.001M HCl λ_{max}	$\varepsilon \times 10^3$	0.01M Tris λ_{max}	$\varepsilon \times 10^3$	0.01M NaOH λ_{max}	$\varepsilon \times 10^3$	Isosbestic point λ	$\varepsilon \times 10^3$
1	360	21.61	362	21.08	468	32.57	396	11.05
2	366	19.31	366	19.43	466	23.93	394	11.69
3	368	19.39	370	18.79	466	25.48	394	13.39
4	352	15.05	352	14.88	400	20.34	368	12.27
5	336	21.03	338	21.66	504	11.33	434	5.42
6a,b	338	19.96	338	20.45	488	10.46	432	6.90

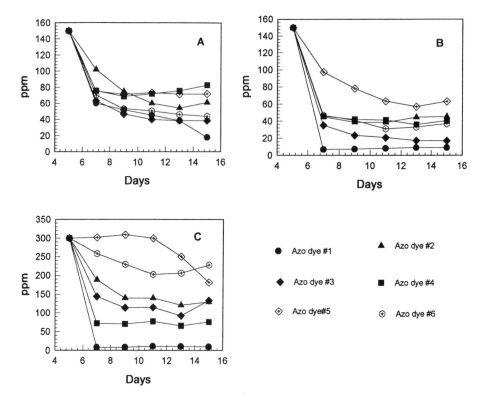

FIGURE 4. Decolorization of azo dyes by submerged cultures of *P. chryso-sporium*. Degradation in stationary conditions, (A) initial concentration 150 ppm. Degradation in agitated conditions, (B) initial concentration 150 ppm; (C) initial concentration 300 ppm.

TABLE 2. Decolorization rate of azo dyes by Mn peroxidase, ligninase, and plate cultures.

Azo dye	Oxidation rate (nmol/min/mg)		Bleaching time by plate cultures (days)
	Mn peroxidase	Ligninase	
1	3,159	1,318	12
2	477	419	8
3	208	690	15
4	220	537	5
5	251	850	8
6a,b	246	96	6

Although the results were less dramatic with ligninase, azo dye 1 was still oxidized the fastest. Dye 6 was extremely slow, and the others were intermediate. With plate cultures, however, azo dye 4 and 6 were oxidized the fastest, and dyes 1 and 3 the slowest (Table 2). Experiments are in progress to investigate the phenomena that underlie these differences.

An oxidizable dye, when coupled to an insoluble support, functioned as a highly effective affinity matrix for purification of LP (Figure 5). Peroxidases were purified 10 × (RZ = A_{408}/A_{280}; RZ = 2.6). Both LP and MP activity were present in the purified preparation. The affinity of other azo dye isomers toward the peroxidases of *P. chrysosporium* now is being investigated.

DISCUSSION

The decolorization rates of 4-hydroxy or 2-hydroxy dimethyl derivatives of azobenzene-4'-sulfonic acid by *P. chrysosporium* were investigated under different culture conditions and with different enzyme preparations. The agitated submerged culture studies (Figure 4B,C) suggested that the most susceptible structure was azo dye 1, followed by azo dye 4. Azo dyes 5 and 6, which showed the least decolorization by agitated submerged cultures (Fig. 4B,C), have an OH group in an *ortho* position relative to the azo linkage. Formation of an intramolecular hydrogen bond may be responsible for much of the recalcitrance of these structures:

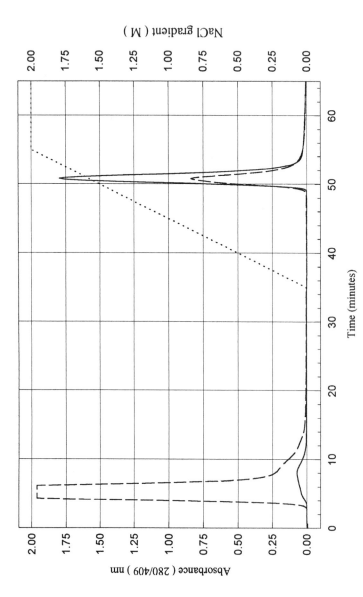

FIGURE 5. Purification of *Phanerochaete chrysosporium* peroxidases on an azo dye 1-EAH-Sepharose column from 6-day-old culture filtrate concentrated 100x (PM-10 Amicon membrane). 100 µL of crude enzyme solution was applied to the column (0.6 × 7 cm). The column was equilibrated with 10 mM sodium succinate buffer (pH 4.5), flowrate 0.3 mL/min. The enzyme was eluted with a 6-mL gradient of NaCl (0 to 2 M) in 20 mM sodium succinate (pH 4.5). The enzyme's absorption ratio of A_{406}/A_{280} increased from 0.27 to 2.6 (purification 10x). This type of affinity substrate may be useful for purification of peroxidases from different sources.

The effects of the dye structures on the decolorization (oxidation) rates and the specific activities of lignin peroxidases towards the various isomers were compared (Table 2). Oxidation by MP was far faster with dye 1 than with any of the others. Oxidation by LP was fastest for 1, slowest for 6a and 6b, and intermediate for the others. Under stationary growth conditions in plate cultures containing 3% malt extract as a low-nitrogen medium, degradation by *P. chrysosporium* was fastest for azo dyes 4 and 6 and slowest for dye 3. These data show that the susceptibility of a particular compound to degradation depends on both structure and culture conditions. Under different culture conditions the fungus may secrete different isozymes of its peroxidases. It is clear that these types of studies may lead to the rational design of biodegradability into many types of industrial chemicals. Our work also has allowed us to devise new affinity-based purification procedures for fungal peroxidases.

REFERENCES

Goszczynski, S., A. Paszczynski, M. B. Pasti-Grigsby, R. L. Crawford, and D. L. Crawford. 1994. "New Pathway for Degradation of Sulfonated Azo Dyes by Microbial Peroxidases of *Phanerochaete chrysosporium and Streptomyces chromofuscus.*" *Journal of Bacteriology* 176:1339-1347.

Hammel, K. E., K. A. Jensen, M. D. Mozuch, L. D. Landucci, M. Tien, and E. A. Pease, E. A. 1993. "Ligninolysis by a Purified Lignin Peroxidase." *Journal of Biological Chemistry* 268: 12274-12281.

Pasti, M. B., A. Paszczynski, S. Goszczynski, D. L. Crawford, and R. L. Crawford. 1992. "Influence of Aromatic Substitution Patterns on Azo Dye Degradability by *Streptomyces* spp. and *Phanerochaete chrysosporium.*" *Applied and Environmental Microbiology* 58:3605-3613.

Paszczynski, A., and R. L. Crawford. 1991. "Degradation of Azo Compounds by Ligninase from *Phanerochaete chrysosporium*: Involvement of Veratryl Alcohol." *Biochemistry and Biophysics Research Communications* 178:1056-1063.

Paszczynski A., and R. L. Crawford. 1995. "The Potential for Bioremediation of Xenobiotic Compounds by the White-Rot Fungus *Phanerochaete chrysosporium.*" *Biotechnology Progress* 19. In press.

Paszczynski, A. M. B. Pasti, S. Goszczynski, D. L. Crawford, and R. L. Crawford. 1991. "New Approach to Improve Degradation of Recalcitrant Azo Dyes by *Streptomyces* spp. and *Phanerochaete chrysosporium.*" *Enzyme Microbial Technology* 13:378-384.

Paszczynski, A., M.B. Pasti, S. Goszczynski, D. L. Crawford, R. L. and Crawford. 1992. "Mineralization of Sulfonated Azo Dyes and Sulfanilic Acid by *Phanerochaete chrysosporium* and *Streptomyces chromofuscus.*" *Applied and Environmental Microbiology* 58:3598-3604.

Pharmacia LKB Biotechnology. 1991. *Affinity Chromatography Principles and Methods* 10-1022-29.

Warishi, H., K. Valli, and M. H. Gold. 1991. "*In Vitro* Depolymerization of Lignin by Manganese Peroxidase of *Phanerochaete chrysosporium.*" *Biochemistry and Biophysics Research Communications* 176: 269-275.

Welinder, K. G. 1992. "Superfamily of Plant, Fungal and Bacterial Peroxidases." *Current Opinion in Structural Biology* 2: 338-393.

Economic Feasibility of Recycling Swine and Poultry Waste: A Case Study from Hong Kong

A. Reza Hoshmand

ABSTRACT

Increasing environmental problems from swine and poultry waste in Hong Kong call for alternative measures of disposing such waste that are effective not only ecologically, but also financially. This paper analyzes the benefit-cost ratio for adopting the "pig-on-litter" and "dry-muckout" method for poultry and swine farms. The results of the analyses indicate that adopting the "pig-on-litter" method is feasible in several scenarios except when the waste collection cost is borne by the farmer and the farmer does not receive any benefits from the sale of the spent sawdust. Dry muckout is feasible for both swine and poultry farms. It is recommended that, in the early stages of adopting pollution-abating strategies, the government should provide incentives to the farming community. Furthermore, a farm waste cooperative, and the establishment of a barter exchange where spent sawdust is exchanged for fresh sawdust, bacterial product, and labor for collection, are recommended.

INTRODUCTION

Environmental degradation from industrial and agricultural sources has become a major problem for many developed and developing nations. As an example, Hong Kong faces major pollution problems not only from industrial sources but also agricultural pollution. Although the territory is not a major agricultural producer, it does have a swine and a poultry industry that accounts for 16% of the total domestic consumption of swine, and 37% of the consumption of poultry (Witt 1993). The potential environmental risk from disposal of waste generated on poultry farms is magnified by the dense confinement of poultry and the decreasing amount of land available for waste disposal (Hodgkiss and Griffiths 1987; Edwards and Daniel 1992).

Previous studies on waste management in the territory have indicated several options for treating waste (Boddington and Heywood 1983). Some of the options are not viable in terms of technical efficiency and economic feasibility. In this study, emphasis is placed on two possible options for handling waste processing.

The first option (dry-muckout) is a collection scheme whereby raw manure is collected daily from each farm, and transported to several processing facilities strategically located among the major pig production areas. Composting, processing, and quality control are performed by these facilities from which the final product is marketed.

The second option is the adoption of the pig-on-litter or POL process which allows in situ composting of the pig manure. This new technique of raising pigs overcomes a number of husbandry problems experienced in the conventional method. The POL method is to overlay the floor of the pig pens with 20 cm of sawdust mixed with a bacterial product that will decompose pig manure on site as it is produced. The success of this approach to raising pigs has been shown in Japan and Taiwan (Hoshmand 1991). One advantage of POL is in reducing the moisture content of the waste, thus reducing the cost of transportation. The additional time required to process the waste, as done in the case of the dry-muckout method, is reduced greatly. In most instances, the spent sawdust could be used immediately as a soil conditioner or as an organic material on vegetable and flower beds (Kwong 1991). This reduction in time in the processing stage saves labor and other costs. After 3 years of trials on research and private farms in Hong Kong, the Agriculture and Fisheries Department's technical findings indicate that the POL method shows a great deal of promise.

To determine the benefit and cost associated with these pollution-abating strategies, the supply of waste and demand for using such waste as a soil conditioner or fertilizer were projected. To project the supply of waste, a trend analysis was performed on the number of pigs and chickens in the territory. From a statistical point of view, the fit of the curve was good, allowing for projections in the near future (see Hoshmand 1991). However, the exact number of animals in the territory in the future depends on many factors including past performance of the industry, favorable price conditions in the market, reduced input costs, and government policy with regard to the agricultural sector. The current projections are the best that could be made from the data available.

BENEFIT-COST ANALYSES

Benefit-cost analysis has been extensively used as a measure of project worth by economists (Gittinger 1982). Formally, the benefit-cost ratio is written as:

$$\text{Benefit-cost ratio} = \frac{\sum_{t=1}^{t=n} \dfrac{B_t}{(1+i)^t}}{\sum_{t=1}^{t=n} \dfrac{C_t}{(1+i)^t}}$$

where B_t is the benefit in each year, C_t is the cost in each year, and i is the interest (discount) rate. To compute the benefit-cost ratio, we have used the 1989 input costs for swine and poultry as the basis for computing the average cost of

TABLE 1. Gross income, input costs, and net income/animal/year of pig and chicken farms in Hong Kong in 1989.

A. Pig Production		HK$/pig
Gross Income		1,126
Input Cost		843
Rent	4	
Purchase price of stock	33	
Feed	711	
Depreciation	25	
Misc.	70	
Net Income		283
B. Chicken Production		HK$/1,000 birds
Gross Income		29,950
Input Cost		23,360
Rent	150	
Purchase price of stock	6,580	
Feed	12,490	
Depreciation	1,100	
Misc.	3,040	
Net Income		6,590

Source: Agriculture and Fisheries Department (1990).

production for subsequent years (Table 1). The average cost of production for sow-porker pigs was Hong Kong (HK) $843 and for poultry it was HK$23.36 (Agriculture and Fisheries Department 1990). The cost of pollution abating strategies are added to the production costs so that its impact could be determined on the net income of the farmers. To better understand the magnitude of the costs involved, different projections and estimates were used to determine the cost of the project (see Table 2). The analyses follow several scenarios to take account of different conditions as well as different enterprises, so as to provide a variety of policy options.

To account for price increases in the future years, the Food and Agriculture Organization (FAO) of the United Nations price index was used to estimate the cost increases of 5% per annum for capital items, 4.5% per annum for operating and maintenance costs, and 5% per annum production cost. On the benefit side, the price increase for livestock products is assumed to be 5% per annum.

RESULTS

Various cost alternatives were taken into account ranging from the farmers paying for the complete cost of pollution abating strategies to those where the

TABLE 2. Estimated input costs, waste treatment costs, collection costs, gross and net income/pig and chicken for years 1991 to 1995 (HK$).

Year	Input Costs[a]	Total Waste Treatment Costs	Waste Collection Cost	Total Cost	Gross Income	Net Income
A. Pig (sow-porker): POL						
1991	929.41	146.0	92.9	1,168.31	1,241.4	73.08
1992	975.88	152.6	97.6	1,228.08	1,303.5	75.41
1993	1,024.67	159.5	102.5	1,286.67	1,368.7	82.02
1994	1,075.90	166.7	107.6	1,350.20	1,437.1	86.89
1995	1,129.70	174.2	113.0	1,416.90	1,509.0	92.09
B. Chicken: Dry Muckout						
1991	25.76	0.93	2.58	29.27	33.02	3.75
1992	27.05	0.98	2.70	30.73	34.67	3.94
1993	28.40	0.017[b]	2.84	31.26	36.40	5.14
1994	29.82	0.018[b]	2.98	32.82	38.22	5.40
1995	31.31	0.019[b]	3.13	34.46	40.13	5.67

(a) The input cost is based on the 1989 data adjusted by the FAO Index of Prices Paid by Farmers which was 4.5% per annum. Gross income is adjusted by 5% per annum.
(b) These costs reflect only the operation and maintenance costs and do not include the capital cost which occurs only in the first 2 years of the project.

government contributed to the cost of removing waste and purchasing the waste from the farmers. Table 3 shows the results of the benefit-cost analysis for the scenario where the cost of adopting the POL is estimated to be HK$76.30 per pig per year. This cost is added to the average cost of production to determine if it is feasible for farmers to adopt this strategy of reducing pollution.

Table 4 summarizes all the scenarios and their feasibility. Each scenario assumes a different technical and economic condition as specified in column 1 of Table 4. All the benefit-cost analyses found that all scenarios are feasible except two (Scenarios 4 and 9). When all the cost of adoption (HK$76.30 per pig per year) and waste collection is borne by the farmers, and the farmers do not receive any further benefits beyond the sale of their animals, the project is not feasible. When the cost of adoption is HK$146 per pig per year, and the farmer pays for the collection of waste and is able to sell the spent sawdust, the project is not feasible. These analyses provide a view of the nature of the costs and benefits and present the policymaker with options as to when they should intervene to make the adoption of a pollution-abating strategy possible.

TABLE 3. Comparing gross benefits to gross costs when the cost/pig/year of adopting the POL method is $76.30 (in millions of HK$).

Year	Project Costs						Total Value of Production(f) (= Gross Benefits)	D.F. 12%	Present Worth 12%
	Capital Cost(a)	O&M Cost(b)	Production Cost(c)	Gross Costs(d)	D.F.(e) 12%	Present Worth 12%			
1991	1.362	7.263	100.452	109.077	0.893	97.406	134.188	0.893	119.830
1992	1.386	7.371	102.449	111.206	0.797	88.631	136.854	0.797	109.073
1993	1.415	7.471	104.331	113.217	0.712	80.610	139.369	0.712	90.231
1994	1.440	7.561	106.088	115.089	0.636	73.197	141.717	0.636	90.132
1995	1.459	7.640	107.720	116.819	0.567	66.236	143.900	0.567	81.591
Total	7.062	37.306	521.040	565.408	3.605	406.080	696.028	3.605	499.857

(a) Capital cost is determined as a product of the estimated capital cost for the respective years and the number of projected animals in each year. For example, the capital cost for the year 1991 is computed as follows: ($12.60)(108,094) = $1,361,984.

(b) O&M is determined as a product of the estimated O&M cost/pig/year and the number of animals that are under the POL for the respective years.

(c) Production cost is a product of the unit cost of production and the projected number of animals on POL in each year. For example, the cost of production is estimated at ($929.3)(108,094) = $100,451,754.

(d) Gross cost is the sum of capital cost, O&M cost, and production cost.

(e) Discount factor.

(f) Total value of production is a product of price of pigs sold and the projected number of animals in each year.

TABLE 4. Feasibility matrix for different scenarios.

Cost/Benefit Conditions	Scenarios								
	1	2	3	4	5	6	7	8	9
<u>Pig</u>									
POL @ $76.30	F[a]	—	—	—	—	—	—	—	—
POL @ $146.00	—	F	—	—	—	—	—	—	—
Dry muckout @ $54.00	—	—	F	—	—	—	—	—	—
POL @ $76.30 + waste collection	—	—	—	NF	—	—	—	—	—
Dry muckout @ $54.99 + waste collection	—	—	—	—	F	—	—	—	—
<u>Poultry</u>									
Dry muckout @ $0.93	—	—	—	—	—	F	—	—	—
Dry muckout @ $0.93 + waste collection	—	—	—	—	—	—	F	—	—
<u>Pig</u>									
POL @ $76.30 + all costs + all benefits	—	—	—	—	—	—	—	F	—
POL @ $146 + all costs + all benefits	—	—	—	—	—	—	—	—	NF

(a) F = Feasible; NF = Not Feasible.

CONCLUSIONS AND POLICY RECOMMENDATIONS

It appears that the solution to the problem of livestock waste can be found in a cohesive and enduring policy that brings together the producers, various interested parties, and the government. The measures that may be drawn upon for implementing a pollution policy are conceived here to serve the needs of the farming community and the requirements set by the government. These measures relate to policy and fiscal issues. Policy measures require a clear statement of what a farmer can and cannot do and how to achieve the objectives of a pollution-free environment. Fiscal measures provide a guideline for the farming community and government alike to find ways of financing these costs.

Given the current attitudes of farmers towards livestock waste disposal, government agencies must find approaches that will induce farmers to voluntarily adopt waste disposal techniques. At present, farmers do not see any financial advantage in adopting any of the waste treatment schemes. Adopting the POL, as practiced on the government experimental research farm, is not what most pig producers have in mind. To make this proposition appealing, economic incentives must be introduced. The government could share in the cost of adopting pollution devices and removal of waste from the farms, at least in the initial stages of implementation. Once all farms have adopted a scheme to deal with farm waste, a licensing fee could be applied toward waste collection costs for the future. How much to charge for a license could be based on the number of animals or birds raised per farm or a fixed annual fee for the pig and poultry farms.

A farm waste cooperative could be established to handle not only waste removal, but also marketing of the composted waste for domestic and export markets. In this study, it is found that some leaders in pig and poultry associations are interested in pursuing a livestock waste cooperative.

Establishment of a barter exchange may provide a convenient mechanism for removing spent sawdust from the farms. At present, farmers consider waste a commodity that has no value and can therefore be washed into the streams. The market demand survey of the nine large landscape contractor companies indicates that there is not only an interest in using spent sawdust as soil conditioner, but a desire on the part of some companies to exchange it for fresh sawdust, bacterial product, and labor for collection. It is therefore possible for farmers to barter for such commodities and services without having to incur a direct cost. The government can play a role in bringing together the interested companies and the farmers.

REFERENCES

Agriculture and Fisheries Department. 1990. *Cost of Production and Returns to Local Farming.* Agricultural Technical Circular, AF-EC. May.

Boddington, M.A.B., and A. Heywood. 1983. *A Strategy for Agricultural Waste Management.* Technical Report prepared for the Environmental Protection Agency of Hong Kong Government, ERL Asia Limited in Association with HFA Hong Kong.

Edwards, D. R., and T. C. Daniel. 1992. "Environmental impacts of on-farm poultry waste disposal — A review." *Bioresource Technology* 41:9-33.

Gittinger, J. P. 1982. *Economic Analysis of Agricultural Projects.* Johns Hopkins University Press, Baltimore, Maryland.

Hodgkiss, I. J., and D. A. Griffiths. 1987. "Need pig waste pollute?" *Hong Kong Engineer* 15(5):43-48.

Hoshmand, A. R. 1991. *Integrated Farming in Hong Kong: A Feasibility Study of Recycling Waste from Swine and Chicken.* Report presented to the Hong Kong Agriculture and Fisheries Department, Kowloon, Hong Kong.

Kwong, E.M.L. 1991. *Recommended Uses of Spent Sawdust Litter from "Pig-on-Litter" System for Crop Farming and Landscape Gardening.* Agricultural Technical Circular, AF-CRP 03/9, IV, February.

Witt, Hugh (Ed.). 1993. *Hong Kong 1993.* Government Printer, Hong Kong.

Biodegradation Potential of a Pentachlorophenol-Degrading Microbial Consortium

Carol Panneton, Juliana Ramsay, Jean-Louis Bertrand,
Raymond Mayer, and Claude Chavarie

ABSTRACT

A pentachlorophenol (PCP)-degrading consortium was obtained from contaminated soil after acclimatization to the three monochlorophenol isomers individually and then to PCP. In batch cultures, the consortium completely degraded PCP up to concentrations of 800 mg/L. In biometric flasks studies, the consortium mineralized 70% of ^{14}C-PCP (50 mg/L) in an aqueous environment (no solids) and 60% in contaminated soil slurries in less than 66 h, whereas noninoculated suspensions required 280 h to achieve the same extent of mineralization. In batch culture, a lag phase was always observed before degradation. When PCP was the sole source of carbon, the lag phase increased with the initial PCP concentration, but the specific degradation rate was constant for concentrations of 50 to 500 mg/L. When the number of cells in the inoculum was increased using a mixture of glucose and PCP, the specific PCP degradation rate was lower than when the inoculum was prepared with PCP as the sole source of carbon.

INTRODUCTION

Flavobacterium (Crawford and Mohn 1985), *Arthrobacter* (Chu and Kirsch 1972; Stanlake and Finn 1982), *Pseudomonas* (Watanabe 1973; Radehaus and Schmidt 1992) and *Rhodococcus* (Häggblom et al. 1988) are some of the pure bacterial cultures that metabolize PCP partially or totally. PCP-degrading microbial consortia have also been isolated from contaminated soils (Rutgers et al. 1991; O'Reilly and Crawford 1989) and from municipal sewage sludge (Kim and Maier 1986).

An acclimatization period of the isolated microorganisms to PCP is essential to induce the enzymes responsible for its degradation (Häggblom et al. 1988; Steiert et al. 1987; Crawford and Mohn 1985). This step is mainly achieved by adapting the culture to PCP or to an analog. It has been shown that when the inoculum of an anaerobic sludge was adapted to each isomer of monochlorophenol individually then to PCP, the latter component was completely metabolized (Mikesell and Boyd 1986).

The objective of this work was to develop a PCP-degrading microbial consortium using the method described by Mikesell and Boyd (1986) and to evaluate its ability to degrade PCP as a sole source of carbon. Moreover, PCP degradation by an inoculum prepared using a more assimilable carbon source such as glucose and PCP to increase cell numbers was compared on the basis of the duration of the lag phase before PCP degradation and the specific and nonspecific degradation rates.

MATERIALS AND METHODS

Acclimatization

The three different isomers of monochlorophenol (2-, 3-, and 4-CP) were fed to three different reactors in which the same sample of PCP-contaminated soil was present in a modified mineral salt (MS) medium described by Edgehill and Finn (1982). The calcium and manganese salts used by these authors were replaced by $Ca(NO_3)_2 \cdot 4H_2O$ (0.03 g/L) and by a trace element solution (1 mL/L) (Ramsay et al. 1990). The effluents of these three chemostats were fed to a fourth one containing the same contaminated soil, but fed with PCP. PCP was added as the sodium salt. Concentrations of the monochlorophenols and PCP were gradually increased as each new steady state was achieved. At the time the consortium was studied, the feed concentrations were 600 mg/L for 2-CP, 200 mg/L for 3- and 4-CP, and 800 mg/L for PCP, but the reactor concentrations were all under 20 mg/L. The total viable bacterial concentration in the PCP chemostat was 1×10^7 cells/mL.

Analytical Methods

Residual PCP concentration was analyzed by ultraviolet (UV) absorption at 319 nm with a Milton Roy Spectronic 1001 Plus after dilution of the samples with 0.1 M NaOH. PCP mineralization was measured as $^{14}CO_2$ released from uniformly labeled ^{14}C-PCP using the technique of Sharabi and Bartha (1993). The concentrations of PCP and certain metabolites were also measured by gas chromatography/mass spectrometry (GC/MS) (Varian Saturn II, 3400) using the extraction technique proposed by the Quebec Environment Ministry based on EPA studies (MENVIQ 1990). Glucose was determined colorimetrically by the DNS method (Miller 1959). The total bacterial concentrations were evaluated by the most-probable-number method (MPN) (Brown et al. 1986; Crawford and Mohn 1985) using five test tubes containing tryptone (5 g/L), yeast extract (2.5 g/L), and glucose (1 g/L).

PCP Degradation Experiments

All experiments were conducted in 500-mL Erlenmeyer flasks containing 100 mL sterile MS medium at room temperature (25°C) and at 180 rpm on a gyro/rotary shaker (New Brunswick Scientific, Edison, New Jersey). The flasks were covered with aluminum foil to avoid PCP photodegradation. A previous experiment showed that PCP did not volatilize. When PCP was the sole source of carbon, 10 mL of the PCP chemostat was used to inoculate the Erlenmeyer flasks containing different PCP concentrations. To study the effect of glucose as a cosubstrate, the inoculum was prepared with 1 g/L of glucose and 50 mg/L of PCP in the MS medium. When glucose and PCP were exhausted, 10 mL of this medium were used to inoculate a flask with only 50 mg/L of PCP in MS medium. The degradation of PCP in this flask was the one studied. All experiments were conducted under aerobic conditions.

RESULTS AND DISCUSSION

PCP Mineralization

For most of the experiments reported in this paper, the degradation of PCP was monitored spectrophotometrically to verify whether PCP was completely mineralized or only partially degraded. GC/MS analyses of the culture supernatant from flasks of initial PCP concentrations of 50, 200, and 500 mg/L showed that the final concentrations of some metabolites were all below 1 mg/L. The metabolites analyzed were phenol, 2-chlorophenol, 2,4- and 2,6-dichlorophenol, 2,3,5-, 2,4,6-, 2,4,5- and 2,3,4-trichlorophenol, and 2,3,5,6- and 2,3,4,5-tetrachlorophenol. In experiments with ^{14}C-PCP, 70% of the initial aqueous ^{14}C-PCP (50 mg/L) was recovered as [14]CO_2; 5% was still present in the medium, either as unmetabolized PCP or metabolites; 20% was incorporated by the biomass; and the remaining 5% probably leaked from the biometric flask. This [14]CO_2 conversion is similar to values reported in the literature (Kirsch and Etzel 1973; Lin and Wang 1991).

Biometric flask studies using approximately 70 mg/L of ^{14}C-PCP diluted in contaminated soil slurries indicated that 60% of the PCP was metabolized in less than 66 h when inoculated with the consortium, whereas noninoculated suspensions required about 280 h to achieve the same extent of mineralization.

Effect of Initial PCP Concentration

PCP as the Sole Source of Carbon. The mixed culture from the PCP chemostat was found to degrade aqueous PCP up to an initial concentration of 800 mg/L when PCP was the sole source of carbon. Figure 1a shows only three of the many concentrations studied (Panneton 1994). At all concentrations, a lag period preceded the degradation. The length of the lag phase increased as the initial PCP concentration increased. For this reason, it took 400 h to degrade all the PCP at 800 mg/L. Most mixed or pure cultures adapted to PCP are reported to degrade and mineralize lower PCP concentration. For example, Radehaus and Schmidt (1992) found that, whereas the growth rate of a species of *Pseudomonas* decreased at PCP concentrations greater than 40 mg/L, PCP was completely mineralized up to concentrations of 160 mg/L, but not at all at concentrations greater than 200 mg/L. PCP was reported to be inhibitory to *Arthrobacter* at concentrations higher than 130 mg/L, but degradation was observed up to concentrations of 300 mg/L (Stanlake and Finn 1982; Edgehill and Finn 1982). The high degradation performance of our consortium is probably due to the acclimatization of the mixed culture to the three different isomers of monochlorophenol and the high PCP concentration fed to the chemostat (800 mg/L).

When PCP was the sole source of carbon, the lag phase increased with the initial PCP concentration (Table 1), but not linearly as observed by Stanlake and Finn (1982). This lag is probably a period of adaptation by the microorganisms to the initial PCP concentration in the fresh batch culture. Even though the chemostat feed concentration was 800 mg/L of PCP, the concentration in the reactor was less than 20 mg/L. When the culture was transferred to higher PCP concentrations in our batch experiments, shock may have occurred such that the microorganisms needed to readapt to the new environment.

PCP-specific (q_s) and nonspecific (r_s) degradation rates were calculated from the data of Figure 1a. The r_s values were evaluated from the slope of PCP versus time neglecting the lag period and $q_s = r_s/X_{average}$ where $X_{average}$ is $((X_{initial} - X_{final})_{during\ degradation})/2$. These values are listed in Table 1.

TABLE 1. Initial and average $((X_{initial} - X_{final})_{during\ degradation}/2)$ biomass concentrations, specific (q_s), and nonspecific (r_s) degradation rates and duration of lag phase for PCP degradation in batch culture using an inoculum prepared with either PCP as the sole source of carbon or a mixture of glucose and PCP.

PCP (mg/L)	$X_{initial}$ (cell/mL)	$X_{average}$ (cell/mL)	r_s (mgPCP. $mL^{-1}.h^{-1}$)	q_s (mgPCP. $cell^{-1}.h^{-1}$)	lag (h)
Inoculum	: PCP				
50	1.1×10^6	7.4×10^7	2.5	3.4×1^{-11}	36
200	1.4×10^6	3.4×10^8	3.1	9.1×10^{-12}	147
500	5.0×10^5	3.4×10^8	11	3.2×1^{-11}	337
Inoculum	: PCP	+ glucose			
50	3.5×10^7	2.6×10^7	0.17	6.5×10^{-12}	43

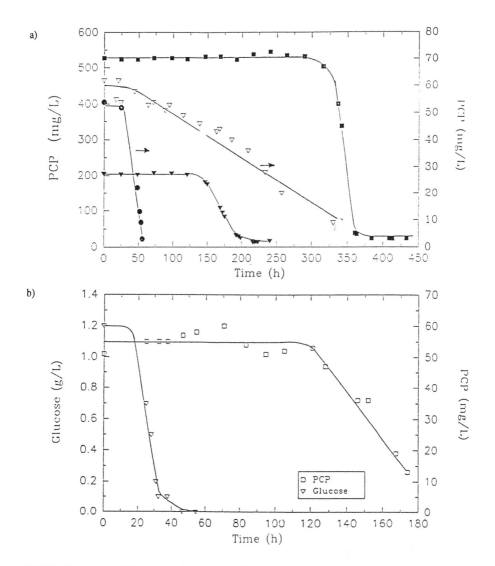

FIGURE 1. a) PCP degradation at different initial concentrations using an inoculum prepared with PCP as the sole source of carbon (solid symbols) or a mixture of glucose and PCP (open symbol) (using different scale, the 50 mg PCP/L experiments (●, ▽) are read on right axis as shown by arrows, and 200 (▼) and 500 (■) on left axis). b) Consumption of glucose and PCP by the consortium.

When PCP was the sole source of carbon, r_s increased with the initial PCP concentration (Table 1). This means that once PCP degradation begins, its degradation is faster at higher PCP concentrations. But, as mentioned previously, the lag period, which is not taken into account in r_s, increased with the initial PCP concentration so that the total time to achieve degradation was longer. When PCP was the only source of carbon, q_s was zero order between 50 and 500 mg/L with an average value of 2.5×10^{-11} mg/cell.h (Table 1). It is difficult to compare our q_s with literature values because of the different methods of measuring the biomass concentration (different units that cannot be easily converted). Nevertheless, Topp et al. (1988) found that the specific degradation rate for *Flavobacterium* was 1.51×10^{-13} mg/cell.h when PCP as the sole carbon source was between 70 and 150 mg/L. For the same PCP range, the calculated q_s for our consortium was ten times higher than the one obtained by Topp et al. (1988).

Effect of Glucose. When a mixture of glucose and PCP was used to prepare the inoculum, glucose was always totally consumed before PCP degradation began (Figure 1b). Using this inoculum did not reduce the lag phase observed when PCP was the sole source of carbon. The length of the lag phase was similar to the one obtained when the inoculum came directly from the chemostat (Figure 1a). The initial biomass concentration in the flask inoculated directly from the PCP chemostat was ten times lower than when the inoculum was prepared with glucose and PCP (Table 1). Upon dilution of the latter to the same initial biomass concentration as the chemostat, no PCP degradation was observed. This may be explained by the fact that the microbial population was different when a cosubstrate was used.

The consortium in the PCP chemostat probably is composed of microorganisms that can degrade PCP or its metabolites as well as those that may use cellular metabolites such as carboxylic acids (e.g., acetate) but do not degrade PCP. The latter microorganisms would be favored by the addition of a more easily assimilable carbon source, such as glucose. So, for the same initial total biomass concentration, the number of microorganisms that specifically degrade PCP may not be the same.

When PCP was the sole source of carbon in the inoculum, the r_s and q_s were higher than when the inoculum was prepared with glucose and PCP. Most authors have demonstrated that the addition of a cosubstrate increases the degradation rates (Topp et al. 1988; Gonzalez and Hu 1991). However, most of these studies were done with pure cultures. If the genes responsible for the degradation of PCP are stable and are not lost during cell division, the addition of a cosubstrate to a pure culture will result in an increase in the number of bacteria that could degrade PCP. This will have a positive effect in reducing the lag period and stimulating the removal rate. However, our mixed culture is not composed entirely of microorganisms that can degrade PCP, and therefore the addition of a cosubstrate such as glucose does not have the same effect as with a pure culture.

Preliminary efforts to characterize the consortium showed that the mixed culture was composed of at least three different bacteria based on colonial morphology. All were gram-negative rods. Growth on eosin-methylene blue agar (EMB) showed that none of these cultures were coliforms. Further studies are needed to identify which one(s) can degrade PCP. However, on a more practical point, our results suggest that the utilization of a cosubstrate to increase the total number of bacterial cells of a mixed culture, as during in situ biostimulation, is not advantageous to improve PCP degradation.

CONCLUSION

Our acclimatized culture both degraded and mineralized PCP up to 800 mg/L in a batch culture in MS medium and in a soil slurry. Increasing the cell numbers in the inoculum using a mixture of glucose and PCP did not result in a reduced lag phase or an increase in the PCP degradation rate. The lag phase was shorter and the specific and nonspecific degradation rates higher when PCP was the sole source of carbon. The addition of a cosubstrate to increase biomass concentration is not recommended in a mixed culture system designed for an in situ biostimulation approach.

ACKNOWLEDGMENTS

The authors thank The Biocapital Biotechnology Fund L.P. (Montreal) and Serrener Consultation Inc. (Montreal) for financial support of this project and F. Roberge for the GC/MS analysis.

REFERENCES

Brown, E.J., J.J. Pignatello, M.M. Martinson, and R.L. Crawford. 1986. "Pentachlorophenol Degradation: A Pure Bacterial Culture and an Epilithic Microbial Consortium." *Appl. Environ. Microbiol.* 52(1): 92-97.
Chu, J.P., and E.J. Kirsch. 1972. "Metabolism of Pentachlorophenol by Anoxic Bacterial Culture." *Appl. Microbiol.* 23(5): 1033-1035.
Crawford, R.L., and W.W. Mohn. 1985. "Microbiological Removal of Pentachlorophenol from Soil Using a *Flavobacterium*." *Enzyme Microbiol. Technol.* 7: 617-620.
Edgehill, E.A., and R.K. Finn. 1982. "Isolation, Characterization and Growth Kinetics of Bacteria Metabolizing Pentachlorophenol." *Eur. J. Appl. Microbiol. Biotechnol.* 16: 179-184.
Gonzalez, J.F., and W.S. Hu. 1991. "Effect of Glutamate on the Degradation of Pentachlorophenol by *Flavobacterium* sp." *Appl. Microbiol. Biotechnol.* 35: 100-104.
Häggblom, M.M., L.J. Nohynek, and M.S. Salkinoja-Salonen. 1988. "Degradation and O-Methylation of Chlorinated Phenolic Compounds by *Rhodococcus* and *Mycobacterium* Strains." *Appl. Environ. Microbiol.* 54(12): 3043-3052.
Kim, C.J., and W.J. Maier. 1986. "Biodegradation of Pentachlorophenol in Soil Environments." *Proceeding 41st Purdue University Industrial Waste Conference*, pp. 303-312.
Kirsch, E.J., and J.E. Etzel. 1973. "Microbial Decomposition of Pentachlorophenol." *J. Water Pollut. Control Fed.* 45(2): 359-363.
Lin, J.E., and H.Y. Wang. 1991. "Degradation of Pentachlorophenol by Non-Immobilized and Co-Immobilized *Arthrobacter* Cells." *J. Ferment. Bioeng.* 72(4): 311-314.

MENVIQ. 1990. "Guide des méthodes de conservation et d'analyses des échantillons d'eau et de sol." Direction des laboratoires, Ministère de l'Environnement du Québec, Québec, Canada.

Mikesell, M.D., and S.A. Boyd. 1986. "Complete Reductive Dechlorination and Mineralization of Pentachlorophenol by Anaerobic Microorganisms." *Appl. Environ. Microbiol.* 52(4): 861-865.

Miller, G.L. 1959. "Use of Dinitrosalicylic Acid Reagent for Determination of Reducing Sugars." *Anal. Chem.* 31: 426-428.

O'Reilly, K., and R.L. Crawford. 1989. "Degradation of Pentachlorophenol by Polyurethane-Immobilized *Flavobacterium* Cells." *Appl. Environ. Microbiol.* 55(9): 2113-2118.

Panneton, C. 1994. "Biodégradation du PCP en milieu aqueux." M.Sc.A. Thesis, École Polytechnique de Montréal, Montréal, Québec, Canada.

Radehaus, P.M., and S.K. Schmidt. 1992. "Characterization of a Novel *Pseudomonas* sp. that Mineralizes High Concentrations of Pentachlorophenol." *Appl. Environ. Microbiol.* 58(9): 2879-2885.

Ramsay, B.A., K. Lomaliza, C. Chavarie, B. Dubé, P. Bataille, and J.A. Ramsay. 1990. "Production of Poly-(β-Hydroxybutyric-Co-β–Hydroxyvaleric) Acids." *Appl. Environ. Microbiol.* 56(7): 2093-2098.

Rutgers, M., J. Bogte, A. Breure. and J. Vanandel. 1991. "Biodegradation of Pentachlorophenol by Aerobic Enrichment Cultures." *International Symposium on Environmental Biotechnology.* Ostend, Belgium, pp. 99-102.

Sharabi, N.E.D., and R. Bartha. 1993. "Testing of Some Assumptions about Biodegradation in Soil as Measured by Carbon Dioxide Evolution." *Appl. Environ. Microbiol.* 59(4): 1201-1205.

Stanlake, G.J., and R.K. Finn. 1982. "Isolation and Characterization of a Pentachlorophenol-Degrading Bacterium." *Appl. Environ. Microbiol.* 44(6): 1421-1427.

Steiert, J.G., J.J. Pignatello, and R.L. Crawford. 1987. "Degradation of Chlorinated Phenols by a Pentachlorophenol-Degrading Bacterium." *Appl. Environ. Microbiol.* 53(5): 907-910.

Topp, E., R.L. Crawford, and R. Hanson. 1988. "Influence of Readily Metabolizable Carbon on Pentachlorophenol Metabolism by a Pentachlorophenol-Degrading *Flavobacterium* sp." *Appl. Environ. Microbiol.* 54(10): 2452-2459.

Watanabe, I. 1973. "Isolation of Pentachlorophenol Decomposing Bacteria from Soil." *Soil Sci. Pl. Nutr.* (19): 109-116.

Biodegradation of Creosote Compounds Coupled with Toxicity Studies

Søren Dyreborg, Erik Arvin, Kim Broholm, and Michael Löfvall

ABSTRACT

The inhibition on the aerobic degradation of toluene by the presence of 14 selected creosote-related compounds was investigated in a reduced factorial experiment over a period of 16 days. A group of heterocyclic aromatic compounds containing nitrogen, sulfur, or oxygen (NSO), i.e., pyrrole, 1-methylpyrrole, thiophene, and benzofuran had a significant inhibiting effect on the degradation of toluene. If the NSO-compounds were present, no or little degradation of toluene was observed during the experimental period, although toluene was completely degraded within the 16 days in batches without the compounds. Concomitant with the chemical analysis of the samples from the degradation study, toxicity was measured with a bioassay. The toxicity of a sample was determined as inhibition to nitrification. A good correlation was obtained between the presence of creosote compounds and the toxicity of a sample. A statistical analysis revealed that the phenolic compounds and the NSO-compounds caused the toxicity to nitrification. The toxicity decreased as the compounds were biodegraded. In some cases, the toxicity to nitrification disappeared completely, indicating a total mineralization of the compounds or an accumulation of nontoxic intermediates, whereas in other cases a remaining inhibition of 10 to 16% was observed. This may indicate that toxic intermediates were accumulated during the degradation of the creosote compounds.

INTRODUCTION

Contaminated groundwater arising from creosote-polluted sites may contain a variety of different organic compounds, including monocyclic aromatic hydrocarbons (MAHs, e.g., toluene, *o*-xylene, and *p*-xylene), polycyclic aromatic hydrocarbons (PAHs, e.g., naphthalene, and 1-methylnaphthalene), phenolic compounds (e.g., phenol, *o*-cresol, and 2,6-dimethylphenol), and heterocyclic aromatic hydrocarbons (NSO-compounds, e.g., pyrrole, 1-methylpyrrole, indole, quinoline, thiophene, benzothiophene, and benzofuran). In evaluating the bioremediation potential of an aquifer, it is important to conduct degradation experiments with complex mixtures. Misleading conclusions of the degradation potential could be drawn by conducting studies with single compounds or simple mixtures because interactions between compounds may be an important factor in the degradation of many compounds.

A biotoxicity test may be used, in addition to the chemical analysis, because the chemical analysis in some cases fails to detect intermediates and/or end products of the degradation. A reduction in toxicity of a complex mixture may not be measured by the parent compound disappearance, because the intermediates/end products can be more toxic than the parent compound.

The purpose of this study was to evaluate the inhibition of 14 creosote compounds on the aerobic degradation of toluene. Concomitant with the chemical analysis, toxicity studies using nitrifying bacteria were conducted to couple the observed degradation with the toxicity of a sample. Nitrifying bacteria were chosen as test organisms, because they are present in many water and wastewater treatment plants. They are also sensitive to many organic chemicals.

METHODS

Degradation Experiment

The degradation experiment is described in detail in a previous paper (Dyreborg et al. 1995), but a short summary is presented here. The aerobic degradation experiments were conducted as batch experiments using 117-mL serum bottles with Mininert valves, with a reduced $\frac{1}{2}*2^6$ factorial design, where the degradation of toluene was the dependent variable. The statistical design is shown in Table 1, and the six independent variables (Factors A through F) are described in Table 2. The factors were divided into different groups of compounds typically found in creosote. By mistake, naphthalene was included in Factor F, but the results showed that this had no influence on the conclusions made in this paper. The factorial experiment was designed so Factor F was confounded with the five-factor interaction $A*B*C*D*E$ (alias relation F = ABCDE). Toluene was added to all bottles at a concentration of 3 mg/L. Each of the 32 combinations in the statistical design was run in duplicate. Three bottles acidified with 8N H_2SO_4 (pH = 1) were used as control for abiotic losses. The inoculum used in the biodegradation experiment was an aerobic enrichment culture originated from a creosote-contaminated aquifer in Fredensborg, Denmark. Samples for chemical and toxicity analysis were withdrawn at Day 0, Day 4, Day 8, and Day 16; 10 mL of sample for chemical analysis and 3 mL for the toxicity test were withdrawn from a serum bottle with a glass syringe and replaced by pure oxygen. The 10-mL sample was extracted with 1 mL of diethylether and 100 μL of pentane containing heptane and undecane as internal standards, and 1 μL of the organic phase was injected into a

TABLE 1. The statistical design of the experiment and the inhibition of samples at Day 0 and Day 16. (See Table 2 for definition of factors.)

Batch No.	A	B	C	D	E	F	Day 0	Day 16
1	—	—	—	—	—	—	29	12
2	+	—	—	—	—	+	54	51
3	—	+	—	—	—	+	34	15
4	+	+	—	—	—	—	49	42
5	—	—	+	—	—	+	53	14
6	+	—	+	—	—	—	48	47
7	—	+	+	—	—	—	30	12
8	+	+	+	—	—	+	59	64
9	—	—	—	+	—	+	34	4
10	+	—	—	+	—	—	46	56
11	—	+	—	+	—	—	28	7
12	+	+	—	+	—	+	63	57
13	—	—	+	+	—	—	25	0
14	+	—	+	+	—	+	62	59
15	—	+	+	+	—	+	36	0
16	+	+	+	+	—	—	43	44
17	—	—	—	—	+	+	29	2
18	+	—	—	—	+	—	55	47
19	—	+	—	—	+	—	30	14
20	+	+	—	—	+	+	62	60
21	—	—	+	—	+	—	25	14
22	+	—	+	—	+	+	61	45
23	—	+	+	—	+	+	33	16
24	+	+	+	—	+	—	56	46
25	—	—	—	+	+	—	27	3
26	+	—	—	+	+	+	61	52
27	—	+	—	+	+	+	42	13
28	+	+	—	+	+	—	58	48
29	—	—	+	+	+	+	37	5
30	+	—	+	+	+	—	55	42
31	—	+	+	+	+	—	30	9
32	+	+	+	+	+	+	60	54
Control	+	+	+	+	+	+	67[c]	75[c]

(a) +: Factor initially present, —: Factor absent,
(b) Average of four samples. (c) Average of six samples.

TABLE 2. Factor description and the initial concentrations of the different compounds.

Factor	Compounds	Initial concentration [mg/L]
A	pyrrole (N-compound)	4.0
	1-methylpyrrole (N-compound)	4.0
	thiophene (S-compound)	4.0
	benzofuran (O-compound)	1.8
B	benzothiophene (S-compound)	1.4
C	indole (N-compound)	3.0
D	quinoline (N-compound)	3.0
E	o-xylene (MAH)	0.9
	p-xylene (MAH)	0.9
	naphthalene (PAH)	0.4
	1-methylnaphthalene (PAH)	0.2
F	phenol (phenolic compound)	6.0
	o-cresol (phenolic compound)	5.0
	2,4-dimethylphenol (phenolic compound)	4.0
	naphthalene (PAH)	0.3

DANI 8520 GC equipped with a flame ionization detector (FID) and a 30-m J&W DB5 capillary column. For further details see Dyreborg et al. (1995). The samples for toxicity testing were frozen (–18°C) until analysis for toxicity.

Toxicity Study

The toxicity of a sample was measured using the MINNTOX-test (Arvin et al. 1994), which is an acute 2-h toxicity test. The toxicity of a sample was measured by the inhibition of nitrification (calculated as a reduction in the rate of ammonia oxidation), using active nitrifying sludge from a wastewater treatment plant as the test organism. For further details see Arvin et al. (1994).

Each sample from the degradation experiment was run as a replicate in the toxicity test. Samples from the control bottles were neutralized (pH = 7) with 8N KOH before they were analyzed for toxicity; 134 data sets were achieved, but data from the control bottles were excluded from the statistical factorial analysis, leaving 128 data sets for each day. When testing for toxicity, samples withdrawn from the degradation experiments were diluted six times according to the analytical procedure (Arvin et al. 1994). Therefore, the final concentrations in the test samples for the toxicity test were one-sixth of the concentrations measured by the chemical analysis.

RESULTS

Degradation Experiment

The results from the degradation experiment are published elsewhere (Dyreborg et al. 1995), but a short summary is presented here, with the focus on the observed degradation of the compounds at Day 16.

Toluene was degraded completely within the 16 days of incubation in all the batches where Factor A was absent. In the presence of Factor A (thiophene, pyrrole, 1-methylpyrrole, and benzofuran), no or a small percent (0 to 30%) degradation of toluene was observed. Phenol, 2,4-dimethylphenol, indole, and 1-methylnaphthalene were completely degraded (>95%) in all the batches after 16 days. Quinoline, *p*-xylene, *o*-xylene, and naphthalene were degraded completely in nearly all the batches, whereas benzothiophene and *o*-cresol were degraded completely in 50% of the batches. Factor A had also a high negative influence on the degradation of *o*-cresol. Only approximately 25% removal of *o*-cresol was observed in the presence of Factor A. Of the four compounds in Factor A, only benzofuran was observed to be degradable; however, no total removal of benzofuran was observed in any of the batches. A statistical analysis showed that Factor A was the main inhibitor to toluene degradation.

Toxicity Study

Table 1 shows the statistical design of the degradation experiment and the results of the toxicity test from Day 0 and Day 16. Data from the toxicity test are averages of four samples (two replicates from two samples), except for the control, which is an average of six samples. In general, the absolute error in toxicity between two replicates was less than 2%. The absolute error in toxicity between two samples was less than 10%, in most cases less than 5%.

The toxicities at Day 0 correlated well with the presence of Factor A. High inhibitions (50 to 60%) were observed when Factor A was initially present, whereas inhibitions were significantly lower (30 to 40%) in the absence of Factor A (Table 1). Also, the three phenolic compounds (Factor F) had an inhibiting effect on the nitrifying bacteria, increasing the inhibition approximately 10%. In the two batches where only toluene was present (batch 1), an inhibition of 30% was observed.

A reduction in inhibition was observed at Day 16 (Table 1). In 16 batches showing an inhibition less than 16% at Day 16, the chemical analysis showed a total removal of all organic compounds initially present (Dyreborg et al. 1995) except from batches 7 and 11 (Table 1), where a small concentration of benzothiophene was left. In 6 out of the 16 batches, no or only a small toxicity (< 5% inhibition) remained after 16 days of incubation, while the last 10 batches showed a toxicity of 5 to 16% inhibition. The remaining toxicities in the 10 batches may indicate that metabolites were formed during the degradation of the compounds. The chemical analysis did not reveal any metabolites, probably because of limitations in the analytical method used.

Statistical Analysis of the Toxicity Study

A small variation in the average percent inhibition of the three control batches was observed from Day 0 to Day 16 (Table 1). The variation between the two days was probably due to changes in the sensitivity of the nitrifying biomass. This variation required a correction of the data, carried out by dividing all observed inhibitions of Day 0 with the average inhibition of the control of Day 0 (67%). The same procedure was done with data from Day 16.

A statistical analysis was run separately on the corrected data from Day 0 and Day 16 with the statistical program package, SAS (1985), using the procedure ANOVA. The statistical model included all main factors, all interactions between two factors, and the interaction between Factors A, E, and F. The significant factors at Day 0 were (significance level $= 0.01$, the asterisk (*) denotes interactions between factors): A, B*C, E, A*E, F, D*F, and E*F, $r^2 = 0.92$. The two main factors were Factor A, contributing 76% of the total variance, and Factor F contributing with 10%. The analysis showed that Factor A had a high inhibiting effect, but the effect of Factor F was much weaker, though it also inhibited nitrification. The other significant factors showed very weak influence on nitrification. For Day 16, the significant factors were A, B, D, A*D, C*D, A*E, B*E, F, A*F, and B*F ($r^2 = 0.97$). Again, the main inhibitor was Factor A, accounting for 91% of the total variance.

The contribution to the total variance from the other significant factors was less than 1%. Factors B and F had a very weak inhibiting effect on the nitrification, whereas Factor D seemed to have a small stimulating effect, except when Factor D was together with Factor A, where the effect was inhibiting. The contribution to the total variance of the phenolic compounds (Factor F) decreased from 10% to less than 1%. This is consistent with the complete degradation of the three phenolic compounds in all the batches, with the exception of the remaining o-cresol in the batches where Factor A was present (Dyreborg et al. 1995). The decrease of toxicity in relation to the degradation of the phenolic compounds and the presence of Factor A are shown in Figure 1. A significant decrease in the toxicity in samples where Factor A was absent was clearly observed. From an inhibition of 42% (Factor F absent) and 54% (Factor

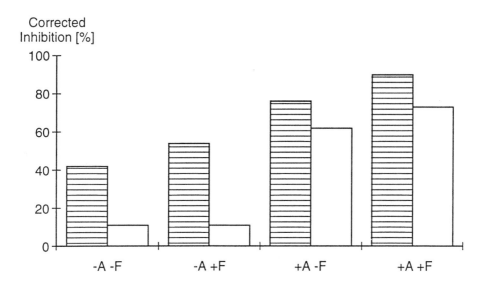

FIGURE 1. The average corrected inhibition in batches, where both Factor A and Factor F were absent (–A –F), Factor A absent and Factor F initially present (–A +F), Factor A present and Factor F absent (+A –F), and Factor A and Factor F present (+A +F). ▤ : Day 0 □ : Day 16.

F initially present), a decrease in inhibition to approximately 11% was observed. However, when Factor A was present, only a small decrease (10 to 16%) in inhibition was observed.

DISCUSSION

A significant correlation was obtained between the content of creosote compounds and the toxicity to nitrification of a sample. Surprisingly, the combination of the four NSO-compounds (thiophene, pyrrole, 1-methylpyrrole, and benzofuran) was the most significant inhibitor. There are indications in the literature that the phenolic compounds are especially toxic to nitrification. A concentration of phenol of 5.6 mg/L has been reported to result in 75% inhibition of ammonia oxidation, whereas 75% inhibition was observed at a concentration of 4.4 mg/L for o-cresol (Stafford 1974). For the NSO-compounds, hardly any data have been reported. For quinoline, 5% inhibition was reported at a concentration of 5 mg/L (Beccari et al. 1980), whereas other studies showed 10% inhibition of ammonia oxidation at a concentration of 0.55 mg/L of pyrrole and 0.6 mg/L of 1-methylpyrrole (Dyreborg 1995). For the S- and O-compounds, studies have shown that the thiophenes and benzofuran are very toxic to ammonium oxidation. An $EC_{50\%}$ value of 0.1 mg/L for thiophene, $EC_{50\%}$ =

3.0 mg/L for benzothiophene, and an $EC_{90\%} = 3.0$ mg/L for benzofuran has been observed (Dyreborg 1995).

Dealing with toxicity of mixtures, three different interactions between compounds can take place: (1) no interactions, which means that the total toxicity is a simple addition of the toxicity of each sample; (2) antagonistic interaction, implying that the total toxicity is less than simple addition; and (3) synergism, implying that the total toxicity is higher than simple addition. The inhibitions reported here showed comparable results to the literature using the no-interactions hypothesis. The insignificance of indole and quinoline at Day 0 is consistent with the fact that these compounds obviously are nontoxic at the concentration levels used in this experiment. At Day 16, quinoline became significant with a very weak stimulating effect, which may indicate that the metabolites formed during the degradation of quinoline were stimulating nitrification. Also, the toxicity of the three phenolic compounds was expected according to the literature; and when the phenolic compounds were degraded, a decrease in the toxicity was observed.

Toluene showed an unexpectedly high inhibition in the batches where it was present as a single compound. At a concentration of 0.5 mg/L, an inhibition of 30% was observed. Keener and Arp (1994) reported that a toluene concentration of 13.8 mg/L reduced the activity of nitrification by *Nitrosomonas europaea* by almost 80%. Blum and Speece (1991) observed a 50% inhibition on *Nitrosomonas*. The reason for this inconsistency has not yet been identified.

The significance of *p*-xylene, *o*-xylene, naphthalene, and 1-methylnaphthalene also is surprising. No inhibition was observed at a naphthalene concentration of 50 mg/L (Richardson 1985). A 95% inhibition was observed at a *p*-xylene concentration of 15.9 mg/L (Keener and Arp 1994). For *o*-xylene an estimate for the concentration causing 100% inhibition was 8.4 mg/L (Dyreborg and Arvin 1995). The concentrations used in this experiment were far below these concentrations, and no significant effect was expected.

In some cases the toxicity disappeared completely during the 16 days of incubation, indicating a total mineralization of the compounds or accumulation of nontoxic intermediates, though the chemical analysis revealed no organic compounds. A remaining inhibition of 5 to 16% was observed in other cases, perhaps indicating that toxic intermediates were formed and accumulated during the degradation of the organic compounds. However, again the chemical analysis detected no traces of organic compounds. The reason for this inconsistency was probably that the intermediates were nondetectable by the chemical analysis used.

The results presented in this paper show that the degradation potential in a contaminated aquifer can be incorrectly estimated, if experiments with only simple mixtures are carried out. The toxicity test can in some cases reveal if any metabolites/end products are formed during the degradation of the organic compounds.

ACKNOWLEDGMENT

This study was financially supported by the Groundwater Research Center at the Technical University of Denmark.

REFERENCES

Arvin, E., S. Dyreborg, C. Menck, and J. Olsen. 1994. "A mini-nitrification test for toxicity screening, MINNTOX." *Water Research* 28(9):2029-2031.

Beccari, M., R. Passino, R. Ramadori, and V. Tandoi. 1980. "Inhibitory effects on nitrification by typical compounds in coke plant wastewaters." *Environmental Technology Letters* 5(1):245-252.

Blum, D.J.W., and R.E. Speece. 1991. "A database of chemical toxicity to environmental bacteria and its use in interspecies comparisons and correlations." *Res. J. Water Pollut. Control Fed. 63* (3):198-207.

Dyreborg, S., and E. Arvin. 1995. "Inhibition of nitrification by creosote-contaminated water." Accepted for publication in *Water Research*.

Dyreborg, S. 1995. Unpublished data.

Dyreborg, S., E. Arvin, and K. Broholm. 1995. "The influence of creosote compounds on the aerobic degradation of toluene." Accepted for publication in *Biodegradation*.

Keener, W. K., and D. J. Arp. 1994. "Transformations of aromatic compounds by *Nitrosomonas europaea.*" *Applied and Environmental Microbiology* 60(6):1914-1920.

Richardson, M. 1985. *Nitrification inhibition in the treatment of sewage.* The Royal Society of Chemistry, Thomas Water reading. Burlington House, London, UK.

SAS. 1985. Version 6, release 6.02. SAS Institute Inc. Cary, NC.

Stafford, D. A. 1974. "The effect of phenols and heterocyclic bases on nitrification in activated sludge." *Journal of Applied Bacteriology* 37:75-82.

Laboratory Biodegradation to Define In Situ Potential for PCE Transformation

Margaret Findlay, Alfred Leonard,
Sam Fogel, and William Mitchell

ABSTRACT

A laboratory biodegradation test for anaerobic dechlorination of tetra-chloroethene (PCE) to ethene was designed to determine the potential for site samples to support rapid PCE transformation by providing a complete mineral supplement and excess electron donor. A 100-mL groundwater sample from one site spiked with 50 μmol PCE produced 17 μmol ethene in 22 days, supporting the proposal that natural attenuation of contamination was occurring. A soil sample from a second site in 100 mL test medium converted 10 μmol PCE completely to ethene in 10 days, providing evidence that the recent spill could be responsible for observed biotransformation products.

INTRODUCTION

Laboratory cultures originating from environmental samples have been developed that are capable of anaerobically transforming PCE via trichloroeth-ylene (TCE), dichloroethylene (DCE), and vinyl chloride (VC) to ethylene within days or hours (DiStefano et al. 1991; deBruin et al. 1992). "Microcosms" constructed from anaerobic subsurface samples, however, typically require months for the appearance of ethene. We describe an anaerobic test procedure that can support rapid biotransformation of high concentrations of PCE to ethene by environmental samples, and can assist in evaluating a site's potential to support natural attenuation or remediation.

METHODS

Media and Test Conditions

Serum bottles (160 mL) contained 100 mL of mineral/yeast extract (YE) medium containing, per liter, 200 mg NH_4Cl, 36 mg $(NH_4)_2SO_4$, 36 mg Na_2SO_4,

240 mg $MgCl_2 \cdot 6H_2O$, 380 mg KCl, 0.3 mg $MnCl_2 \cdot 4H_2O$, 0.6 mg $CoCl_2 \cdot 6H_2O$, 0.3 mg H_3BO_3, 2.7 mg $CaCl_2 \cdot 2H_2O$, $Na_2MoO_4 \cdot 2H_2O$, 0.5 mg $ZnCl_2$, 1.0 mg $NiCl_2$, 36 mg $FeCl_2 \cdot 4H_2O$, 2.64 g $NaHCO_3$, 71 mg YE, 1 mg resazurin ($NaHCO_3$ powder added after sterilizing). While adding soil, bottles were purged via cannula with 70/30 N_2/CO_2. Bottles were then sealed with Teflon™-faced silicone septa and crimped aluminum caps, and were incubated cap side down at 35°C.

Analytical Procedures

Chlorinated ethenes were determined by analyzing 10-μL aqueous samples according to U.S. Environmental Protection Agency (EPA) Method 601. Methane and ethene were determined by analyzing 100-μL samples of headspace by direct injection into an HP5880 gas chromatograph with flame ionization detection, operated isothermally at 220°C. The column was packed glass, 6 ft × 4 mm i.d., Suppelco Carbosieve II support. Calibration for CH_4 and ethene was by external standards. The total amount of ethene per bottle (headspace + aqueous) was calculated using dimensionless Henry's law constants for chlorinated ethenes at 35°C (Gossett 1987), 7.6 for ethene, and 23.3 for methane.

Site #1 History

Site #1 is the location of a manufacturing facility having small releases from both a PCE and a TCE degreaser to the subsurface between 1950 and 1980. Wells downgradient from these spots yielded anaerobic biotransformation products suggesting that natural treatment of the contaminants might be occurring on site. Because groundwater moving at 6 ft/day (1.8 m/day) appeared to have the potential for carrying the VC plume off site, a laboratory biodegradation experiment was planned to demonstrate the potential of site samples to biodegrade VC to ethylene.

For the TCE degreaser plume, no aqueous samples were available from the source area, but soil gas analysis detected 7 ppm TCE. Table 1 presents typical data from three wells located 350 ft, 650 ft, and 1,350 ft (107 m, 198 m, and 411 m) directly downgradient from the TCE source. Well #1 in 1992 had no TCE, but contained 16 ppm cDCE and 77 ppm VC. Well #2 had only trace quantities

TABLE 1. Site #1 wells downgradient from TCE source; TCE transformation products, μg/L.

Well #	Feet from source	Days from source	11/92 TCE	cDCE	VC	10/93 TCE	cDCE	VC
1	350	60	<1	16	77	<1	22	24
2	650	110	<1	1.1	<1	<1	<1	<1
3	1,350	225	<1	<1	<1	<1	<1	<1

of *c*DCE, and no VC. Water from the farthest well, having a travel time of 225 days from the source, contained no chlorinated ethenes. It was postulated that the VC was being anaerobically dechlorinated to ethene downgradient from well #1.

Table 2 presents data for wells downgradient from the PCE source. Well #4 is 75 ft from the PCE source and contains PCE and TCE, but no DCE or VC. Well #5 is 250 ft (76 m) downgradient and contains only DCE and a trace of VC. Although no wells are located further downgradient, it was suggested that the *c*DCE in well #5 would be converted to VC and ethene downgradient.

Site #1 Biodegradation Test

To obtain evidence in support of the idea that the anaerobic biotransformation is proceeding to ethylene on this site, a sample of groundwater was obtained from well #1 and tested for the presence of anaerobic organisms capable of carrying out those processes in the laboratory. A 100-mL sample, which contained TCE, *c*DCE, and VC at 1, 16, and 35 ppb, respectively, was amended with concentrated mineral solutions and yeast extract to provide the same concentrations as the mineral medium described above, and placed in a 160-mL serum bottle with 250 μmol of methanol. After 5 days, vigorous methane production was measured. At this point, the bottle was flushed with N_2/CO_2 to remove any remaining original chlorinated ethenes, and spiked with 50 μmol PCE and additional 250 μmol methanol.

Table 3 presents the results of analysis of samples from this bottle. After 4 days, TCE appeared, but by 12 days no conversion to DCE or VC had occurred. An additional 500 μmol of methanol was then added, and by 15 days *c*DCE appeared and VC began to accumulate. On Day 18 ethylene was detected, and on Day 22 accounted for 34% of the spiked PCE. It was concluded that well #1 contained a microbial population with the potential to transform PCE to ethylene.

Site #2 History

A PCE spill occurred when a pressure line ruptured during delivery of PCE to a dry cleaning establishment, releasing 5 gal (19 L). Although visibly contaminated soil was removed immediately, within 11 weeks a groundwater

TABLE 2. Site #1 wells downgradient from PCE source PCE and transformation products, μg/L.

Well #	Feet from source	9/92				12/93			
		PCE	TCE	DCE	VC	PCE	TCE	DCE	VC
4	75	13	4	<1	<1	33	2.4	<1	<1
5	250	<1	1.1	27	<1	<1	<1	24	1

TABLE 3. Site #1, well #1 test bottle: μmol per bottle (aqueous + headspace); 100 mL site water + 60 cc headspace; Day 0, 50 μmol PCE, 250 μmol methanol, minerals, 7 mg YE.

Day	PCE	TCE	cDCE	VC	Ethene
0	50	bdl[a]	bdl	bdl	bdl
4	12	31	bdl	bdl	bdl
12	bdl	38	bdl	bdl	bdl
		add 500 μmol methanol			
15	bdl	29	16	17	bdl
18	bdl	bdl	bdl	61	6
22	bdl	bdl	bdl	29	17

(a) bdl is below detection limit. For detection limits, see Table 4.

sample taken 10 ft from the spill site contained 130 mg/L PCE, as well as 9 mg/L TCE and 5 mg/L VC, anaerobic biotransformation products of PCE. It was critical to obtain evidence that these products could have arisen from the observed spill, and not from some previous unreported spill. Rapid dechlorination was a logical possibility for this site, because the contaminated area was adjacent to a septic leach field. Therefore, a laboratory experiment was carried out to determine the potential of site soils to rapidly mediate the anaerobic dechlorination of PCE.

Site #2 Biodegradation Tests

A soil sample was collected 10 months after the spill from the immediate spill area by hand auger from 13 cm above to 13 cm below the water table at 2 m. Analysis for purgable organics indicated the presence of at least 27 mg/kg PCE, 12 mg/kg TCE, 50 mg/kg cDCE, 9 mg/kg tDCE, and 19 mg/kg VC. Four 5- to 11-g subsamples were immediately transferred (without purging the soil to remove chlorinated ethenes) to serum bottles containing the mineral/YE medium described above, and fed 250 μmol of methanol. Methane production in 10 days was about 0.05 μmol, demonstrating that a methanogenic community was functioning. At this point, the bottles were purged to remove chlorinated ethenes associated with the original samples. Then 9.8 μmol of PCE and 250 μmol of methanol were added to the bottles, and they were incubated as before.

The biodegradation test data are presented in Table 4. On Day 0, although 9.8 μmol PCE was added per bottle, somewhat less was detected by analyzing aqueous samples, probably due to PCE adsorption to soil. In Bottle B, 2 μmol of DCE were present on Day 0, indicating insufficient purging of the original contamination. In all bottles, within 1 to 2 weeks, PCE was extensively depleted and was replaced by DCE and/or VC. In B, 8.3 μmol DCE were present on

TABLE 4. Site #2 test bottles: μmol per bottle (aqueous plus headspace); 100 mL mineral medium/soil + 60 mL headspace; Day 0, 9.8 μmol PCE, 250 μmol methanol, 7 mg YE.

Bottle	Day	PCE	TCE	DCE	VC	ethylene	methane
B	0	5.6	bdl	2.0	bdl	—	—
	6	0.5	bdl	8.3	bdl	—	—
	10	—	—	—	—	0.5	43
C	0	8.2	bdl	bdl	bdl	—	—
	6	1.0	0.5	bdl	11.0	—	—
	10	—	—	—	—	14	0.1
A	0	2.6[a]	bdl[b]	bdl[b]	bdl[b]	—	—
	7	bdl[b]	bdl	12.4	bdl	—	—
	15	1.0	2.1	bdl	6.2	—	—
	19	—	—	—	—	3.0	7.1
D	0	2.0[a]	bdl[b]	bdl[b]	bdl[b]	—	—
	7	bdl	bdl	11.2	bdl	—	—
	15	0.5	bdl	bdl	bdl	—	—
	19	—	—	—	—	1.9	220
Detection Limits		0.4	0.4	0.6	2.4	0.1	0.05

(a) Day 0 samples for bottles A and D were frozen and analyzed after 1 week.
(b) bdl means not detected.   "—" means not analyzed.

Day 6, and ethylene (accounting for 5% of the added PCE) was present on Day 10. In C, 11 μmol of VC were measured on Day 6, and on Day 10, ethylene accounted for 140% of the spiked PCE. Bottles A and D had mostly VC by Day 15, and 4 days later contained ethylene accounting for 30% and 19% of the PCE. In summary, 3 of the 4 bottles showed significant conversion of PCE to ethylene in 10 to 19 days. It was concluded that soil from Site #2 did have the potential to bring about production of biotransformation products on a time frame of weeks, rather than years.

DISCUSSION

Laboratory Degradation of High Concentrations of PCE to Ethene at High Rates

DiStefano et al. (1991) developed an anaerobic batch culture from a sewage treatment inoculum by gradually increasing the concentration of PCE and electron donors in successive subcultures. Repeated additions of 50 μmol PCE plus

1,600 μmol methanol and 2 mg yeast extract to 100 mL of culture resulted in the degradation of the PCE to a mixture of VC and ethylene in 2 days, obtaining 88% reducing equivalents from methanol and 12% from yeast extract. PCE dechlorination accounted for about one-third of the reduced product, and acetate formation for two-thirds. If these cultures were allowed to proceed for 4 days, 99% of the PCE was converted to ethylene. In another study, DeBruin et al. (1992) developed a laboratory column culture from river sediment and anaerobic digester sludge by gradually increasing the PCE input and flow until 9 μM PCE could be completely converted to ethene and ethane in less than 2 h. These authors stressed the importance of providing a high ratio of reducing equivalents (150-fold) to PCE to achieve complete conversion to ethene or ethane. It should be possible to achieve rates of dechlorination in situ comparable to those reported.

In Situ Association of Ethene with Anaerobic Dechlorination

Several sites have been studied in which the presence of ethene has been associated with either natural or remedial anaerobic dechlorination of ethenes. Beeman et al. (1994) stimulated treatment of PCE, TCE, and DCE by groundwater recirculation and addition of benzoate and sulfate, observing ethene production in the treated area but not in a control area. Fiorenza et al. (1994) reported ethene associated with a plume of DCE, VC, and 1,1-DCA produced from anaerobic dechlorination of disposed PCE, TCE, and 1,1,1-TCA, whereas no ethene was detected upgradient. Major et al. (1991) compared two areas of a PCE spill, one also having methanol contamination able to act as an electron donor for anaerobic processes. The methanol area exhibited more extensive PCE transformation, as well as higher ethylene concentrations. Anaerobic microcosms created from site soil/groundwater slurries produced significant quantities of ethene after 83 days at 10°C, substantiating the proposed mechanism of contaminant attenuation.

Design of Laboratory Biodegradation Tests

The purpose of the laboratory biodegradation tests reported here is to demonstrate the potential of site samples to support rapid dechlorination of high concentrations of contaminants. In this regard, it is reasoned that "microcosms" constructed from unamended subsurface samples do not correctly represent in situ situations in which microbes immobilized on soil may be continuously provided with mineral nutrients and electron donors via moving groundwater. The biodegradation test bottles used in this study provided subsurface samples with complete mineral supplement as well as excess electron donor. The results for Sites #1 and #2 show conversion of PCE to ethylene in about 20 days, indicating that the test can provide a relatively rapid assessment of a site's potential for remediation.

The test for Site #2 (bottles A, C, and D) resulted in the conversion of 10 μmol PCE, amended with complete minerals, 250 μmol MeOH and 7 mg YE, to between 20% and 100% ethene in 10 to 19 days at 35°C. This is a faster conversion than that reported by Major et al. (1991) for 1.8 μmol of PCE added to a 60-mL site slurry amended with 50 μmol acetate and 94 μmol methanol to 28% ethene in 83 days at 10°C. The major differences in the design to these two biodegradation tests are the incubation temperature and the lack of amendment with mineral nutrients in the Major et al. test. It is possible that temperature may not be a large factor, because deBruin et al. (1991) reported the same rate of dechlorination after acclimation at 10°C as obtained at 20°C.

Because our tests were carried out directly with site samples, rather than with laboratory cultures, our results may indicate that the rate of anaerobic dechlorination of PCE in the subsurface environment can be, at sites characterized by high dissolved organic carbon and mineral nutrients, faster than the time frame of years usually assumed for natural attenuation.

REFERENCES

Beeman, R. E., J. E. Howell, S. H. Shoemaker, E. A.Salazar, and J. R. Buttram. 1994. "A field evaluation of in situ microbial reductive dehaleogenation by the biotransformation of chlorinated ethenes." In R.E. Hinchee et al. (Eds.), *Bioremediation of Chlorinated and Polycyclic Aromatic Hydrocarbon Compounds*, pp. 14-27. Lewis, Chelsea, MI.

deBruin, W. P., M. Kotterman, M. A. Posthumus, G. Schraa, and A. Zehnder. 1992. "Complete biological reductive transformation of tetrachloroethene to ethane." *Appl. Environ. Microbiol. 58*: 1996-2000.

DiStefano, T. D., J. M. Gossett, and S. H. Zinder. 1991. "Reductive dechlorination of high concentrations of tetrachloroethylene to ethene by an anaerobic enrichment culture in the absence of methanogenesis." *Appl. Environ. Microbiol. 57*: 2287-2292.

Fiorenza, S., E. L. Hockman, S. Szojka, R. M. Woeller, and J. W. Wigger. 1994. "Natural anaerobic degradation of chlorinated solvents at a canadian manufacturing plant." In R. E. Hinchee et al. (Eds.), *Bioremediation of Chlorinated and Polycyclic Aromatic Hydrocarbon Compounds*, pp. 277-286. Lewis, Chelsea, MI.

Gossett, J. M. 1987. "Measurement of Henry's Law Constants for C1 and C2 Chlorinated Hydrocarbons." *Environ. Sci. Technol. 21*(2): 202-208.

Major, D. W., E. W. Hodgins, and B. J. Butler. 1991. "Field and laboratory evidence of in situ biotransformation of tetrachloroethene to ethene and ethane at a chemical transfer facility in North Toronto." In R.E. Hinchee and R.F. Olfenbuttel (Eds.), *On-Site Bioreclamation Processes for Xenobiotic and Hydrocarbon Treatment*, pp. 147-171. Butterworth-Heinemann, Stoneham, MA.

Thermophilic Slurry-Phase Treatment of Petroleum Hydrocarbon Waste Sludges

Frank J. Castaldi, Karl J. Bombaugh,
and Beverly McFarland

ABSTRACT

Chemoheterotrophic thermophilic bacteria were used to achieve enhanced hydrocarbon degradation during slurry-phase treatment of oily waste sludges from petroleum refinery operations. Aerobic and anaerobic bacterial cultures were examined under thermophilic conditions to assess the effects of mode of metabolism on the potential for petroleum hydrocarbon degradation. The study determined that both aerobic and anaerobic thermophilic bacteria are capable of growth on petroleum hydrocarbons. Thermophilic methanogenesis is feasible during the degradation of hydrocarbons when a strict anaerobic condition is achieved in a slurry bioreactor. Aerobic thermophilic bacteria achieved the largest apparent reduction in chemical oxygen demand, freon extractable oil, total and volatile solids, and polycyclic aromatic hydrocarbons (PAHs) when treating oily waste sludges. The observed shift with time in the molecular weight distribution of hydrocarbon material was more pronounced under aerobic metabolic conditions than under strict anaerobic conditions. The changes in the hydrocarbon molecular weight distribution, infrared spectra, and PAH concentrations during slurry-phase treatment indicate that the aerobic thermophilic bioslurry achieved a higher degree of hydrocarbon degradation than the anaerobic thermophilic bioslurry during the same time period.

INTRODUCTION

Thermophilic microorganisms capable of growth on hydrocarbons have been isolated from natural mesophilic and thermophilic environments contaminated with hydrocarbons (Zarilla & Perry 1987). The isolates were obtained from hot springs, estuarine muds, and wastewater treatment activated sludges and include strains of obligately thermophilic bacteria which use *n*-alkanes as growth substrates. The microbes were aerobic, endospore-forming rods capable of growth at temperatures ranging from 42 to 75°C, with optimum growth between 55 and 65°C. Optimum pH for growth was between 6.2 and 7.5 (Zarilla & Perry 1987). Thermophilic bacteria capable of growth on hydrocarbons also were isolated from oil-covered lake muds. These microbes grew at temperatures between 45 and 70°C, with optimum growth between 55 and 60°C (Klug & Markovetz 1967).

Bacterial isolates capable of using selected *n*-alkanes and *l*-alkenes were obtained from natural hot springs (60 to 65°C) in Yellowstone National Park (Merkel et al. 1978). These obligate thermophilic bacteria had the capacity to grow at the expense of selected long-chain saturated and unsaturated hydrocarbon substrates.

A hydrocarbon-utilizing, obligately aerobic thermophilic bacterium, *Thermomicrobium fosteri*, was isolated from a littoral area of North Carolina (Phillips and Perry 1976). This microorganism can use various long-chain *n*-alkanes, *l*-alkenes, primary alcohols, or ketones as substrates for growth. Optimum growth occurs at a temperature of 60°C.

Relatively few data exist for thermophilic petroleum degraders in soil, although the activity of such microorganisms may be significant around Middle Eastern desert oil wells and petroleum refinery environments. However, evidence suggests that hydrocarbon-utilizing thermophiles may be isolated from both mesophilic and thermophilic sources.

This study investigated the feasibility of using thermophilic bacteria to achieve enhanced oil degradation during slurry bioreactor treatment of oily waste sludge from refinery operations. The objectives of this proof-of-concept biotreatability study were (1) to develop chemoheterotrophic cultures of thermophilic bacteria that can sustain growth on petroleum hydrocarbons, (2) to achieve a microbial treatment process that can effectively reduce waste mass through petroleum hydrocarbon degradation, (3) to evaluate the effects of mode of thermophilic bacterial metabolism on the potential for petroleum hydrocarbon degradation, and (4) to evaluate the potential for thermophilic methanogenesis during hydrocarbon degradation.

EXPERIMENTAL APPROACH

Inoculum

The thermophilic bacteria examined in the study were isolated from chemical industry waste compost, petroleum refinery waste compost, anaerobic digester sewage sludge, activated sludges from three individual chemical industry wastewater treatment facilities, and activated sludge from a pulp and paper manufacturing wastewater treatment facility. These materials were obtained from their respective sources and used as inoculum for both aerobic and anaerobic thermophilic bacteria enrichment experiments. The inoculum screening studies were conducted in triplicate sealed serum bottles using batch culturing techniques (Owen et al. 1979). The enrichments were carried out at 55°C, and microbial activity was monitored by measuring gas evolved during processing. All serum bottles in the thermophilic anaerobic screening study received an inoculum of anaerobic digester sludge to provide methanogenic bacteria to the test population.

Anaerobic Thermophilic Bacteria Enrichments

The procedure for cultivating the chemoheterotrophic and methanogenic anaerobic thermophilic bacteria used in the experiment was originally described by Owen et al. (1979). The nutrient formulation of the culture medium incorporates modifications described by Shelton and Tiedje (1984) and is presented in Table 1. The medium was prepared by mixing stock solutions of macronutrients, trace nutrients, vitamins, and a mixture of mineral oil and diesel oil , at approximately 1,000 mg/L, in distilled water. The mineral oil was a mineral type nonPCB transformer oil (Univolt-60), and the diesel oil was a fuel for diesel engines obtained from distillation of petroleum.

Gas production in the serum bottles was measured in triplicate using a syringe in the manner described by Owen et al. (1979), and the extent of gas generation was compared between controls and treatment vials. The average cumulative gas productions of the thermophilic anaerobic bacteria examined in the screening study are presented in Figure 1. The measured methane and carbon dioxide levels in the generated gas from each test inoculum are presented in Table 2. These data indicate that the petroleum refinery waste compost (Inoculum Source B) and the pulp and paper manufacturing wastewater-activated sludge (Inoculum Source E) samples had the most active thermophilic anaerobic microorganism cultures. Although at lower levels of gas production, both the chemical waste compost (Inoculum Source A) and the chemical wastewater #1-activated sludge bacterial culture (Inoculum Source D) were active during the anaerobic screening test. Methane and carbon dioxide levels in the generated gas were 47 to 53% and 33 to 39%, respectively for the four most active inocula. These levels indicate that healthy methanogenic conditions developed with all of these cultures in the presence of the diesel oil and mineral oil substrates. The other inocula sources were generally similar to the controls in their behavior.

A mixture of the most active sources of thermophilic inoculum was selected as the seed bacteria for the treatment of oily waste sludges from petroleum refining operations. The majority of this thermophilic anaerobic inoculum consisted of the petroleum refinery waste compost (59 wt%) and the pulp and paper wastewater-activated sludge materials (22 wt%) , with smaller portions of chemical waste compost (10 wt%) and activated sludge (9 wt%) added for enhanced population diversity.

Aerobic Thermophilic Bacteria Enrichments

Aerobic thermophilic bacteria enrichments were performed in a similar manner as the anaerobic enrichments, except that oxygen was added to the

TABLE 1. Composition of anaerobic mineral medium.

Compound [a]	Concentration (mg/L)
KH_2PO_4	270
$K_2HPO_4 \cdot 3H_2O$	460
NH_4Cl	530
$CaCl_2 \cdot 2H_2O$	75
$MgCl \cdot 6H_2O$	100
$FeCl_2 \cdot 4H_2O$	20
$MnCl_2 \cdot 4H_2O$	0.5
H_3BO_3	0.05
$ZnCl_2$	0.05
$CuCl_2 \cdot 2H_2O$	0.038
$NaMo_4 \cdot 2H_2O$	0.01
$CoCl_2 \cdot 6H_2O$	0.5
$NiCl_2 \cdot 6H_2O$	0.05
Na_2SeO_3	0.05

(a) In addition to the above components, the following were added to the medium: 1.0 mg resazurin/L to indicate oxidation/reduction potential; 1.2 g $NaHCO_3$/L as a source of CO_2 for the methanogenic bacteria; and $Na_2S \cdot 9H_2O$ and $FeCl_2 \cdot 4H_2O$ to generate 0.05 millimole FeS with an excess of sodium sulfide to serve as a reducing agent.

serum bottles as air at the time that gas displacement measurements were made. Inocula sources that showed positive responses during aerobic thermophilic culturing included the chemical waste compost, the petroleum refinery waste compost, and the pulp and paper wastewater-activated sludge materials. These inocula showed levels of gas production above the controls that ranged from 160% for Inoculum Source E to 464% for Inoculum Source B, and a higher percent of carbon dioxide in the produced gas that ranged from

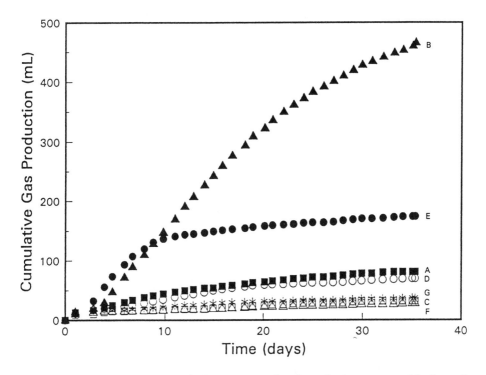

FIGURE 1. Average cumulative gas production during anaerobic inocula screening.

17.2% to 28.0% by volume, while also indicating a substantive depletion of oxygen level in the headspace. These data suggest that a similar mixture of thermophilic bacteria sources as used for anaerobic inoculum would provide a viable seed culture for aerobic thermophilic processing of oily waste sludges from petroleum refining operations.

Biodegradation Testing

The experimental thermophilic bacteria slurry bioreactor system is illustrated in Figure 2. It is a continuously stirred tank reactor that can be operated in the thermophilic temperature range (55 to 75°C) under either aerobic or anaerobic conditions. The reactor has a 4.0-L glass reaction chamber and a Teflon™ stirrer that provides contact between the oily waste sludge and the thermophilic bacteria inoculum developed for the study. A gas sparge system is provided with the aerobic process configuration to deliver air or oxygen to the reactor. Temperature and pH are monitored daily in the bioreactor slurry. Reactor temperature is maintained at levels consistent with thermophilic processing by adding heat from a heating mantle designed to control temperature in the proper range. When operated under anaerobic conditions, the bioreactor was provided with a wet test meter for continuous gas volume monitoring.

TABLE 2. Thermophilic anaerobic inocula screening results.

	Inoculum Source [a]	Average Total Cumulative Gas Production (mL)	Percent CO_2	Percent CH_4	Percent N_2 [b]
A	Chemical Waste Compost	80.8 ± 8.3	34.8 ± 4.4	47.1 ± 1.8	6.7 ± 11.6
B	Petroleum Refinery Waste Compost	464.9 ± 87.5	38.9 ± 6.0	47.7 ± 1.3	0
C	Anaerobic Digester Sludge	32.0 ± 1.3	35.2 ± 9.4	17.9 ± 1.4	51.2 ± 1.3
D	Chemical Wastewater No. 1 Activated Sludge	69.8 ± 3.7	38.0 ± 13.7	50.2 ± 2.4	5.6 ± 9.7
E	Pulp and Paper Wastewater Activated Sludge	173.8 ± 15.5	32.8 ± 1.0	53.0 ± 0.3	0
F	Chemical Wastewater No. 2 Activated Sludge	27.7 ± 1.9	33.8 ± 4.6	9.4 ± 6.4	54.6 ± 7.4
G	Chemical Wastewater No. 3 Activated Sludge	37.5 ± 0.5	38.9 ± 17.1	21.4 ± 7.3	46.4 ± 4.6

(a) All serum bottles were incubated at 55°C (131°F). Each inoculum source was evaluated in triplicate. Reported values are averages of three measurements.

(b) All serum bottles had a headspace filled initially with nitrogen. Where the total percentage of measured gases is less than 100, the difference is probably equal to the amount of nonmethane hydrocarbon volatilized from the waste during treatment.

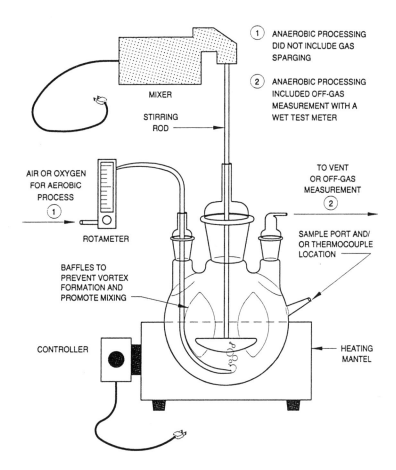

FIGURE 2. Schematic of bench-scale thermophilic slurry bioreactor.

The flask reactors (Figure 1) were operated at an average temperature of 65°C under either strict anaerobic or strict aerobic thermophilic conditions to assess the feasibility of achieving hydrocarbon biodegradation of petroleum refining waste sludges. The reactors were operated under batch conditions except during periods when nutrients were added to the slurry to stimulate the rate of biodegradation.

The thermophilic bacteria consortium used to seed the flask reactors was initially cultured on a mixture of diesel oil and mineral oil under either anaerobic or aerobic conditions. Thereafter, oily waste sludge was added to the individual test bioreactors in increments of 6,000 mg/kg of solids until approximately 12 to 14 wt% slurry was achieved in each reactor. Nitrogen and phosphorus were added as a combination of ammonium phosphate and ammonium sulfate to the slurry bioreactors with each new oily waste sludge increment blended into the test reaction vessels. These nutrients were added with each waste charge on the basis of a carbon:nitrogen:phosphorus ratio of 10:0.45:0.07, respectively. The thermophilic bacteria consortium, including the thermophilic methanogenic bacteria (anaerobic digester sludge) added to the strict anaerobic bioreactor, comprised approximately 3% of the final target slurry mass at the time when performance testing began on the slurry reactors. This incremental loading of the oily waste sludge required approximately 20 days of consecutive feedings for each reactor.

Each of the bioreactors was monitored monthly for freon extractable hydrocarbons (EPA Method 413.1), total and soluble chemical oxygen demand (COD) (EPA Method 410.1), total solids (EPA Method 160.3), total volatile solids (EPA Method 160.4), alkalinity (EPA Method 310.1), ammonia-nitrogen (EPA Method 350.2), total Kjeldahl nitrogen (EPA Method 351.3), orthophosphate (EPA Method 365.2), specific gravity (ASTM Method D-1429), and viscosity (ASTM Method D-2983). Analytical methods used to determine organic compound biodegradation were PAHs analyzed by high-performance liquid chromatography (HPLC), petroleum products using simulated distillation by gas chromatography (GC), and tetrahydrofuran (THF) extractable hydrocarbon analyzed by infrared (IR) spectrophotometry. The analytical methods for analysis of PAHs and petroleum products employed modifications to EPA Method 8310 and ASTM Method D-2887, respectively.

The bioreactors also were monitored daily for temperature (EPA Method 170.1), pH (EPA Method 150.1), dissolved oxygen (aerobic reactor) by EPA Method 360.1, and oxidation-reduction potential (anaerobic reactor) by ASTM D-1498-76. Frequent measurements of the rate of dissolved oxygen consumption (SM 2710B) by the thermophilic microorganisms were made in the aerobic bioreactor during the study. Biological gas production in the anaerobic reactor was measured using ASTM E-1196-87.

RESULTS

The waste material treated in the thermophilic biodegradation study was a K-waste (i.e., petroleum refinery hazardous sludge) composed of a combination of American Petroleum Institute (API) oil-water separator sludge and float from a dissolved air flotation (DAF) oil-water separator. The wastestreams were combined at the point of generation and a sample was obtained of the composite material. The grain-size distribution indicates that the waste-contained solids that were composed of 48.8 wt% sand, 35.8 wt% silt, and 15.4 wt% clay. The sand fraction was 98% fine and the silt was a gradation from coarse to fine material. Approximately 96.9 wt.% of the grain sizes were less than 0.075 mm, whereas only 26.2 wt% of the solids had grain sizes less than 0.038 mm.

The measurements for apparent viscosity (4,095 cps) and specific gravity (1.12) were typical of most K-waste. The material behaved as a pseudoplastic, and the apparent viscosity was a function of both temperature and shear rate.

The raw waste had high concentrations of calcium and iron at 29,000 mg/kg and 8,600 mg/kg, respectively. It also contained high levels of chromium and zinc at 550 mg/kg and 390 mg/kg, respectively. These concentrations were for a solids level of 25.3 wt% total solids with 49% volatile matter.

Conventional Pollutant Evaluation

The results of conventional pollutant parameter analyses for the anaerobic and aerobic bioreactors are presented in Table 3. These data are for three sampling events over a period of 60 days of batch operation and provide information for the changes in the concentration of conventional pollutants during the test. Each thermophilic bioreactor was actually in operation for a period of 90 days because of the incremental loading strategy employed to introduce the oily waste sludge to the bacterial cultures.

In addition, two identical nonsterile unseeded control reactors also were operated in parallel with the test biodegradation reactors to monitor physical and chemical changes in the waste for comparison with the biochemical changes that occurred in the seeded reactors during the test.

The anaerobic thermophilic bioreactor experienced smaller changes in total chemical oxygen demand (COD) and oil and grease during the test period than did the aerobic reactor. However, soluble COD values did not change noticeably in the anaerobic reactor during the test period. The aerobic thermophilic bioreactor experienced large apparent reductions in total COD, oil and grease, total solids, and total volatile solids during the test period. However, it was determined that a separation had occurred in this reactor, which resulted in a viscous tar material that settled out during the period prior to the last sampling. This tarry residue had a solids concentration of 69.1 wt% total solids with 50.6% volatile matter. Both bioreactors were maintained between pH 6.8 and 7.9 during the first 90 days of testing.

Gas production in the anaerobic bioreactor is presented in Figure 3 in terms of cumulative liters of gas over the test period. The apparent dips in the graph are associated with sample collection and waste or nutrient loadings during the initial portion of the study. A culture of methanogenic bacteria was introduced to the anaerobic reactor on the 20th day of operation. Three gas samples were collected from the anaerobic bioreactor for analysis of methane and carbon dioxide during the study period. The bioreactor off-gases between day 98 and 105 were found to contain 54% carbon dioxide and 38% methane, with the remainder possibly a composite of hydrocarbon constituents found in the waste but not identified in the analysis. The pH in the anaerobic bioreactor dropped below pH 6.0 after day 105, which resulted in a cessation of gas production.

The microbial respiration rates measured with a membrane electrode according to the American Water Works Association (AWWA) Standard Method 2710B during the operation of the aerobic bioreactor are presented in Figure 4. The highest bioreactor respiration rates were experienced during the periods of incremental waste loading and/or the addition of nutrients to supplement the

TABLE 3. Change in conventional pollutants during thermophilic treatment of oily waste sludge.

Parameter	Anaerobic Bioreactor			Aerobic Bioreactor		
	Time 0 day	Time 30 days	Time 60 days	Time 0 day	Time 30 days	Time 60 days [a]
Total Solids, mg/kg	121,130	131,090	147,670	146,970	158,250	40,720
Total Volatile Solids, mg/kg	57,030	66,950	83,900	62,380	73,350	11,750
Total COD, mg/kg	195,330	156,010	113,950	190,630	162,430	23,950
Soluble COD, mg/L	1,220	1,090	1,300	3,380	5,010	5,140
Ammonia-Nitrogen, mg/L	191	203	230	180 [b]	17 [c]	11 [c]
Total Kjeldahl Nitrogen, mg/kg	730	530	565	620	440	169
Orthophosphate, mg/L	0.90	0.05 [d]	0.11	0.06 [e]	0.10 [d]	0.11 [d]
Alkalinity as $CaCO_3$, mg/L	1,085	1,860	2,130	260	250	260
Oil and Grease, mg/kg	79,300	71,160	64,060	84,900	65,720	11,110

(a) A viscous heavy tar material separated from the mixed liquor of the aerobic reactor and settled out during this period. This material accounted for 55% of the original waste charged to the bioreactor, on a dry-weight basis.
(b) Adjusted to 205 mg/L NH_3-N.
(c) Adjusted to 150 mg/L NH_3-N.
(d) Adjusted to 20 mg/L PO_4-P.
(e) Adjusted to 10 mg/L PO_4-P.

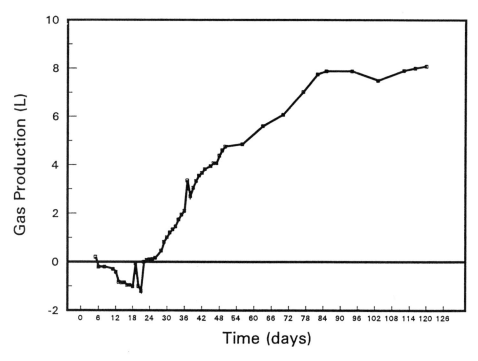

FIGURE 3. Anaerobic bioreactor gas production.

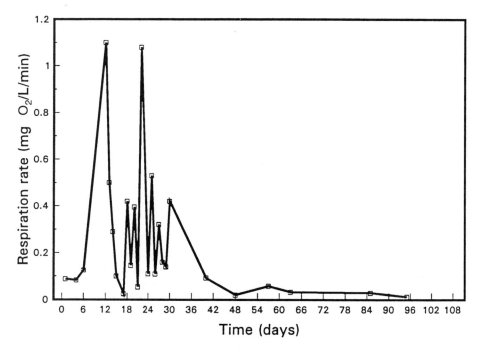

FIGURE 4. Aerobic bioreactor respiration rates.

measured deficiencies during the biodegradation test. There also was a significant increase in soluble fraction COD in the aerobic bioreactor during the test period. However, foam was not present in the slurry during the operation of the aerobic reactor.

The two nonsterile, unseeded control reactors showed a change in COD and oil concentration between the day 30 and day 60 samplings. These changes occurred concurrently with a rise in apparent indigenous microbial activity (i.e., oxygen uptake and carbon dioxide production) in the control reactors during the test. This was expected because the unseeded controls were not sterilized.

Polycyclic Aromatic Hydrocarbon Degradations

The results of the PAH analyses for each of the test reactors are presented in Table 4. These data are presented as dry weights. The results are for two sampling events over a period 60 days of batch operation. The results of the initial sampling of the anaerobic bioreactor are reasonably consistent with the PAH levels measured in the raw waste. However, the lower PAH concentrations measured in the initial sampling of the aerobic bioreactor can be attributed to the biodegradation that probably occurred during the period of incremental waste loading.

The anaerobic thermophilic bioreactor was less effective than the aerobic thermophilic bioreactor for the removal of PAHs in the oily waste sludge. The apparent PAH removal potential under thermophilic conditions appears to increase with increasing oxygenation of the slurry reactor. Strict anaerobic conditions were not conducive to the biodegradation of PAHs, and any apparent removal may have resulted from volatilization of the low-molecular-weight polycyclic aromatics.

The apparent removal of PAHs during aerobic thermophilic treatment was influenced by the separation of a viscous tarry residue during the test period. The fate of the PAHs and other pollutants in the aerobic bioreactor slurry was determined by a mass balance of materials in the reactor (Table 5). The concentrations of pollutants remaining in both the aerobic bioreactor slurry and separated residue at the conclusion of the test period were compared to pollutant levels in the raw waste at the start of treatment. The pollutants examined in this analysis included oily waste sludge total solids, total volatile solids, total COD, oil and grease, and PAHs. The HPLC chromatograms of PAHs detected in the raw oily waste sludge, the aerobic bioreactor treated slurry, and the separated residue are presented in Figure 5. These data indicate that substantial PAH degradation had occurred in both the bioreactor-treated slurry and the separated residue fractions during aerobic thermophilic treatment despite the apparent separation of the viscous tar material. The overall removal of PAHs during aerobic thermophilic treatment was approximately 58 wt%, while the apparent PAH removal during anaerobic thermophilic treatment was 15 wt% for the same time period.

TABLE 4. Change in PAHs during thermophilic treatment of oily waste sludge.

Analytes [a]	Raw Waste	Anaerobic Bioreactor		Aerobic Bioreactor	
		Time 0 day	Time 60 days	Time 0 day	Time 60 days [b]
Naphthalene	681.69	478.82	271.62	125.2	ND
Acenaphthylene	ND[c]	201.44	185.48	ND	ND
Acenaphthene	ND	ND	ND	ND	ND
Fluorene	127.72	118.88	100.02	65.99	ND
Phenanthrene	317.52	293.07	251.78	207.53	ND
Anthracene	30.05	20.64	17.47	12.93	ND
Fluoranthene	426.26	378.93	387.55	342.93	74.66
Pyrene	1,074.34	994.8	908.99	769.55	315.82
Benzo(a)anthracene	357.45	340.13	272.23	210.25	63.61
Chrysene	141.95	66.05	81.06	64.64	ND
Benzo(b)fluoranthene	39.94	36.33	8.94	14.29	ND
Benzo(k)fluoranthene	32.82	42.93	41.17	23.13	ND
Benzo(a)pyrene	21.75	20.64	9.35	11.57	ND
Dibenzo(a,h)anthracene	ND	26.42	25.19	2.04	16.70
Benzo(g,h,i)perylene	ND	ND	ND	ND	ND
Indeno(1,2,3-cd)pyrene	ND	ND	ND	4.76	ND
Total PAHs	3,251.49	3,019.08	2,560.85	1,854.81	470.79

(a) All data as $\mu g/g$ dry-weight basis.
(b) A viscous heavy tar material separated from the mixed liquor of the aerobic reactor and settled out during this period. This material accounted for 55% of the original waste charged to the bioreactor, on a dry weight basis.
(c) ND is not detected for values less than 0.2 $\mu g/g$.

Hydrocarbon Distribution Changes

Hydrocarbon distributions using a simulated distillation by wide-bore capillary column gas chromatography were measured for slurries in the test reactors at three points during the batch treatments. The results of these analyses on samples from the anaerobic and aerobic thermophilic bioreactors are presented in Figures 6 and 7, respectively. These data indicate that the aerobic bioreactor achieved higher treatment levels than that obtained during anaerobic thermophilic treatment when compared over the same time period. However, changes with time are still apparent in the anaerobic thermophilic bioreactor for the measured hydrocarbon distribution, which suggests a slower

TABLE 5. Fate of pollutants in a thermophilic aerobic bioreactor.

Parameter	Raw Waste [a]		Reactor Slurry [b]		Reactor Residue [c]		Mass Removed from Reactor for Analysis [e]	Percent Removal
	Concentration [d]	Mass [e]	Concentration [d]	Mass [e]	Concentration [d]	Mass [e]		
Total Solids	252,900	587.88	40,720	51.19	690,900	323.41	86.51	25.29%
Total Volatile Solids	124,000	288.25	11,750	14.77	349,600	163.65	36.88	29.02%
Total COD	202,500	470.72	23,950	30.11	280,080	131.11	94.26	57.18%
Oil and Grease	150,600	350.08	11,110	13.97	304,800	142.68	40.44	49.41%
PAHs	3,251.49 [f]	1.529	470.79 [f]	0.024	1,708.79 [f]	0.554	0.168	57.53%

(a) K-waste at 25.3% solids.
(b) Slurry remaining in reactor at Time 60 days.
(c) Tarry solids settled in reactor at Time 60 days.
(d) All data as mg/kg, as received (wet weight basis), except as noted.
(e) All data as grams, dry weight of a given parameter in the raw waste or treated residue (except as noted).
(f) Data as µg/g dry-weight basis.

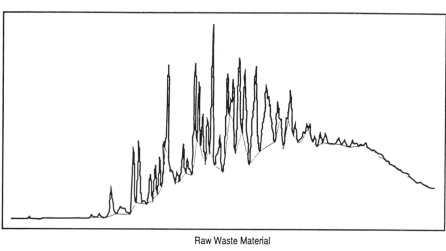

Raw Waste Material

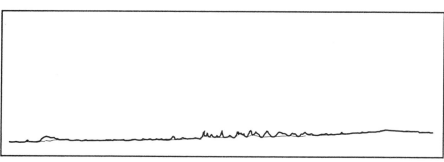

Treated Reactor Slurry

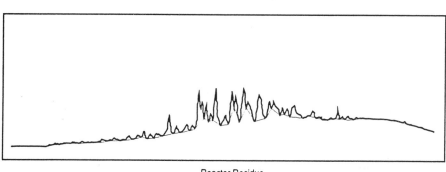

Reactor Residue

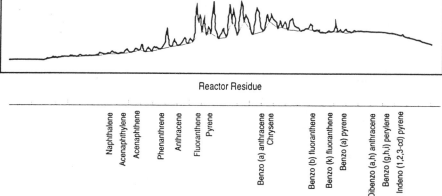

Naphthalene
Acenaphthylene
Acenaphthene
Phenanthrene
Anthracene
Fluoranthene
Pyrene
Benzo (a) anthracene
Chrysene
Benzo (b) fluoranthene
Benzo (k) fluoranthene
Benzo (a) pyrene
Dibenzo (a,h) anthracene
Benzo (g,h,i) perylene
Indeno (1,2,3-cd) pyrene

FIGURE 5. HPLC chromatograms of polycyclic aromatic hydrocarbons detected in aerobic bioreactor materials.

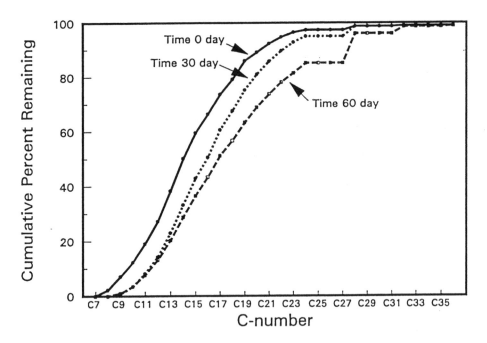

FIGURE 6. Hydrocarbon distribution for anaerobic bioreactor slurry.

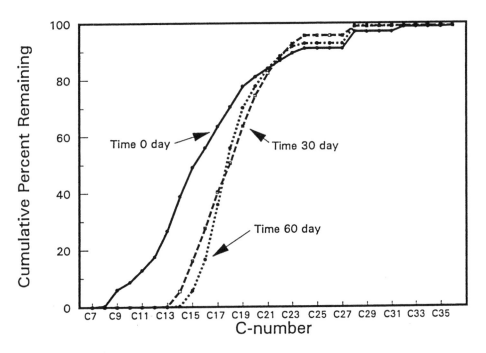

FIGURE 7. Hydrocarbon distribution for aerobic bioreactor slurry.

metabolic process that was nonetheless productive because methanogenesis was occurring.

Figure 8 presents a comparison of the hydrocarbon distributions of the raw waste material with the aerobic bioreactor-treated residue. The treated residue sample has a larger amount of higher-molecular-weight hydrocarbon material than does the raw waste. When this comparison is extended to include the data from the aerobic bioreactor slurry molecular weight distribution, both the aerobic bioreactor slurry (Figure 7) and residue (Figure 8) hydrocarbon distributions appear similar, suggesting that the untreated hydrocarbon fractions for these treated materials are also similar.

Infrared Spectra Evaluation

IR spectra of the tetrahydrofuran (THF) soluble and insoluble fractions of slurry samples from the anaerobic and aerobic bioreactors collected at day 60 of the test are shown overlain in Figure 9. When the spectra of the THF extracts from the anaerobic and aerobic bioreactors are compared, a strong band in the 1,000 to 1,100 K region is evident in the spectrum of the THF soluble fraction of the aerobic bioreactor sample. A significant difference in the symmetrical methyl bending vibration is also in evidence between the spectrum of the anaerobic bioreactor and the aerobic bioreactor, indicating that the slurry of the

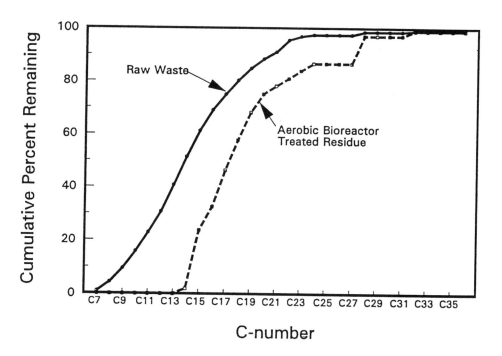

FIGURE 8. Comparison of raw waste and aerobic bioreactor residue hydrocarbon distributions.

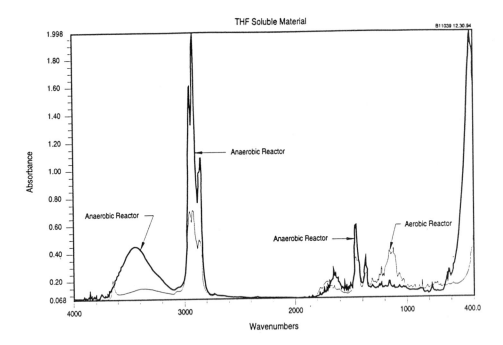

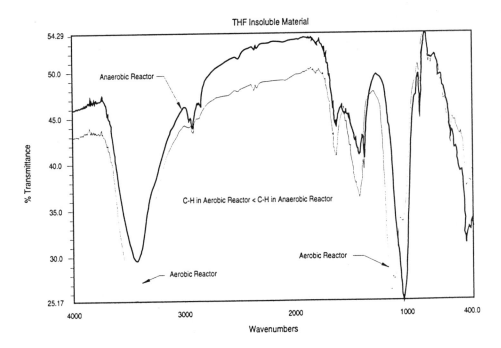

FIGURE 9. Comparison of anaerobic and aerobic bioreactor infrared spectra for THF soluble and insoluble materials.

aerobic bioreactor has significantly more gem-dimethyl groups than the anaerobic bioreactor material. The relative intensities of the methyl to methylene asymmetrical C-H stretching vibrations indicate that significantly fewer methylene carbons are present in the THF extract of the aerobic bioreactor than that of the anaerobic bioreactor.

The THF insoluble materials of the anaerobic and aerobic bioreactors are also shown overlain in Figure 9. This comparison shows that insoluble material from the aerobic bioreactor contains significantly less alkyl organic substance than does material from the anaerobic bioreactor. The spectra also show that the 1,030 K band is significantly smaller in the aerobic bioreactor sample, and the 1,080 K band is significantly stronger. If the 1,030 K band is due to primary alcohol functionality and the 1,080 K band to secondary alcohol functionality, it may follow that, during biodegradation, primary alcohols disappear in parallel with methylene carbons, while secondary alcohols remain with the isopropyl and gem dimethyl functionality.

CONCLUSIONS

The study of thermophilic bacteria treatment of petroleum refinery wastes indicated that cultures of chemoheterotrophic thermophilic bacteria can be developed from wastewater and waste treatment sources. Both anaerobic and aerobic thermophilic bacteria are capable of growth on petroleum hydrocarbons. The study also concluded that thermophilic methanogenesis is feasible during the degradation of petroleum hydrocarbons when a strict anaerobic condition is achieved in a slurry bioreactor. Aerobic thermophilic bacteria cultured in a slurry bioreactor achieved the largest apparent reduction in total COD, oil and grease, total solids, and total volatile solids when treating oily waste sludges.

A shift in the hydrocarbon distribution occurred during the period of biodegradation treatment to a distribution of higher-molecular-weight hydrocarbon material under all thermophilic metabolic conditions. However, this change in the distribution of petroleum hydrocarbons was more pronounced under an aerobic metabolic condition than under a strict anaerobic metabolic condition.

The apparent biodegradation of PAHs in petroleum refinery wastes under thermophilic conditions was best achieved by aerobic bacteria. Strict anaerobic conditions may not have been conducive to biodegradation of PAHs in the oily waste sludges examined by the study.

REFERENCES

Klug, M.J. and A.J. Markovetz. 1967. "Thermophilic Bacterium Isolated on *n*-Tetradecane." *Nature. 215*: 1082-1083.

Merkel, G.J., W.H. Underwood, and J.J. Perry. 1978. "Isolation of Thermophilic Bacteria Capable of Growth Solely on Long-Chain Hydrocarbons." *FEMS Microbiology Letters.* 3: 81-83.

Owen, W.F., D.C. Stuckey, J.B. Healy, Jr., L.Y. Young, and P.L. McCarty. 1979. "Bioassay for Monitoring Biochemical Methane Potential and Anaerobic Toxicity." *Water Research.* 13: 485-492.

Phillips, W.E. and J.J. Perry. 1976. "*Thermomicrobium fosteri* sp. nov., Hydrocarbon-Utilizing Obligate Thermophile." *Inter. Journal of Systematic Bacteriology.* 26(2): 220-225.

Shelton, D.R. and J.M. Tiedje. 1984. "General Method for Determining Anaerobic Biodegradation Potential." *Applied and Environmental Microbiology.* 47: 850-857.

Zarilla, K.A. and J.J. Perry. 1987. "Bacillus Thermoleovorans, sp. nov., Species of Obligately Thermophilic Hydrocarbon Utilizing Endospore-Forming Bacteria." *System Appl. Microbial.* 9: 258-264.

Toxicity and Biodegradability of Selected N-Substituted Phenols Under Anaerobic Conditions

Brian Donlon, Elias Razo-Flores,
Ching Shyung Hwu, Jim Field, and Gatze Lettinga

ABSTRACT

The anaerobic toxicity and biodegradability of N-substituted aromatics were evaluated in order to obtain information on their ultimate biotreatment. The toxicity of selected N-substituted aromatic compounds toward acetoclastic methanogens in granular sludge was measured in batch assays. This toxicity was highly correlated with compound hydrophobicity (log P octanol/water), indicating that partitioning into the bacterial membranes was an important factor in the toxicity. However, other factors, such as chemical interactions with key cell components, were suggested to be playing an important role. Nitroaromatic compounds were, on the average, over 300-fold more toxic than their amino-substituted counterparts. This finding suggests that the facile reduction of nitro-groups known to occur in anaerobic environments would result in a high level of detoxification. To test this hypothesis, continuous lab-scale upward-flow anaerobic sludge bed reactors treating 2-nitrophenol and 4-nitrophenol were established. The 4-nitrophenol was readily converted to the corresponding 4-aminophenol, whereas complete mineralization of 2-nitrophenol via intermediate formation of 2-aminophenol was obtained. These conversions led to a dramatic detoxification of the nitrophenols, because it was feasible to treat the highly toxic nitrophenolics at high organic loading rates.

INTRODUCTION

Nitrosubstituted aromatic compounds are widely used in the manufacture of azo dyes, explosives, pharmaceuticals, and pesticides. The presence of these aromatic xenobiotics in the environment creates serious public health and environmental problems (Kriek 1979). Of particular concern is the fact that the nitrosubstituted compounds are generally considered to persist in the environment. The aerobic metabolism of N-substituted aromatics has been well documented (Haigler & Spain 1991; Marvin-Sikkema & de Bont 1994), whereas the anaerobic biodegradation and toxicity of these compounds have only recently been addressed (O'Connor & Young 1993; Haghighi Podeh et al. 1995). Consequently, it was of interest to evaluate the anaerobic toxicity and biodegradability of these xenobiotic compounds to gain insights on how to establish protocols for their ultimate biotreatment.

EXPERIMENTAL PROCEDURES AND MATERIALS

Biomass

Elutriated methanogenic granular sludge from the full-scale upward-flow anaerobic sludge blanket (UASB) reactor of Shell Nederland Chemie (SNC) was used as inoculum.

Basal Medium

The basal macromedia and micromedia used in the anaerobic toxicity assay and the continuous experiments were as described previously (Sierra & Lettinga 1991).

Anaerobic Toxicity Assay

The specific acetoclastic methanogenic activity test was performed in duplicate in 120-mL vials sealed with butyl rubber septa and aluminium caps. Measurements of methane were performed according to a method modified from that of Sierra and Lettinga (1991). In the present article, the acetoclastic activity of the granular sludge (2 g [VSS]/L) was determined after 3 days of exposure to the toxicants. The toxicants studied included all monoisomers of aminophenol and nitrophenol, 2,4-dinitrophenol, 2-4-diaminophenol, 2-nitroaniline, and 2-phenylenediamine at concentrations ranging from nontoxic to completely inhibitory concentrations. This involved direct measurement of the methane production in the gas headspace by gas chromatography.

Continuous Upflow Anaerobic Sludge Blanket Experiments

The continuous experiments were performed in glass upflow anaerobic sludge blanket (UASB) reactors with a liquid volume of 0.16 L placed in a temperature-controlled room at 30 +2°C. All reactors were inoculated with 20 g VSS/L of anaerobic granular sludge. The continuous UASB reactors were operated at hydraulic retention times of 6 h. The reactors were started up with partially neutralized (pH = 6.0) volatile fatty acid (VFA) solutions (acetate:propionate:butyrate, 23:34:41 on a chemical oxygen demand [COD] basis) at a concentration of 4 g COD/L for 15 days. Thereafter, the nitrophenolic compounds were added at subtoxic concentrations in addition to the VFA substrate (8 and 12 mg/L for 2-nitrophenol and 4-nitrophenol, respectively). Nitrophenol concentrations were increased every two weeks, usually by increments of 25 mg toxicant/L, and the two reactors were treating 300 mg/L of 2-nitrophenol and 260 mg/L of 4-nitrophenol, respectively, at the time of sampling.

Analysis

Methane and VFAs were determined as described previously (Sierra & Lettinga 1991). Ultraviolet (UV) light absorption was measured with a Perkin-Elmer 550 A spectrophotometer. UV analyses of samples at less than 280 nm gave information on the breakdown in the aromatic structure of the parent compound. In addition, reduction of the nitro group to the corresponding amino group could be followed by monitoring color at 370 and 400 nm for 2-nitrophenol and 4-nitrophenol, respectively. Reversed-phase high-performance liquid chromatography (HPLC) (C18 Chromosphere) chromatography was used to quantify the N-substituted phenols. The solvent phase was methanol/acetic acid (20/80). The solvent flow was 0.5 mL per min. UV absorbance was detected at 280 nm. Aminophenols were also determined chemically using a modification of the method described by Oren et al. (1991). VSS determinations were performed according to Standard Methods (American Public Health Association 1985).

RESULTS

The 50% inhibitory concentration (50% IC) of N-substituted phenols against the activity of acetoclastic methanogenic bacteria was evaluated in this study. The 50% IC values of aminophenols and nitrophenols are plotted in Figure 1 as a function of compound hydrophobicity based on the log P octanol/water. A high correlation between toxicity to acetoclastic methanogens and compound hydrophobicity ($r^2 = 0.988$, p <0.001) was determined. It should be noted that aminophenols with a lower log P value were less toxic than the hydrophobic nitrophenols. Figure 1 also contains the regression lines for alkylphenols and chlorophenols to acetoclastic methanogens from a previous report (Sierra & Lettinga 1991). The comparison with alkylphenols and chlorophenols indicates that, at any given log P, the N-substituted phenols were much more toxic. The type of N-substitution had a strong effect on toxicity. Figure 2 demonstrates that nitroaromatics (NO_2 group) were consistently several orders of magnitude more toxic than their amine counterparts. Because NO_2 groups on aromatic structures are often reduced under anaerobic environments, the findings illustrated in Figure 2 suggest that a high level of detoxification can be expected in anaerobic environments. Consequently, the conversion of nitrophenols in continuous anaerobic reactors was tested. Figure 3 shows ultraviolet/visual (UV/VIS) scans of influent and effluent samples from the reactors treating

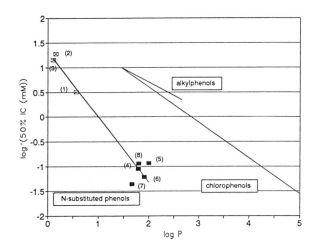

FIGURE 1. The effect of log P on the 50% inhibitory concentration of pheno-
lics to methanogens. N-substituted phenols this study; aminophenols are
indicated by the squares with ×, nitrophenols are indicated by the filled
squares: (1) 2-aminophenol, (2) 3-aminophenol, (3) 4-aminophenol, (4) 2-
nitrophenol, (5) 3-nitrophenol, (6) 4-nitrophenol, (7) 2,4-dinitrophenol,
and (8) 2,5-dinitrophenol. Alkylphenols and chlorophenols are adapted
from Sierra and Lettinga (1991).

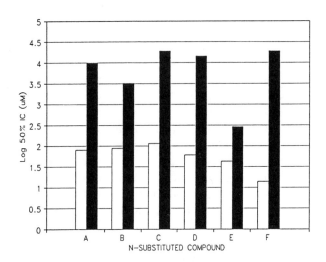

FIGURE 2. The 50% inhibitory concentration of nitroaromatic compounds to
methanogens with their corresponding aromatic amine. A = nitroben-
zene/aniline, B = 2-nitrophenol/2-aminophenol, C = 3-nitrophenol/3-
aminophenol, D = 4-nitrophenol/4-aminophenol, E = 2,4-dinitrophenol/
2,4-diaminophenol, and F = 2-nitroaniline/2- phenylenediamine.

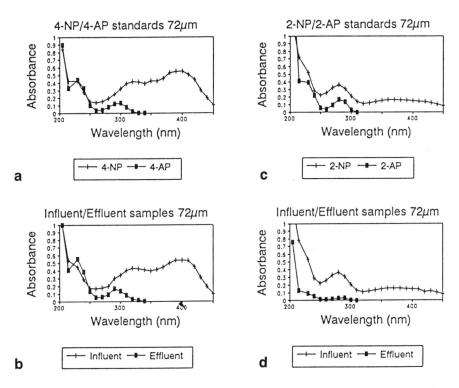

FIGURE 3. UV scans of (a) authentic 4-nitrophenol and 4-aminophenol standards, (b influent and effluent samples from reactor treating 4-nitrophe-
nol, (c) authentic 2-nitrophenol and 2-aminophenol standards, (d)
influent and effluent samples from reactor treating 2-nitrophenol. All
samples were diluted to give a concentration of 72 μM in 0.2M phosphate
buffer pH 7.0.

2-nitrophenol and 4-nitrophenol. When comparing these scans to similar scans
made for nitrophenol and aminophenol standards, 4-nitrophenol was clearly
converted to the corresponding aromatic amine. This was confirmed with
HPLC. On the other hand, 2-nitrophenol was mineralized via the intermediate
production of 2-aminophenol. The UV scan of effluent from the reactor treating 2-nitrophenol (Figure 3d) closely resembles a UV scan from a reactor
treating VFA only (results not shown). Further, the mineralization of 2-amino-
phenol was confirmed by its absence in the chemical assay for aminophenols.
The reactors were able to convert VFAs to methane at a high organic loading
rate of 15 g (COD) . $L^{-1}.d^{-1}$ (of which there was 1 g COD nitrophenol.l-$L.d^{-1}$),
with greater than 90% COD removal efficiency (results not shown), with an
influent concentration of nitrophenol greatly in excess of the 50% IC. This efficiency may possibly be attributed to the fact that in this case we have a reactor
system with an adapted sludge population whereas in determination of the

50% IC, a nonadapted sludge under batch conditions was employed. However, these results indicate that the anaerobic processing of nitrophenols can yield a high level of detoxification.

DISCUSSION

The experimental data suggest that some general relationships exist between the nature of the ring substitution and the inhibitory effect of N-substituted aromatics on acetoclastic methanogens. In particular, the nitro group was found to be the most toxic ring substituent. The higher toxicity of the nitro group was due in part to the high hydrophobicity of nitroaromatics compared to the other aromatics tested. The 50% IC values evidenced by the mono- and di-nitrophenols were similar (Figure 1), which indicates that hydrophobicity rather the oxidation state of the compounds was a more important factor related to the toxicity of these zenobiotics. Increasing hydrophobicity leads to easier passage of the organic compound into the lipophilic membranes, eventually disturbing the membrane structure and functions (Sikkema et al. 1994). Therefore, the greater the hydrophobicity of a compound (log P), the greater the toxicity and, subsequently, a lower 50% IC concentration would be expected for the organic compound in question.

Indeed, a high correlation was found between the toxicity and the hydrophobicity (log P octanol/water) of N-substituted phenols. However, at any given log P, the N-substituted phenols had a higher toxicity than alkylphenols and chlorophenols, indicating that their toxicity is not due solely to membrane toxicity. In this case, other factors such as the high chemical reactivity of nitroaromatics (Manahan 1993) could damage key enzymes and account for the elevated toxicity.

It is widely reported that, under anaerobic conditions, nitroaromatics are reduced to aminoaromatics (Gorontzy et al. 1993; O'Connor & Young 1993). Consequently, it was of interest to compare the relative toxic effects of nitroaromatic compounds and their corresponding amino compound in batch toxicity assays. The aromatic amines were on the average over 300-fold less toxic than the parent nitroaromatics. Therefore, a high level of detoxification would be expected to occur during continuous anaerobic treatment.

It is also evident from this study that, under continuous anaerobic conditions, nitrophenolics can be reduced to their corresponding aromatic amines. This biotransformation made it feasible to treat these highly toxic aromatics at high organic (nitrophenolic) loading rates. The ability of anaerobic consortia to remove and detoxify the nitro group makes anaerobic processes a useful treatment system for dealing with nitroaromatic wastewaters.

ACKNOWLEDGMENTS

Support from the Human Capital and Mobility program of the EC is gratefully acknowledged. Funding from the Consejo Nacional de Ciencia y Tecnologia and the Instituto Mexicano del Petroleo from Mexico also is appreciated.

REFERENCES

American Public Health Association. 1985. *Standard Methods for Examination of Water and Wastewater*, 16th ed. American Public Health Association, Washington, DC.

Gorontzy, T., J. Kuver, and K. H. Blotevogel. 1993. "Microbial transformation of nitroaromatic compounds under anaerobic conditions." *J. Gen. Microbiol. 139*: 1331-1336.

Haigler, B. E., and J. C. Spain. 1991. "Biotransformation of nitrobenzene by bacteria containing toluene degradative pathways." *Appl. Environ. Microbiol. 57*: 3156-3162.

Haghighi Podeh, M. R., S. K. Bhattacharya, and M. Qu. 1995. "Effects of nitrophenols on acetate utilizing methanogenic systems." *Wat. Res. 29*: 391-400.

Kriek, E. 1979. "Aromatic amines and related compounds as carcinogenic hazards to man." In P. Emmelot and E. Kriek (Eds), *Environmental Carcinogenesis*, pp. 143-164. Elsevier, Amsterdam.

Manahan, S. E. 1993. *Fundamentals of Environmental Chemistry.* Lewis Publishers, Inc. Chelsea, MI.

Marvin-Sikkema, F. D., and J. A. M. de Bont. 1994. "Degradation of nitroaromatic compounds by microorganisms." *Appl. Microbiol. Biotechnol. 42*: 499-507.

O'Connor, O. A., and L. Y. Young. 1993. "Effect of nitrogen limitation on the biodegradability and toxicity of nitro- and aminophenol isomers to methanogenesis." *Arch. Environ. Contam. Toxicol. 25*: 285-291.

Oren, A., P. Gurevich, and Y. Henis. 1991. "Reduction of nitrosubstituted aromatic compounds by the halophilic anaerobic eubacteria *Haloanaerobium praevalens* and *Sporohalobacter marismortui.*" *Appl. Environ. Microbiol. 57*: 3367-3370.

Sierra R., and G. Lettinga. 1991. "The effect of aromatic structure on the inhibition of acetoclastic methanogens in granular sludge." *Appl. Microbiol. Biotechnol. 34*: 544-550.

Sikkema, J., J. A. M. de Bont, and B. Poolman. 1994. "Interactions of cyclic hydrocarbons with biological membranes." *J. Biol. Chem. 269*: 8022-8028.

Ecotoxicological Assessment of Bioremediation of a Petroleum Hydrocarbon-Contaminated Soil

Agnès Y. Renoux, Rajeswar D. Tyagi,
Yves Roy, and Réjean Samson

ABSTRACT

A battery of bioassays [barley seed germination, barley plant growth, lettuce seed germination, worm mortality, Microtox®, lettuce root elongation, algae *Selenastrum capricornutum* growth, *Daphnia magna* mortality, and SOS Chromotest (±S9)] was used to assess an above-ground heap pile treatment of a soil contaminated with aliphatic petroleum hydrocarbons (12 to 24 carbons). Despite an initial oil and grease concentration of 2,000 mg/kg, no significant (geno)toxicity was apparent in the soil sample before treatment. During the treatment, which decreased oil and grease concentrations to 800 mg/kg, slight toxicity was revealed by three bioassays (barley seed germination, worm mortality, *Daphnia magna* mortality), and a significant increase in genotoxicity was measured with the SOS Chromotest (±S9). It appears that ecotoxicological evaluation revealed harmful condition(s) that were not detected by chemical assessment. This suggests that the remediation had ceased before complete detoxification occurred. This phenomenon must be further investigated, however, to furnish solid conclusions on the toxicological effectiveness of the biotreatment.

INTRODUCTION

To assess the environmental risk associated with contaminated soils, governmental agencies and the scientific community are now promoting the ecotoxicological evaluation of soils using bioassays in addition to the chemical characterization (Porcella 1983; Athey et al. 1987). If properly used, a battery of bioassays is intended to act as a direct indicator of ecological effects. Because soil is a complex ecological system, two major pathways leading to exposure have to be considered: (1) the direct contact of organisms living in soil, and (2) the indirect contact via runoff water. The second element involves the use of a leaching procedure to simulate contaminant runoff. The objective was to evaluate, with a battery of bioassays, the effectiveness of a biological process of soil remediation. For that purpose, four soil samples were collected during the biorestoration of a petroleum refinery soil that originally contained 2,000 mg/kg of oil and grease. The toxic effects were evaluated and compared to those of four "noncontaminated" soils.

MATERIALS AND METHODS

Soils Studied

Four soil samples were taken during the course of a bioremediation performed at a petroleum refinery site. More details are given in Samson et al. (1994). In this study, 1,500 m^3 of soil contaminated with an average concentration of 2,000 mg/kg of petroleum hydrocarbons having an average chain length of 12 to 24 carbons, characterized by gas chromatography/mass spectrocopy (GC/MS), were treated in an aboveground bioremediation cell. Three samples were collected in the bioreactor at various times in the remediation process: one shortly after the beginning of the bioprocess (December 1991), a second 5 months later (April 1992), and the last one (August 1992) at the end. A 100-m^3 control pile (Ctl), which received no particular treatment, also was sampled in December 1991. Each sample represents a mix of 8 subfractions, 2-mm sieved and well homogenized.

Four "noncontaminated" soils (urban soil, agricultural soil, from an urban area, strictly agricultural soil, and forestal soil) were sampled at specific areas where no point source contamination had been reported. These noncontaminated samples were collected from the topsoil horizon, thoroughly mixed, and 2 mm-sieved. Chemical analyses for mineral oil and grease, polycyclic aromatic hydrocarbons (PAHs), pentachlorophenol, and heavy metals showed that none of these contaminants were present at levels of concern.

Ecotoxicological Characterization

A battery of nine bioassays [barley (*Hordeum vulgare*) seed germination, barley (*Hordeum vulgare*) plant growth, lettuce (*Lactuca sativa*) seed germination, worm (*Eisenia foetida*) mortality, Microtox® (*Photobacterium phosphoreum*), lettuce (*Lactuca sativa*) root elongation, algae (*Selenastrum capricornutum*) growth, *Daphnia magna* mortality, and SOS Chromotest (±S9)] was performed on each soil. Each soil sample was split into two fractions; one to be used in direct-contact bioassays (solid phase) and the other to produce a leachate for liquid-phase bioassays. The persistence of the toxic effect upon dilution was given by the toxic threshold; for each bioassay, the NOEC (no observed adverse effect concentration in % v/v for leachates and % w/w for soils) and LOEC (lowest observed adverse effect concentration) were determined. The magnitude of the toxic effect (MT) was then given by the maximum observed effect for a sample in terms of percent inhibition or percent mortality, regardless of the soil or leachate concentration. For both SOS Chromotests (±S9), the maximum induction factor corrected for cell viability (IFCV)) was given as a genotoxic end point.

RESULTS AND DISCUSSION

Results for chemical analysis and biological activity studies of the bioremediated soils have been published elsewhere (Samson et al. 1994). The results demonstrated that the decontamination criterion (below 1,000 mg/kg) for mineral oils and grease has been reached. The hydrocarbon contamination decreased during the bioremediation process from 2,112 ± 615 mg/kg (n = 8) in the Dec91 sample (1,782 mg/kg (n = 1) detected in the Ctl sample), to 1,512 (n = 1) and 688 ± 228 mg/kg (n = 8) for Apr92 and Aug92, respectively (Samson et al. 1994).

For each soil sample, bioassays exhibiting significant (geno)toxicity, are shown in Table 1. In four bioassays, no detectable response was observed for any of the tested soils; i.e., lettuce seed germination (solid phase), Microtox®, algal growth inhibition, and root elongation (leachate). The absence of effect of the water-soluble fraction of petroleum hydrocarbons on *Photobacterium phosphoreum* bioluminescence has already been pointed out and has been attributed to a very low solubility or a poor toxicity of these compounds (Eisman et al. 1991). The other results, which showed that bioassays were sensitive to soils, will be discussed test by test.

TABLE 1. Results of the bioassay exhibiting significant toxicity.

Soil	Ctl	Dec91	Apr92	Aug92
Barley germination				
NOEC/LOEC (% w/w)	100/>100	56/100	56/100	100/>100
MT (%)	0	28	18.6	0
Worm mortality				
NOEC/LOEC (% w/w)	100/>100		56/100	100/>100
MT (%)	0		100	0
Cladoceran mortality				
NOEC/LOEC (% v/v)	100/>100	100/>100	100/>100	18/32
MT (%)	0	0	0	40
SOS Chromotest				
NOEC/LOEC (% v/v)	50/>50	50/>50	1/2	1/2
IFCV	1.7	1.7	2.5[a]	2.5[a]
SOS Chromotest +S9				
NOEC/LOEC (% v/v)	50/>50	50/>50	10/50	10/50
IFCV	1.0	1.5	1.9[a]	1.9[a]

(a) Significant genotoxicity (a = 0.01).

Solid-Phase Bioassays

Plant Growth and Seed Germination of Barley **(Hordeum vulgare).** The control soil Ctl and biotreated soils stimulated plant growth: 40% for Ctl, 100% for samples Dec91, Apr92, and Aug92. This increase can be explained partly by the addition of nutrients to biotreated soil. Sheppard et al. (1993) indicated that nutrient supply was a key factor when studying the variables influencing the plant life-cycle bioassay using *Brassica rapa*. However, barley seed germination was significantly inhibited by the Dec91 and the Apr92 samples (Table 1). This slight effect disappeared by the end of the treatment.

Worm Mortality. A toxic effect was detected when worms (*Eisenia foetida*) were put in contact with the Apr92 sample and a steep dose-response was exhibited; all animals died in the replicates of the 100% concentration. Worm mortality was not observed in the Ctl and Aug92 samples (the Dec91 sample was not analyzed). This result may be associated with the transitory appearance of toxicity during the biotreatment or with poor sample selection, although special care was taken to randomize and homogenize the samples.

Bioassays on Soil Leachate

Cladoceran Mortality. Sample Aug92 significantly contributed to the mortality of *Daphnia magna* (Table 1). Although previous studies have shown that crude oil is toxic to *Daphnia* species (Wong et al. 1981; Wong et al. 1983), the observed toxicity to *D. magna* cannot be directly attributed to the present compounds, which have been weathered. The agricultural "noncontaminated" soils used as a negative control also gave an effect indicating the sensitivity of this species to soil leachate (NOEC = 32% v/v). However, the toxic effect observed in Aug92 was greater (NOEC = 18% v/v) and cannot be attributed to soil leachate characteristics, because no toxicity was observed with Ctl, Dec91, and Apr92 samples.

SOS Chromotests. The SOS Chromotests performed with and without the liver activation fraction (S9) were the most sensitive bioassays used in this study (Table 1). The contaminated soils tested at the beginning of the treatment gave different results from the soil sampled at the end. The genotoxicities of Ctl and Dec91 samples were not statistically significant, although most IFCV values were greater than 1.5. On the other hand, the SOS Chromotests ($\pm$ S9) detected an increased genotoxic activity for both the Apr92 and Aug92 samples. If the induction factors (IFCV) are plotted against time, an obvious increase of genotoxicity during the biotreatment is observed (Fig. 1). The leachate of the "noncontaminated" soils also produced detectable genotoxic effects, indicating the potential presence of "naturally" occurring genotoxic compounds. However, when compared to the "noncontaminated" soils, the genotoxicity of the Apr92 and Aug92 samples was greater in terms of both the toxic thresholds (LOEC/NOEC) and magnitude of toxicities (IFCV).

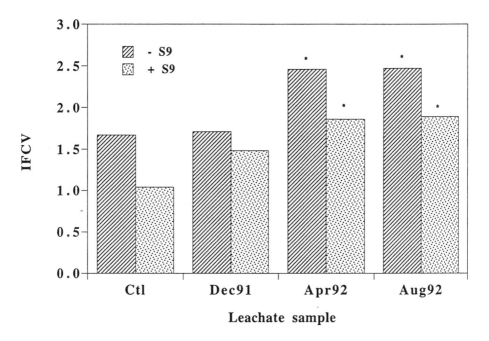

FIGURE 1. Genotoxicity of contaminated soil leachates during biotreatment as measured with the SOS Chromotest in the absence and the presence of liver activation fraction (S9). *: Significant genotoxic activity.

Although these soils were contaminated with petroleum hydrocarbons, PAHs were not detected (Samson et al. 1994). However, PAHs present in minute quantities could account for the detected genotoxicity, because PAHs in combination with aliphatic compounds such as dodecane have been shown to exhibit increased carcinogenic activity (Goldschmidt 1981; Kipling 1984). Moreover, the obvious increase in mutagenicity of soil during treatment suggests the appearance of one or more mutagenic compound(s) or an increase in the bioavailability of compounds already present. During the evaluation of a petroleum refinery waste, Donnelly et al. (1985) also observed that, although chemical analysis failed to identify any genotoxic constituent of the waste, biological analysis detected transitory mutagenic activity. Moreover, after a 1-year biodegradation of this petroleum-sludge-amended soil, the mutagenicity was increased (Brown et al. 1986; Barbee et al. 1992).

CONCLUSION

The ecotoxicological monitoring of an aboveground bioprocess tended to indicate that there was an increase in toxicity during treatment, although the soil contamination before treatment and the nutrient amendments did not seem to be harmful to the environment. This sometimes transitory appearance of toxicity, mainly associated with genotoxicity and acute animal toxicity, may be due to different factors such as (1) the formation of toxic by-products (not detected by chemical analysis), (2) an increase in bioavailability, or (3) a long-term negative effect of amendments. The appearance of toxicity can be assessed only by ecotoxicological characterization. This, with concomitant chemical assessment, would provide a basis to manage a biotreatment. Further research is necessary to confirm the previous hypotheses and to determine which factors could trigger bioassay response. Also, the development of specific bioassays for soil is needed because existing methods for water and sediments have limited ecological relevance and do not take into account the physicochemical characteristics of the soil. Nevertheless, the applicability of ecotoxicological soil assessment was demonstrated.

REFERENCES

Athey, L. A., J. N. Thomas, J. R. Skalski, and W. E. Miller (Eds.). 1987. *Role of Acute Toxicity Bioassays in the Remedial Action Process at Hazardous Waste Sites.* U.S. Environmental Protection Agency, EPA/600/8-87/044.
Barbee, G. C., K. W. Brown, and K. C. Donnelly. 1992. "Fate of mutagenic chemicals in soils amended with petroleum and wood preserving sludges." *Waste Management & Research* 10: 73-85.
Brown, K. W., J. C. Donnelly, J. C. Thomas, and P. Davol. 1986. "Mutagenic activity of soils amended with two refinery wastes." *Water, Air, and Soil Pollution* 29: 1-13.
Donnelly, K. C., K. W. Brown, J. C. Thomas, P. Davol, B. R. Scott, and D. Kampbell. 1985. "Evaluation of the hazardous characteristics of two petroleum wastes." *Hazardous Waste & Hazardous Materials* 2(2): 191-208.
Eisman, M. P., S. Landon-Arnold, and C. M. Swidoll. 1991. "Determination of petroleum hydrocarbon toxicity with Microtox." *Bulletin of Environmental Contamination and Toxicology* 47: 811-816.
Goldschmidt, B. M. 1981. "Nonnitrogenous carcinogenic industrial chemicals." In J. M. Sontag (Ed.), *Carcinogens in Industry and the Environment*, pp. 283-343. Marcel Dekker, New York, NY.
Kipling, M. D. 1984. "Soots, tars, and oils as causes of occupational cancer." In C. E. Searle (Ed.), *Chemical carcinogens*, Vol. 1, 2nd ed., pp. 165-174. ACS Monograph 182, Washington, DC.
Porcella, D. B. (Ed.) 1983. *Protocol for Bioassessment of Hazardous Waste Sites.* U.S. Environmental Protection Agency, EPA-600/2-83-054.
Samson, R., C. W. Greer, T. Hawkes, R. Desrochers, C. H. Nelson, and M. St-Cyr. 1994. "Monitoring an aboveground bioreactor at a petroleum refinery site using radio-respirometry and gene probes: Effects of winter conditions and clayey soil." In R. E. Hinchee, et al. (Eds.), *Hydrocarbon Bioremediation*, pp. 329-333. Lewis, Boca Raton, FL.
Sheppard, S. C., W. G. Evenden, S. A. Abboud, and M. Stephenson. 1993. "A plant life-cycle bioassay for contaminated soil, with comparison to other bioassays: mercury and zinc." *Archives of Environmental Contamination and Toxicology* 25(1): 27-35.
Wong, C. K., F. R. Engelhardt, and J. R. Strickler. 1981. "Survival and fecondity of *Daphnia pulex* on exposure to particulate oil." *Bulletin of Environmental Contamination and Toxicology* 26: 606-612.
Wong, C. K., J. R. Strickler, and F. R. Engelhardt. 1983. "Feeding behavior of *Daphnia pulex* in crude oil dispersions." *Bulletin of Environmental Contamination and Toxicology* 31: 152-157.

Degradation of High Concentrations of Glycols, Antifreeze, and Deicing Fluids

Janet M. Strong-Gunderson, Susan Wheelis,
Susan L. Carroll, Michael D. Waltz,
and Anthony V. Palumbo

ABSTRACT

A microbial consortium (EG-c) capable of degrading high concentrations of glycol-based waste was isolated from soil enrichments. The isolate primarily responsible for glycol degradation was a gram-negative rod (EG-y) that produced a water-soluble pigment. Initial laboratory experiments measured the degradation of ethylene glycol (EG) and propylene glycol (PG) at 25°C. Cell biomass optical densities (660 nm) increased from 0.01 to 0.5 within 48 h and reached a maximum of 0.73 at 72 h. Respirometry experiments showed oxygen consumption rates of ca. 1,000 mg/L/day with glycol degraded at a rate of ca. 2,000 mg/L/day (confirmed with gas chromatography). Laboratory tests were expanded to evaluate the degradation of a commercial antifreeze (primarily EG-based) and two deicing fluids (one PG- and one EG-based) at both 25 and 4°C (a more realistic winter temperature).

INTRODUCTION

Glycol-based products are used in numerous industries. EG is a primary component of equipment coolants and aircraft- and runway-deicing fluids and is used in the pharmaceutical-manufacturing industry. Approximately 4.93 billion pounds (2.2 billion kg) were produced in 1991, making it the 30th most-produced chemical in the United States (American Chemical Society 1992). PG is an antifreeze additive, as well as a preservative and emulsifier in food and bath products. More than 745 million pounds (335 million kg) were produced in 1991 (American Chemical Society 1992).

One significant source of glycol waste is generated from spent deicing fluids. Millions of gallons of deicing fluids (EG/PG mixtures) are used each year at northern aircraft facilities, with estimates greater than 25,000 gal (95,000 L)/y for one small military air base and 1.5 million gal (5.7 million L) for one commercial airline (1992-1993) (Airline representative, Minnesota, personal communication). Use of these deicing compounds results in large amounts of spent fluid discharged into sewer systems or collected for treatment at off-site facilities (Anon 1989).

On-site degradation of high concentrations of deicing fluids and anti-freezes may prove to be a cost-effective method for glycol disposal. Gycol waste is currently diluted to <10% before municipal facilities will accept it for treatment (Airline representative, Minnesota, personal communication). Wastewater treatment facilities usually specify 1 to 5% glycol as the maximum concentration for efficient microbial degradation and acceptable oxygen demands. Thus, the degradation of glycol to concentrations <5% can significantly reduce the volume of material discarded to municipal facilities.

This work focused on the isolation of bacteria capable of degrading a >10% concentration of glycol to <5% glycol. Preliminary results also showed that contaminant-degradation rates at 4°C were only slightly slower than at 25°C.

EXPERIMENTAL PROCEDURES AND MATERIALS

Soil Enrichment

Soil samples were received from a site with a long history of glycol contamination. Soils (5 g) were established in 100 mL minimal salt medium (Little et al. 1988) in 250-mL milk bottles with varying carbon contamination. PG and EG were analytical grade (J.T. Baker, Inc., Phillipsburg, New Jersey), the commercial antifreeze was EG-based, and the deicing fluids were a concentrated, unused EG-based product (NDEG) and a dilute, spent PG-based product (PPG).

All solutions were filter sterilized (0.2 μm filter) and kept at 4°C prior to use. The EG, PG, commercial antifreeze, and concentrated deicing fluid were handled as pure solutions (100%). PG-based deicing fluid was a spent solution with a concentration of almost 7% PPG in rain water. Enrichments were established in duplicate at 1, 5, and 10% solutions with the exception of the spent deicing fluid (1, 5, and 7%). Fresh transfers were made weekly (100 μL supernatant: 100 mL minimal salts medium). Bottles were incubated at room temperature on an orbital shaker.

Culture Identification

Supernatant samples (100 μL) were plated onto nonspecific nutrient agar (Difco) plates and Noble agar plates containing 1% EG or 1% PG. Plates were incubated at room temperature for 3 to 7 days. Pure cultures were isolated, and gram stains were performed.

Analytical Assay

Quantitative degradation was verified using gas chromatography. Supernatant samples were filtered using a 0.8/0.2-μm stacked Acrodisc filter.

Liquid injections of 1 μL were analyzed using a Perkin-Elmer Sigma 2000 gas chromatograph equipped with a flame ionization detector using a 30-m J & W Scientific DBWAX column (0.53 megabore, 1-μ film). Method conditions were oven temperature 160°C, injector temperature 250°C, and detector 230°C, with a nitrogen gas carrier flow of 5 mL/min.

RESULTS AND DISCUSSION

The microbial consortium (EG-c) isolated from soil enrichments initially consisted of 5 to 10 phenotypically different organisms and was able to degrade a variety of glycol-based products at high concentrations. The consortium was reduced to 3 isolates by maintaining colonies on plates containing only EG or PG as the sole carbon source. The dominant isolate (EG-y) produced a water-soluble yellow pigment; however, biochemical tests and lipid analyses have not provided conclusive identification for any of the gram-negative isolates.

Microbial growth and respiration indicated different patterns of glycol metabolism. The greatest increase in optical density occurred in cultures grown on PG, followed by cultures grown on PPG, EG, antifreeze, and NDEG (Figure 1). However, respirometry results (Figure 2) showed that growth on

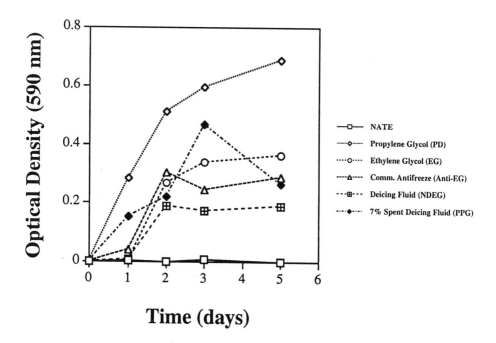

FIGURE 1. Qualitative measure of glycol degradation by increased optical densities. All carbon compounds were at 5% glycol in NATE (minimal salts media).

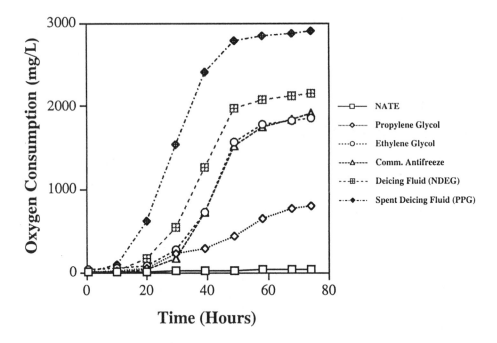

FIGURE 2. **Oxygen-consumption rates for various carbon compounds at 5% glycol.**

the spent deicing fluid (PPG) resulted in the greatest amount of oxygen consumed, followed by the EG deicing fluid (NDEG), pure EG, the commercial antifreeze, and, finally, pure PG. Quantification for all test compounds was confirmed using gas chromatography (Figure 3). EG was reduced from 10% to <6% within 7 days, and the commercial antifreeze was reduced from 5% to <2%. PG was the most recalcitrant. Typically, 10 to 40% of the initial concentration was degraded over a 7-day period, which corresponded to an average glycol metabolism of 1,000 to 4,000 mg/L/7 days. Overall, the degradation of 1 to 5% glycol was not as significant as that of the 10% solutions.

The on-site biodegradation of glycol wastes may be a cost-effective disposal method for a variety of industries. However, to have cost-effective degradation, other operating conditions must be evaluated and optimized. For example, we have begun to measure the degradation of deicing fluids at reduced ambient temperatures (≥4°C). This is important for the potential to bioremediate spent deicing fluids at northern glycol-waste-generating facilities during winter months.

Degradation of antifreeze and deicing fluids has the potential to significantly reduce the glycol concentration released to treatment plants, where it is the single most common compound during the winter months at numerous northern wastewater treatment facilities. Municipal airports could significantly

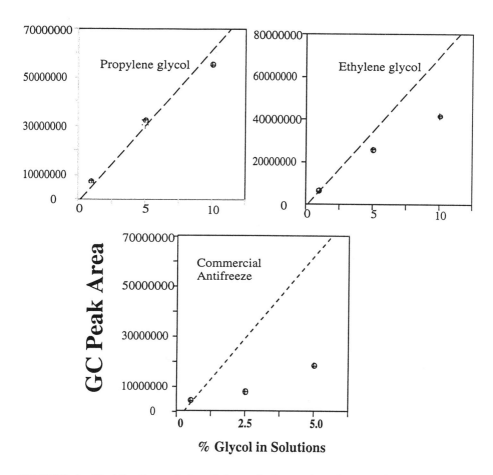

FIGURE 3. Verification of glycol degradation using the gas chromatograph. The dotted line represents the standard curve. Individual data points (n = 2) represent the glycol concentration after a 7-day incubation at 25°C.

reduce their waste-disposal costs by reducing the concentration of glycol fed to a treatment facility. Information from this work will be applied to a field-scale demonstration addressing on-site degradation of spent deicing fluids.

ACKNOWLEDGMENTS

We would like to thank Bret Summers for excellent technical assistance and Tina Anderson, Center for Environmental Biotechnology, The University of Tennessee, for analysis of fatty acid methyl esters using the Microbial Identification System, Microbial ID, Inc. This research was sponsored in part by the In-Situ Remediation Technology Development Program of the Office of Technology Development (Jef Walker), U.S. Department of Energy. The Oak Ridge

National Laboratory is managed by Martin Marietta Energy Systems, Inc., under contract DE-AC05-84OR21400 with the U.S. Department of Energy.

REFERENCES

American Chemical Society. 1992. "Facts and Figures for the Chemical Industry." *Chemical & Engineering News 70*: 32-75.

Anon. 1989. "Control of Storm Water Runoff Containing Aircraft Deicing Fluids." *Water Environment and Technology 1*: 12.

Little, C. D., A. V. Palumbo, S. E. Herbes, S. E., M. E. Lindstrom, R. L Tyndall, R. L., and P. J. Gilmer. 1988. "Trichloroethylene Biodegradation by Pure Cultures of a Methane-Oxidizing Bacterium." *Appl. Environ. Microbiol. 54*: 951-956 .

Microbial Activity in Subsurface Samples Before and During Nitrate-Enhanced Bioremediation

J. Michele Thomas, Virginia R. Gordy, Cristin L. Bruce,
Stephen R. Hutchins, James L. Sinclair, and C. Herb Ward

ABSTRACT

A study was conducted to determine the microbial activity at a site contaminated with JP-4 jet fuel before and during nitrate-enhanced bioremediation. Samples at three depths from six different locations were collected aseptically under anaerobic conditions before and during treatment. Cores were located in or close to the source of contamination, downgradient of the source, or outside the zone of contamination. Parameters for microbial characterization included (1) viable counts of aerobic heterotrophic, JP-4 degrading, and oligotrophic bacteria; (2) the most probable number (MPN) of aerobic and anaerobic protozoa; (3) the MPN of total denitrifiers; and (4) the MPN of denitrifiers in hydrocarbon-amended microcosms. The results indicate that the total number of denitrifiers increased by an order of magnitude during nitrate-enhanced bioremediation in most samples. The number of total heterotrophs and JP-4-degrading microorganisms growing aerobically also increased. In addition, the first anaerobic protozoa associated with hydrocarbon-contaminated subsurface materials were detected.

INTRODUCTION

The effects of in situ bioremedial processes on microbial communities are relatively unknown. At best, viable cell counts are the only indicators of microbial activity generally measured during in situ bioremediation operations. Although viable cell counts may indicate gross changes in population size, changes within the microbial community will not be detected. Of importance will be the selection pressure by the bioremedial process for contaminant-degrading organisms, and the fitness traits that allow these microorganisms to survive and effectively compete with other organisms for nutrients.

Knowledge of the effects of bioremediation on the microbial community may provide information that can be used to develop refined methods for process design and to enhance the bioremedial process.

The present study investigated the effect of nitrate-enhanced bioremediation on microbial populations at a site contaminated with JP-4 jet fuel. We assessed changes in the microbial ecology of a site by determining aerobic viable counts, the MPN of total denitrifiers and JP-4-degrading microorganisms with nitrate as the electron acceptor, and counts of aerobic and anaerobic protozoa. The study was designed so that samples would be collected before, during, and after treatment. In this paper, data from the before and interim sampling periods are compared.

EXPERIMENTAL PROCEDURES AND MATERIALS

Site Characterization

In 1984, a leak in an underground jet-fuel pipeline was discovered at Eglin Air Force base, Florida (Hinchee et al. 1989). An estimated 20,000 gal (75,708 L) of JP-4 jet fuel leaked into a sand and gravel aquifer. The aquifer consisted mainly of very fine to coarse quartz sand, but it also contained gravel and clay lenses and stringers. The hydraulic conductivity of the sands was 0.0212 cm/s; the hydraulic gradient and saturated thickness of the aquifer were estimated to be 0.015 ft/ft and 40 ft (12.5 m), respectively.

In 1986, a 2-year project was undertaken to evaluate bioremediation of the JP-4 using hydrogen peroxide as the oxygen source (Hinchee et al. 1989). The field test was only marginally successful, in part because of problems with iron precipitation and gas evolution encountered when aerobic processes were imposed in the anaerobic aquifer. Because these problems were expected to be minor during anaerobic treatment with nitrate, the site was considered for nitrate-enhanced bioremediation.

For use in the pilot demonstration of nitrate-enhanced bioremediation, two treatment cells were constructed; each measured 100 ft by 100 ft (30.5 m by 30.5 m). One cell received recharge amended with 10 to 20 mg/L NO_3^-, and the second received the same treatment without nitrate (Figure 1). The recharge was continuously applied through surface sprinklers at a rate of 11 gpm/cell (Hutchins et al. 1995). The leak of jet fuel had occurred close to the northwest side of the nitrate treatment cell.

Core Materials

Core material was collected before treatment at three depths from the following six boreholes: 80AA, 80BA, 80DA, 80EB, 80JB, and 80KB (Figure 1). Boreholes 80AA, 80BA, 80DA, and 80EB were to represent zones where residual

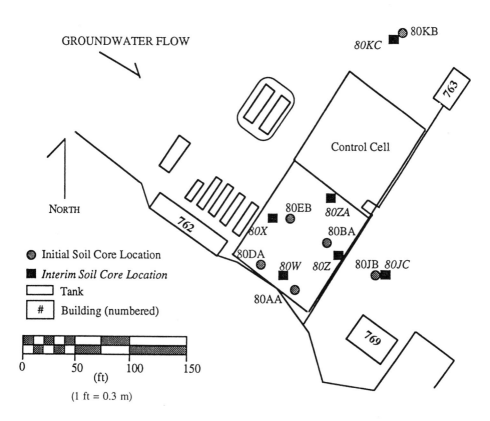

FIGURE 1. Soil core locations—Eglin Air Force Base.

JP-4 was located, 80JB was to represent a zone downgradient but influenced by the contamination, and 80KB was to represent an uncontaminated zone (Table 1). Interim core samples that corresponded in depth and location to the initial samples were collected after 5 months of treatment from boreholes 80Z, 80ZA, 80W, 80X, 80JC, and 80KC. Boreholes 80Z, 80ZA, 80W2, and 80X were supposed to represent zones containing residual contamination; 80JC borehole was downgradient but influenced by the contamination and remedial treatment; 80KC represented uncontaminated material that would not be impacted by treatment. Generally, the top samples in each core were unsaturated and the bottom two samples saturated; however, the water table fluctuated throughout the site, depending on rainfall.

The initial samples were collected in March 1993, and the demonstration was initiated in March 1994. The samples were collected using steam-cleaned drilling equipment and were prepared aseptically under anaerobic conditions (Hutchins et al. 1991). Core material was subsampled in an anaerobic glovebox. The samples were kept on ice in the field and during shipping, and then stored at 5°C in the laboratory until used.

TABLE 1. Physical characteristics of initial and interim samples.

Initial Sample	Initial Sample Depth (ft)	pH	JP-4 Conc. mg/kg	Interim Sample	Interim Sample Depth (ft)	pH	JP-4 Conc. mg/kg
80AA2	2.3-3.4	5.46	214	80ZA2	2.3-3.4	7.81	138.0
80AA1	3.4-4.5	5.49	1260	80ZA1	3.4-4.5	8.08	2630.0
80AA7	4.5-5.6	6.77	276	80ZA4	4.8-5.9	8.11	55.4
80BA3	1.0-2.2	4.88	2.8	80Z2	1.3-2.4	7.81	1.1
80BA2	2.2-3.4	6.23	355	80Z1	2.4-3.5	7.13	3750.0
80BA5	4.5-5.6	6.95	8.3	80Z4	4.9-6.0	8.32	34.1
80DA1	2.5-3.2	5.26	34.6	80W2	2.3-3.4	7.77	10.0
80DA5	4.0-5.0	5.77	377	80W1	3.4-4.5	7.76	1310.0
80DA8	6.0-6.8	6.82	54.7	80W4	4.8-5.9	7.75	4.1
80EB2	3.2-4.2	5.29	1160	80X2	2.5-3.8	8.75	4560.0
80EB1	4.2-5.2	5.46	1600	80X1	3.8-5.0	8.46	2620.0
80EB5	6.5-7.5	7.18	6.8	80X4	5.3-6.4	8.38	5780.0
80JB2	2.5-3.5	6.69	8.5	80JC2	2.3-3.4	6.87	11.1
80JB1	3.5-4.5	6.87	4.0	80JC1	3.4-4.5	6.90	1.7
80JB5	6.0-7.0	6.59	ND[a]	80JC3	5.9-7.0	7.87	0.0
80KB2	3.2-4.4	5.07	6.4	80KC2	5.0-6.0	5.63	0.20
80KB1	4.4-5.5	5.80	3.3	80KC1	6.0-7.5	7.01	0.00
80KB6	5.5-6.7	5.98	3.6	80KC4	7.8-8.9	7.55	0.30

(a) ND means not detected.

Hydrocarbon Biodegradation Under Denitrifying Conditions

Subsurface material collected before treatment was used to assess the denitrification potential of compounds found in JP-4 jet fuel. Microcosms were constructed with the Eglin core samples collected before treatment to evaluate benzene, toluene, ethylbenzene, and xylenes (BTEX), and trimethylbenzene removal under strictly denitrifying conditions as described previously (Hutchins 1991). Selected alkylbenzenes were degraded under denitrifying conditions by indigenous aquifer microorganisms. The mean zero-order rate constants were 1.2 ± 0.5 mg/L/d alkylbenzene biodegradation, and 2.6 ± 1.3 mg/L/d NO_3^--N removal.

Media and Culture Conditions

Serial dilutions of each sample were prepared in triplicate under aerobic conditions by aseptically adding 10 g of subsurface material to dilution bottles that contained 95 mL of 0.1% $Na_4P_2O_7 \cdot 10H_2O$. The bottles were shaken on a wrist-action shaker (Burrell Corporation, Pittsburgh, Pennsylvania) at a setting of 10 for 1 h, after which the rest of the dilution series was prepared using 0.1% $Na_4P_2O_7 \cdot 10H_2O$ as the diluent. The dilution series was used to determine the number of total heterotrophs, JP-4 degraders, oligotrophs, total denitrifiers, microorganisms that denitrify with JP-4 as the sole carbon source, and aerobic and anaerobic protozoa in each sample. The number of total heterotrophs, JP-4 degraders, and oligotrophs in each sample was determined under aerobic conditions on R2A medium (Difco Industries, Detroit, Michigan), a mineral salts medium incubated in the presence of JP-4 vapors, and a mineral salts medium incubated without JP-4 vapors, respectively. The colonies growing on R2A medium were counted after 1 to 1.5 weeks of incubation, whereas colonies growing on the other media were counted after 2 weeks of incubation. Counts of aerobic microorganisms are important because most denitrifiers are aerobic and switch to anaerobic respiration only in the absence of oxygen.

The mineral salts medium (pH 7) contained per liter of deionized water: 0.8 g KH_2PO_4, 5.58 g Na_2HPO_4 or 6.99 $Na_2HPO_4 \cdot 2H_2O$, 1.8 g $(NH_4)_2SO_4$, 0.017 g $CaSO_4 \cdot 2H_2O$, 0.123 g $MgSO_4 \cdot 7H_2O$, 0.5 mg $FeSO_4 \cdot 7H_2O$, 1.54 mg $MnSO_4 \cdot H_2O$, 2.86 mg H_3BO_3, 0.039 mg $CuSO_4 \cdot 5H_2O$, 0.021 mg $ZnCl_2$, 0.041 mg $CoCl_2 \cdot 6H_2O$, and 0.025 mg $Na_2MoO_4 \cdot 2H_2O$. The total number of denitrifiers was determined using Nitrate Broth (Difco Industries). The number of organisms that denitrify with JP-4 as the sole carbon source was determined in the mineral salts medium amended with 1 g/L KNO_3 and 200 μL of JP-4 jet fuel. Vials (40-mL volume) containing 20 mL of sterile aerobic mineral salts medium were amended aseptically with 200 μL of filter-sterilized JP-4, inoculated with serial dilutions of the samples, and sealed. Equal numbers of samples were incubated in the same medium without JP-4 to determine the effect of ambient carbon on denitrification potential. Because some oxygen was present, the denitrification detected in these samples could be the result of biodegradation of JP-4 or JP-4 degradation intermediates produced during the initial aerobic phase of incubation. Denitrifying activity was measured colorimetrically by testing for NO_2^- using sulfanilic acid and N,N dimethyl-1-naphthylamine (Blazevic et al. 1973). The total number of denitrifiers and the number of JP-4 degraders that use NO_3^- as the terminal electron acceptor were determined after 3 and 6 weeks, respectively.

The number of aerobic and anaerobic protozoa was determined (Sinclair and Ghiorse 1987) using subsurface sediment or dilutions of the sediment. Plates containing the protozoan enrichments were incubated aerobically or anaerobically in an anaerobic glovebox. The aerobic enrichments were observed at 2 weeks, 1 month, and 2 months. The anaerobic enrichments were observed every 3 weeks for 3 months.

Physical Analysis of Samples

The pH was determined with U. S Environmental Protection Agency method 9045 (U. S. Environmental Protection Agency 1986). Texture analysis was conducted by Law Engineering, Houston, Texas. The initial samples were found to consist of at least 92% sand and the rest silt. The interim samples have not been analyzed yet.

Samples were analyzed for JP-4 at the R.S. Kerr Environmental Research Laboratory, U. S. Environmental Protection Agency, Ada, Oklahoma, using the standard operating procedure, RSKSOP-72 (U.S. Environmental Protection Agency 1991).

Statistics

The data were compared using the t-test for equal or unequal variances, depending on the samples (95% confidence). The samples were compared in different ways. The entire depth intervals of initial and interim samples within the treatment cell (all cores except for JB, JC, KB, and KC) were compared to determine the overall effect of remediation. Proximate cores that could be paired by depth and compared statistically were 80EB and 80X, 80BA and 80Z, 80JB and 80JC, and 80KB and 80KC.

RESULTS

Denitrification Potential

The average total denitrifier population in the treatment zone (all samples except the J and K cores) increased from $\log_{10}$ 5.8 to 6.9 during the 5-month period of nitrate-enhanced bioremediation (Table 2). The number of total denitrifiers in the entire interval of the K (control) region did not increase,

TABLE 2. Denitrification potential in initial and interim samples.

Initial Sample	Log₁₀ Denitrifiers MPN/g dry wt (SD)			Interim Sample	Log₁₀ Denitrifiers MPN/g dry wt (SD)		
	Total	JP-4	No JP-4		Total	JP-4	No JP-4
80AA2	7.1 (0.4)	6.8 (0.2)	3.4 (0)	80ZA2	6.8 (0.6)	2.0, 1.7, <2	2.7 (0.3)
80AA1	7.2 (0.6)	6.4 (0.1)	<1	80ZA1	6.5 (0.9)	2.0, <2, <2	<1
80AA7	4.2 (0.2)	3.2 (0.5)	<1	80ZA4	6.5 (0.2)	<2, 1.7, 2.0	0.7, <1, <1
80BA3	5.2 (0.7)	1.6 (0.3)	1.9 (0.5)	80Z2	7.2 (0.2)	6.3 (0.2)	2.1 (0.1)
80BA2	6.0 (0.4	4.5 (0.2)	3.2 (0.1)	80Z1	7.9 (0.2)	3.6 (0.4)	2.2 (0.4)
80BA5	4.3 (0.4)	2.9 (0.3)	<1	80Z4	7.5 (0.2)	1.7, <2, <2	<1
80DA1	6.5 (0.2)	3.9 (0.2)	2.5, 2.9, <1	80W2	6.0 (0.5)	3.8 (0.5)	1.2 (0.3)
80DA5	6.1 (0.1)	4.1 (1.2)	1.6 (0.2)	80W1	7.1 (0.4)	<2	<1
80DA8	6.3 (0.4)	5.6 (0.3)	<2	80W4	6.6 (0.9)	4.0 (1.0)	0.7 (0)
80EB2	6.6 (0.5)	5.5 (0.9)	<1	80X2	8.4 (0.3)	3.6 (1.1)	0.6, <1, <1
80EB1	4.4 (0.9)	2.3 (0.5)	<1	80X1	7.0 (0.8)	2.3 (0.5)	<1
80EB5	6.4 (0.2)	5.1 (0.3)	<1	80X4	4.8 (0.2)	<2	0.7, <1, <1
80JB2	4.7 (0)	3.7 (0.3)	<1	80JC2	7.5 (0)	3.1 (0.4)	0.7, 0.7, <1
80JB1	7.1 (0.4)	6.0 (0.4)	<1	80JC1	6.9 (0.2)	2.9 (0.5)	<1
80JB5	4.4 (0.2)	3.0 (0.2)	1.5, 1.5, <1	80JC3	5.2 (0.1)	2.6 (0.2)	<1
80KB2	4.8 (0.2)	0.9 (0.2)	<1	80KC2	5.6 (0.2)	1.0 (0.3)	<1
80KB1	5.7 (0.4)	1.4 (0.7)	<1	80KC1	4.7 (0.6)	1.7 (0.4)	<1
80KB6	5.3 (0.2)	2.0 (0.2)	<1	80KC4	5.6 (0.4)	2.7 (0.3)	<1

suggesting that the treatment stimulated denitrification potential within the nitrate cell.

A comparison of the average number of denitrifiers in the entire depth interval of initial BA core with the average number in interim Z core indicated that numbers were higher after treatment. When individual samples within the initial BA core were paired by depth to those in interim core Z, counts were also higher after treatment. The average number of total denitrifiers in the entire interval of initial EB and interim X cores was not different. When individual samples within BA and Z or EB and X cores were paired by depth and compared, the data indicated that there was a significant increase in total denitrifiers after treatment in all samples except for the lowest depth, X4, in which there was a significant decline.

Comparison of the entire depth interval of initial JB and interim JC samples indicated numbers of total denitrifiers were not different; however, when individual samples within cores were paired by depth and compared, stimulation was observed in the upper and lower depths but the middle depth was not different.

Because the individual samples in cores KB and KC did not correspond directly with depth (Table 1), the number of total denitrifiers in initial sample KB6 was compared with the average in interim samples KC2 and KC1 combined. Numbers of total denitrifiers did not increase at this depth interval.

In contrast to an increase in the number of total denitrifiers, numbers of denitrifiers growing on JP-4 or JP-4 degradation products were lower after 5 months of treatment (Table 2). The MPNs of many samples incubated in the presence of JP-4 could not be calculated because they were below the detection limit of the assay ($\log_{10}$ 2 cells/g dry wt). The results of control samples (no added JP-4) indicated that ambient carbon could be responsible for some of the denitrification detected in some samples.

Aerobic Viable Counts

When the entire depth interval of every core in the treatment zone was considered, there was an overall increase in heterotrophic, JP-4-degrading, and oligotrophic microorganisms after treatment (Table 3). Numbers of heterotrophs, JP-4 degraders, and oligotrophs in the treatment zone increased from (cells/g dry wt) $\log_{10}$ 5.8 to 6.9, 4.7 to 5.4, and 4.3 to 5.5, respectively; however, numbers of these organisms in the K cores (control) increased from 5.6 to 5.8, 4.0 to 4.8, and 4.1 to 5.0, respectively, suggesting that some of the growth resulted from other factors, such as seasonal influences. Analysis of KB and KC soil samples and ground water from wells a few feet from the K region indicated that the nitrate injected into the nitrate cell had not influenced the K region. There were no consistent trends when pairs of samples were compared by depth.

Protozoa

Numbers of aerobic protozoa in the initial samples ranged from $\log_{10}$ 6.2 in shallow samples from the contaminated zone to less than detection limit

TABLE 3. Aerobic viable counts in initial and interim samples[a].

Initial Sample	Log₁₀ Viable Counts/g dry wt (SD)			Interim Sample	Log₁₀ Viable Counts/g dry wt (SD)		
	R2A	JP-4	no JP-4		R2A	JP-4	no JP-4
80AA2	6.73 (0.05)	6.57 (0.08)	6.59 (0.07)	80ZA2	6.97 (0.09)	4.88 (0.12)	4.99 (0.14)
80AA1	6.84 (0.16)	5.48 (0.1)	5.45 (0.15)	80ZA1	6.97 (0.07)	5.00 (0.17)	5.04 (0.1)
80AA7	4.68 (0.22)	2.43 (0.13)		80ZA4	6.84 (0.08)	4.95 (0.14)	5.72 (0.17)
80BA3	5.69 (0.05)	4.53 (0.14)	3.64 (0.09)	80Z2	6.12 (0.04)	5.12 (0.1)	5.47 (0.14)
80BA2	6.01 (0.08)	4.38 (0.18)	3.94 (0.08)	80Z1	7.59 (0.05)	6.61 (0.2)	6.99 (0.05)
80BA5	4.42 (0.08)	3.57 (0.08)	3.61 (0.13)	80Z4	7.11 (0.08)	6.05 (0.07)	6.09 (0.07)
80DA1	5.86 (0.04)	4.87 (0.13)	5.07 (0.1)	80W2	6.55 (0.07)	5.16 (0.26)	5.04 (0.05)
80DA5	5.91 (0.03)	3.8 (0.12)	3.39 (0.1)	80W1	6.42 (0.04)	5.84 (0.17)	5.93 (0.1)
80DA8	5.76 (0.07)	5.23 (0.08)	5.15 (0.1)	80W4	7.21 (0.13)	5.70 (0.04)	5.76 (0.13)
80EB2	6.80 (0.06)	5.65 (0.33)	6.18 (0.03)	80X2	7.57 (0.14)	6.93 (0.04)	6.79 (0.1)
80EB1	4.55 (0.06)	4.16 (0.10)	3.78 (0.09)	80X1	7.61 (0.09)	5.19 (0.56)	4.67 (0.37)
80EB5	5.61 (0.12)	5.65 (0.08)	5.8 (0.09)	80X4	6.17 (0.04)	3.46 (0.13)	3.46 (0.17)
80JB2	5.58 (0.07)	3.33 (0.09)	3.32 (0.11)	80JC2	6.92 (0.03)	6.00 (0.22)	5.99 (0.17)
80JB1	6.64 (0.06)	6.26 (0.08)	6.29 (0.05)	80JC1	6.90 (0.04)	5.28 (0.14)	5.71 (0.16)
80JB5	4.75 (0.08)	4.15 (0.04)	4.15 (0.04)	80JC3	5.54 (0.06)	4.17 (0.41)	4.55 (0.17)
80KB2	5.8 (0.1)	4.2 (0.1)	4.2 (0.1)	80KC2	5.6 (0.1)	5.3 (0.1)	5.1 (0.1)
80KB1	5.2 (0.02)	4.4 (0.03)	4.9 (0.1)	80KC1	6.0 (0.1)	4.8 (0.2)	5.5 (0.03)
80KB6	5.9 (0.2)	3.5 (0.1)	3.2 (0.1)	80KC4	5.8 (0.3)	4.1 (0.1)	4.0 (0.1)

(a) Samples were plated on low nutrient agar (R2A), and on a mineral nutrient agar with or without JP-4 vapors. The detection limit for the assay is log₁₀ 2.

(<10 cells/g) for some KB (control) samples (Table 4). Of interest is that anaerobic protozoa were detected in low to moderate numbers (not greater than 10,000 cells/g). Data sets for the interim sampling of anaerobic protozoa were not available when this paper was written.

CONCLUSIONS

Denitrification potential was stimulated in the treatment zone after 5 months of nitrate-enhanced bioremediation. The number of total denitrifiers was higher in the interim than initial samples and did not increase in the control core. The increase in the number of viable counts may have been the result of treatment; however, other factors may have affected growth.

Numbers of viable microorganisms also increased in the control core, which could have been a seasonal effect. In general, operation of the pilot system resulted in increased pH, nitrate, ammonia, and orthophosphate levels throughout the nitrate cell (data not shown). Total Kjeldahl nitrogen and total

TABLE 4. Aerobic protozoa in initial and interim samples[a].

Initial Sample	Log10 no.cells/g dry wt (SD)	Interim Sample	Log10 no.cells/g dry wt (SD)
80AA2	4.4 (0.2)	80ZA2	5.7 (0.5)
80AA1	3.0, <1, <1 [a, b]	80ZA1	2.7 (0.1)
80AA7	2.9 (0.1)	80ZA4	3 (0.1)
80BA3	6.2 (0.1)	80Z2	5.9 (0.2)
80BA2	5.8 (0.4)	80Z1	3.8 (0.2)
80BA5	2.9 (0.5)	80Z4	2.4 (0.1)
80DA1	5.7 (0.4)	80W2	2 (0)
80DA5	6.0 (0.3)	80W1	<1, 3.7, 3.0
80DA8	>6	80W4	2.9 (0.1)
80EB2	2.3 (0.3)	80X2	3.9 (0.3)
80EB1	2.7 (0.2)	80X1	2.7 (0)
80EB5	3.5 (0.1)	80X4	2.7 (0)
80JB2	3.6 (0.7)	80JC2	>6.2
80JB1	2.5 (0.2)	80JC1	3.5 (0.2)
80JB5	2.8 (0.4)	80JC3	<1, 2.3, 2.3
80KB2	3.2 (0.2)	80KC2	3.3 (0.2)
80KB1	<1, <1, <1	80KC1	3.3 (0.1)
80KB6	2.5 (0.2)	80KC4	>6.2

(a) Detection limit for the assay, $\log_{10} 1$.
(b) Replicate subsamples were averaged unless a replicate was greater or less than the detection limit.

phosphate levels generally decreased. This probably resulted from the combined effects of nitrate assimilation and decomposition, denitrification, and leaching of minerals. This also has resulted in slightly higher cell counts, as determined by phospholipid fatty acids. Other parameters (total organic carbon, benzene, toluene, the xylenes, the trimethylbenzenes, JP-4) are too variable in concentration to generalize. In summary, these data suggest that the microbial activity at the site has increased as a result of the pilot operation.

REFERENCES

Blazevic, D.J., M.H. Koepcke, and J.M. Matsen. 1973. "Incidence and Identification of *Pseudomonas fluorescens* and *Pseudomonas putida* in the Clinical Laboratory." *Appl. Microbiol.* 25:(1): 107-110.

Hinchee, R.E., D.C. Downey, J.K. Slaughter, D.A. Selby, M.S. Westray, and G. M. Long. 1989. *Enhanced Bioreclamation of Jet Fuels: A Full-Scale Test at Eglin AFB FL.* Air Force Engineering & Services Center Technical Report, ESL-TR-88-78, Engineering & Services Laboratory, Tyndall Air Force Base, FL.

Hutchins, S.R. 1991. "Optimizing BTEX Biodegradation Under Denitrifying Conditions." *Environ. Toxicol. Chem.* 10:1437-1448.

Hutchins, S. R., W. C. Downs, J. T. Wilson, G. B. Smith, D. A. Kovaks, D. D. Fine, R. H. Douglass, and D. J. Hendrix. 1991. "Effect of Nitrate Addition on Biorestoration of Fuel-Contaminated Aquifer: Field Demonstration." *Groundwater* 29:571-580.

Hutchins, S. R., D. E. Miller, F. P. Beck, A. Thomas, S. E. Williams, and G. D. Smith. 1995. "Nitrate-Based Bioremediation of JP-4 Jet Fuel: Pilot-Scale Demonstration." In R. E. Hinchee, J. A. Kittel, and H. J. Reisinger (Eds.), *Applied Bioremediation of Petroleum Hydrocarbons*. Battelle Press, Columbus, OH. (In Press).

Sinclair, J. L., and W. C. Ghiorse. 1987. "Distribution of Protozoa in Subsurface Sediments of a Pristine Groundwater Study Site in Oklahoma." *Appl. Environ. Microbiol.* 53:1157-1163.

U.S. Environmental Protection Agency. 1986. *Testing Methods for Evaluating Solid Waste*. Laboratory Manual. Office of Solid Waste and Emergency Response, U.S. Environmental Protection Agency, Washington, DC.

U.S. Environmental Protection Agency. 1991. *Quantitative Analysis of Aviation Gasoline and JP-4 Jet Fuel in Coarse and Medium Textured Soils by Gas Chromatography*. R. S. Kerr Environmental Research Laboratory, U.S. Environmental Protection Agency, Ada, OK.

Bacterial Degradation of an Organomercurial Micropollutant in Natural Sediments

Elise Marcandella, Corinne Bicheron, and Michel A. Buès

ABSTRACT

Phenyl mercuric acetate (PMA) is a dangerous fungicide found in the water and sediments of the Rhine River. To avoid PMA migration from sediments to groundwater, the bacterial degradation mechanisms of *Pseudomonas fluorescens* were studied. A mercury-resistant strain of *P. fluorescens* (PL IV strain) was isolated from Rhine River water. The strain can break the carbon-mercury link of organomercurial compounds. The resulting mercuric ion (Hg^{2+}) is reduced into volatile metallic mercury Hg. With static bioassays, we studied the influence of different parameters on bacterial growth and PMA degradation. We showed that biodegradation is effective and takes place during the early logarithmic phase of growth. To determine if the strain is able to express its resistance when the medium is not supplemented with sulfur source (enzymatic processes need thiol groups to be active), we replaced it with a porous medium (sand). The results will be used to model the processes occurring during the transport of PMA through a porous medium.

INTRODUCTION

Agricultural and industrial expansion involves the intensive use of pesticides, fertilizers, and other chemicals on the land surface. In the upper Rhine basin, the top of the aquifer is very close to the soil surface and many exchanges exist with surface waters. Because this groundwater layer is an important drinking water resource, a major priority is to fight the pollution that imperils water quality. Mechanisms taking place during contaminant transport need to be understood to forecast pollutant spreading. Knowledge of the dynamics of pollutant transfer is still fragmentary at the field scale. It is necessary to carry out laboratory experiments to identify and estimate exchange mechanisms during transport. For this study, the experimental tool was a one-dimensional porous medium model. The potential for organomercurial degradation by environmental bacteria was studied to determine if a biological term must be considered for this modeling (Marcandella et al. 1994).

EXPERIMENTAL PROCEDURES AND MATERIALS

Analysis Methods

Phenyl mercuric acetate (PMA) is an organomercurial compound for which the empirical formula is $C_8H_8HgO_2$. This pesticide was used mainly in the treatment of cereal seed. Some chemical and physical properties are summarized in Royal Society of Chemistry (1990). PMA is analyzed by high-performance liquid chromatography (HPLC) associated with an ultraviolet (UV) detector (Parkin 1986). The threshold of sensitivity is 50 ng/mL for this method. The chromatographic conditions used were Ultraspher column ODS C18 Waters, 240 × 4.6 cm, particle size 5 μm; mobile phase—methanol:acetonitryl: water in proportions 1:4:5 containing $5.10^{-4}\%$ (w/v) of 6 mercaptopurine mono-hydrate and 5 μM of KH_2PO_4; injection flowrate, 1.5 mL/min; injected volume, 40 μL. A phenyl mercuric-6-mercaptopurine complex was formed, with specific absorption properties at 280 nm.

Because this experimental process is very expensive, Marcandella and Buès (1994) developed a sample preparation method using the Hatch-Ott procedure (1968). This procedure involves acid hydrolysis with concentrated HNO_3-H_2SO_4, oxidation by $KMnO_4$, destruction of excess permanganate with hydrox-ylamine, and, finally, reduction of mercury to metallic Hg° by addition of $SnCl_2$. The Hg° is sparged into the mercury vapor cell of a flameless atomic absorption spectrometer (Perkin-Elmer, Model MAS-50B).

Bacteria and Media

Complete Medium. *Pseudomonas fluorescens* strain (PL IV), PMA-resistant, was cultivated at ambient temperature in a semisynthetic liquid medium (KH_2PO_4, K_2HPO_4, $(NH_4)_2SO_4$, $MgSO_4$, trace elements) supplemented with a solution of casamino acids (0.1%) as the source of sulfur and with sodium citrate (0.1%) as the sole source of carbon. This culture medium is a complete medium. When the culture reached an optical density (OD = 0.1 at 640 nm), $HgCl_2$ (0.8 μg/mL) was added as an inductor to reduce the lag phase. The $HgCl_2$ concentration in the medium should remain subtoxic. After exposure to the $HgCl_2$ for 30 min, the cells were harvested by centrifugation at 6,000 rpm for 20 min and resuspended in the culture medium without mercury.

Influence of PMA Concentration. Bacterial suspensions were prepared by the method described previously. Then, different PMA concentrations (0; 1; 1.5; 2; 3 and 5 μg/mL) were added to the culture medium. Bacterial growth was followed by measuring the suspension's optical density with a spectropho-tometer (Seconam S 250).

Estimation of the Degradation Constant. Both induced and noninduced bacterial suspensions, prepared as previously described, were blended with an aqueous solution (100 mL) of PMA (initial concentration equal to 1 μg/mL). The decrease of PMA concentration in the supernatant was determined by the HPLC method. To measure the amount of PMA bound to the cells, 1.5-mL samples were collected at regular intervals in the medium. Then the cells were harvested by centrifugation at 4°C (15,000 rpm for 10 min), and then organo-mercurial content remaining in the supernatant was determined by HPLC. A control without PMA was incubated along with the PMA-treated batches to determine abiotic losses. Disappearance of PMA was not observed up to 12 h (Bicheron 1992).

Capacity of Sediments to Provide Thiol Groups. Bacterial resistance to mer-cury and organomercurials is determined by two enzymes: organomercurial-lyase and mercury-reductase. The enzyme activity directly depends on the presence of a sulfhydryl compound (R-SH) (Clark et al. 1977; Robinson and Tuovinen 1984). Carrying out batch reactor tests, we investigated the capacity of sediments to provide these thiol groups. The sand used was composed of about 99% quartz and 1% clay (Bicheron 1992).

Bacterial suspensions were prepared by the method previously described, but the semisynthetic liquid medium was not supplemented with a solution of casamino acids. Aqueous solution (25 mL) of PMA at 1 μg/mL and solid phase (25 g of nontreated sand) were blended in a 50-mL Erlenmeyer flask. When pseudoequilibrium was reached, 1 mL of suspension of induced bacteria (ini-tial OD = 0.05) was injected in the batch reactor. The same batch, but without sand, was used as the reference test, along with two other batches, whose con-tent is given in Table 1. The decrease of PMA concentration in both batches (with and without sand) was determined by HPLC.

Every test was repeated two or three times under the same conditions, and the results obtained were reproducible and reliable.

RESULTS

Figure 1 shows growth curves of strain PL IV according to the initial PMA concentration. In the case of no PMA in the culture medium, we observed a lag phase before the bacteria's exponential growth. This lag phase did not appear in the range of PMA concentrations included between 1 and 3 μg/mL. For a concentration equal to 5 μg/mL, we did not observe any growth after 7 h.

Figure 2 shows a PMA concentration that allowed optimal growth of the PL IV strain. Growth was optimal for the PMA concentrations between 1 and 2 μg/mL. Therefore, most of the later experiments were carried out with an ini-tial concentration equal to 1 μg/mL.

An estimation of the degradation constant was derived using batch reactor tests. Figure 3 shows the decrease of PMA in the complete medium when

TABLE 1. Capacity of sediments to provide thiol groups (content of batches and results).

	Batch a	Batch b	Batch c	Batch d
Sand (g)	0	0	25	25
Growth medium (mL)	25	25	25	25
Casamino acids (g/L)	1	0	1	0
[PMA]i (μg/mL)	1	1	1	1
[PMA]f (μg/mL)	0	1	0	0
Growth after 60h	+ +	—	+ +	+ +

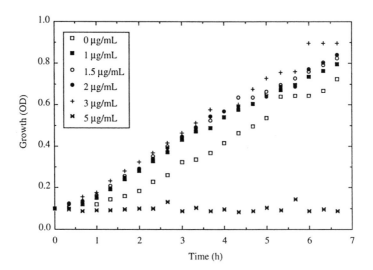

FIGURE 1. Influence of PMA concentration on bacterial growth.

bacteria were exposed (induced bacteria) or not exposed (noninduced bacteria) to $HgCl_2$ (0.8 μg/mL) at t = 0. In both cases (induced and noninduced bacteria), the decrease of PMA can be defined by an exponential equation. Using the least-squares method, the degradation constant (noted λ) can be estimated.

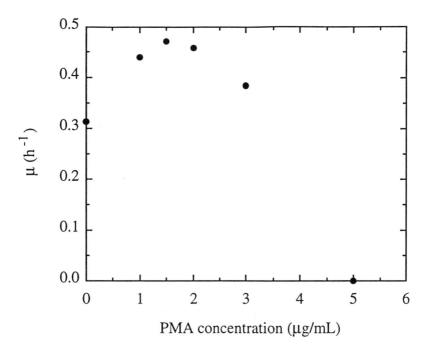

FIGURE 2. Growth rate according to PMA concentration.

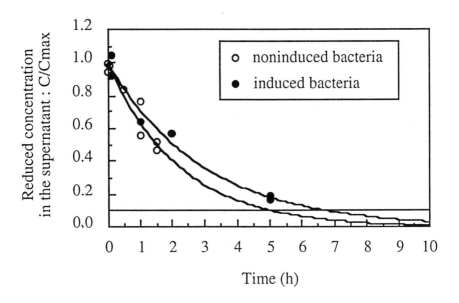

FIGURE 3. Decrease of PMA concentration (Co = 1 μg/mL) in the complete medium.

The values of λ obtained are 0.453 h^{-1} for noninduced bacteria and 0.341 h^{-1} for induced bacteria.

The estimation of PMA bound had shown that about 15 to 35% of the PMA present was immediatly found associated with the cells and that the cell-bound PMA decreased roughly in parallel to the decrease in total PMA present so that always the same percentage was bound and the remainder free in solution. This PMA can be subsequently volatilized after dissociation of the cell surface and after making contact with the responsible enzymes.

Table 1 presents the results of the experiments on the capacity of sediments to provide thiol groups. After 35 to 60 h, no more PMA was found in the liquid phase of the reactors with sand (batches c and d); whereas, in the batch without sand or casamino acids (batch b), the organomercurial concentration in solution remained stable. Adding casamino acids (batch a), we observed disappearance of PMA and bacterial growth. Microscopic observations of supernatants showed that, in the batch where sand is the sole sulfur source, the bacterial population is very important and very active, whereas in the batch without sand or casamino acids, there are many dead bacteria. Only some bacteria appeared to survive.

DISCUSSION

PMA bacterial degradation can occur in the complete medium and begins in the exponential growth phase. The degradation constant obtained should be considered with caution because these experiments were carried out in batch reactors. To obtain greater accuracy and a representative value, dynamic assays are required. Dynamic tests in sand columns will allow observation of bacterial behavior under more natural dynamic conditions.

On the other hand, the positive results of the PMA bacterial degradation in a natural porous medium (without adding compounds R-SH) tend to show that the bacterial populations are able to use the compounds present in the medium to detoxicate it. According to these results, as well as the results previously obtained (Marcandella & Buès 1994), this biological degradation process could be used for the in situ treatment of a contaminated site. Nevertheless, the natural medium must be amended with nutrient because the PMA is degraded by cometabolism (i.e., the PMA degradation provides neither energy nor carbon to the bacteria).

REFERENCES

Bicheron, C. 1992. "Transport d'un micropolluant organomercuriel à travers un milieu poreux saturé: réactivite physico-chimique et biodégradation bactérienne du phényl acétate de mercure." Ph.D., Université Louis Pasteur, Strasbourg, France.

Clark, D.L., A.A. Weiss, and S. Silver. 1977. "Mercury and organomercurial resistance determined by plasmids in *Pseudomonas*." *J. Bacteriol.* 132 : 186-196.

Hatch, W.R., and W.L. Ott. 1968. "Determination of sub-microgram quantities of mercury by atomic absorption spectrophotometry." *Anal. Chem. 40* : 2085-2087.

Marcandella, E., and M.A. Buès. 1994. "Caractérisation par des essais en réacteur fermé de la réactivité physico-chimique et de la biodégradation bactérienne d'un organomercuriel. Application a la dépollution de sediments contaminés." In *Gestion active des aquifères.* I.A.H., Paris, France.

Marcandella, E., C. Bicheron, and M.A. Buès. 1994. "Biodegradation of an organomercurial by a bacterial resistant strain and transport of a micropollutant through an homogeneous porous medium." In T. Dracos and F. Stauffer (Eds.), *Transport and Reactive Processes in Aquifers,* pp. 53-58. A.A. BALKEMA, Rotterdam.

Parkin, J.E. 1986. "Assay of phenyl mercury acetate and nitrate in pharmaceutical products by high performance liquid chromatography with indirect photometric detection." *J. Chromatograph.:* 210-213.

Robinson, J.B., and O.H. Tuovinen. 1984. "Mechanisms of microbial resistance and detoxification of mercury and organomercury compounds: physiological, biochemical, and genetic analysis." *Microbiol. Rev. 48* : 95-124.

Royal Society of Chemistry (Ed.). 1990. *The Agrochemicals Handbook.* 2nd ed., England.

Temperature Effects on Kinetics and Economics of Slurry-Phase Biological Treatment

Patrick M. Woodhull and Douglas E. Jerger

ABSTRACT

OHM Remediation Services Corp. conducted bench-, pilot-, and full-scale slurry-phase biological treatment of polycyclic aromatic hydrocarbons (PAHs) present in Resource Conservation and Recovery Act (RCRA)-listed K001 wood-preserving wastes at a Superfund site in Mississippi. As part of an optimization program for the full-scale treatment system, several bench-scale experiments were conducted to determine the effect of operating temperature on biodegradation of the PAHs present in the wastes. Experiments were conducted in 6.5-L bench-scale slurry reactors to determine the effect of various operating temperatures on PAH biodegradation. The bench-scale operating temperatures (15, 25, 35, and 42°C) were selected based on the range of operating temperatures observed in the full-scale treatment system. PAH degradation in the reactors was monitored by both total PAH concentration and by individual PAH concentrations in the slurry. Biodegradation rates at various temperatures from the 750,000-L slurry reactors in the full-scale treatment system were compared to the data obtained from the bench-scale reactors. An engineering-economic analysis compared the cost of heating the full-scale reactors during periods of low ambient temperatures versus extending operating time to achieve treatment criteria.

INTRODUCTION

The Southeastern Wood Preserving Superfund site is an abandoned wood-preserving facility that was operated from 1928 to 1979. The U.S. Environmental Protection Agency (U.S. EPA) initiated an emergency response at the site in 1986, excavating approximately 10,500 yd³ (8,050 m³) (14,100 tons) of contaminated soils from lagoons, treatment facilities, and storage areas. The lagoon material was classified as RCRA-listed waste number K001. The excavated material was stabilized with kiln dust and stockpiled on site until further treatment.

PAH concentrations in the stockpiled material ranged from 8,000 mg/kg (dry weight) to 15,000 mg/kg (dry weight) for total PAHs, and from 1,000 mg/kg (dry weight) to 2,500 mg/kg (dry weight) for carcinogenic PAHs. Treatment criteria for the project were 950 mg/kg (dry weight) for total PAHs, and 180 mg/kg (dry weight) for benzo(a)pyrene-equivalent carcinogenic PAHs (Jerger et al. 1994).

OHM conducted numerous bench- and pilot-scale treatability studies to obtain site-specific information necessary to select and implement slurry-phase biological treatment at the site (Jerger et al. 1994; Jerger et al. 1993). The results of these studies were used to develop the process design and operating conditions for the full-scale treatment system. One set of studies was designed to determine the effects of operating temperature on the biodegradation kinetics of PAHs in slurry-phase reactors.

MATERIALS AND METHODS

Bench-Scale Slurry Reactors

The bench-scale slurry reactors were 6.5-L glass vessels with an agitator and aeration system. The configuration and operation of the bench-scale slurry reactors have been described in detail elsewhere (Jerger et al. 1993). Samples were collected from the reactor side port and the solids extracted according to U.S. EPA SW-846 Method 3540. The instrumental method used for PAH quantification was gas chromatography with a mass spectrometer detector (SW-846 Method 8270). All results were reported on a dry-weight basis.

Based on the range of temperatures observed in the full-scale reactors during field operation, the bench-scale slurry reactors were operated at temperatures of 15, 25, 35, and 42°C (±2°C). The reactors were heated with electric heat tape and a thermostat and cooled by circulating water through a chiller and an internal coil of tubing placed in the reactors.

The contaminated material for the bench-scale experiments was collected from the full-scale slurry preparation system in 5-gal (19-L) containers and shipped overnight to the laboratory. The full-scale slurry preparation system produced a 20% solids (weight/weight) slurry of –200-mesh material.

Full-Scale Slurry Reactors

The full-scale slurry-treatment system consisted of materials handling and screening, soil washing, slurry-phase biological treatment, and slurry dewatering. The full-scale process has been described in detail in previous publications (Jerger et al. 1994; Woodhull and Jerger 1994; Woodhull et al. 1993). Slurry samples were collected from the reactor using a bomb-type sampler to collect discrete samples that were composited in the field into a single slurry sample. The composite sample was shipped to an off-site laboratory for analysis. The samples were extracted and analyzed according to SW-846 Method 3540 and SW-846 Method 8270, respectively. All results were reported on a dry-weight basis.

Based on the data collected from the bench-scale reactors, two of the four full-scale slurry reactors were operated November 1993 through April 1994, when average ambient temperatures were less than 15 to 20°C. The reactors were heated by a diesel-powered hot-water boiler, circulating water pumps, and submerged heat exchangers. Each of the reactors was fitted with two submerged heat exchangers located near the reactor walls and oriented radially from the tank center. A single boiler and circulation pump heated both reactors. The slurry temperature in each reactor was controlled by adjusting the flowrate of the hot water through each set of submerged heat exchangers.

RESULTS AND DISCUSSION

Bench-Scale Slurry Reactors

The reduction in total PAH concentration in the bench-scale reactors versus time for each operating temperature is plotted in Figure 1. The first-order

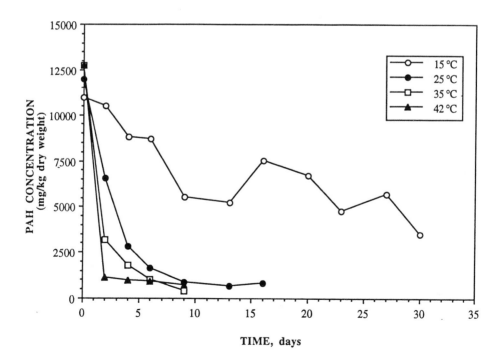

FIGURE 1. Effect of temperature on biodegradation rate of PAHs in K001 wastes (6.5-L bench-scale slurry reactors).

biodegradation rate constants (k) for the operating temperatures of 15, 25, 35, and 42°C were 0.029, 0.175, 0.357, and 1.21, respectively, and are plotted versus temperature in Figure 2. The rate constants are linear with temperature in the range of 15 to 35°C. The biodegradation rate constant for 42°C is higher than expected for a based linear relationship between temperature and rate constants. The increase in biodegradation kinetics at higher temperatures may be related to the increase in solubility of the PAHs or increased desorption of the PAHs from the matrix as temperatures increase. As an example, the solubility of phenanthrene in water at 8.5°C is 0.423 mg/L; it increases to 0.816 mg/L at 21°C and to 1.23 mg/L at 30°C (Verschueren 1983). Although data for the other PAHs was not available, a similar pattern is expected.

In addition to an increase in the rate constant for total PAH biodegradation, a temperature increase has a significant impact on biodegradation of the higher-ring PAHs. Using the four-ring PAHs as an example, Figure 3 shows the impact of operating temperature on removal of the four-ring compounds. Biodegradation of the four-ring PAHs at 15°C is almost nonexistent, whereas at 42°C degradation is complete by day 9. Again, the increase of PAH solubility or desorption with increasing temperature may have a significant effect on PAH biodegradation.

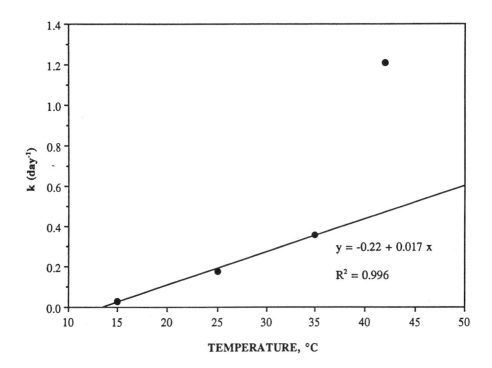

FIGURE 2. Effect of temperature on biodegradation kinetic rate constants for PAH biodegradation.

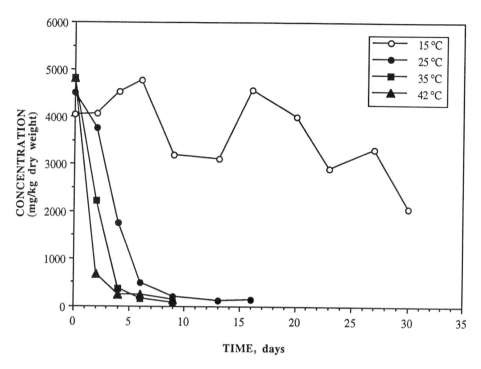

FIGURE 3. **Effect of temperature on biodegradation rate of four-ring PAHs in K001 wastes (6.5-L bench-scale slurry reactors).**

Full-Scale Slurry Reactors

The effect of temperature on the biodegradation of PAHs in the full-scale slurry reactors is shown in Figure 4. The kinetic rate constants were 0.27 and 0.43 for an average operating temperature of approximately 25 to 27°C in the unheated reactor and approximately 35 to 37°C in the heated reactor. The field kinetic constants are slightly higher than the bench-scale kinetic constants; however, the operating temperature in the field reactors varied by approximately 5 to 10°C compared to ±2°C in the bench-scale slurry reactors. Depending on ambient conditions, the boiler system usually required approximately 2 to 3 days to raise the operating temperature of the slurry to the optimal range.   However, once optimal slurry temperatures in the reactors were achieved, the slurry temperature was generally within ± 5°C of the operating temperature.

Economics

The total capital and operating costs for the boiler system were approximately $150 per day per reactor. Using an average solids loading of approximately 170 tons per reactor (Woodhull et al. 1993), the cost of heating the

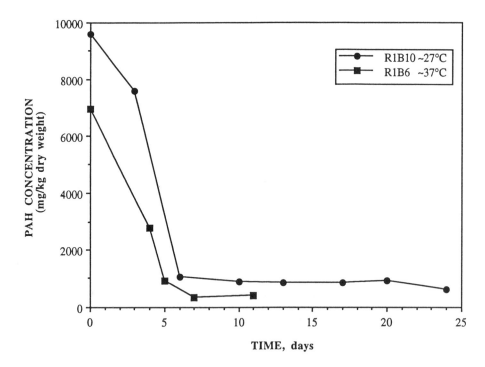

FIGURE 4. **Effect of temperature on biodegradation rate of PAHs in K001 wastes (750,000-L full-scale slurry reactors).**

reactors equates to approximately $0.86 per ton per day. Based on the data collected from the field reactors, heating the reactors from approximately 25 to 35°C increases the kinetics for PAH removal by a factor of 1.6. For an initial concentration of 10,000 mg/kg PAHs, slurry-phase biological treatment at 25°C requires approximately 9 days to achieve the treatment criterion of 950 mg/kg, whereas biodegradation at 35°C requires approximately 6 days. The cost to operate the unheated treatment system, adjusted for a 4-month shutdown during the winter, is approximately $205 per ton for a 9-day batch time (Woodhull and Jerger 1994; Woodhull et al. 1993). Operation of the treatment system without a winter shutdown at a 6-day batch time yields a total cost of approximately $180 per ton. In addition, heating the reactors for a period of 6 days requires an additional $5 per ton in operating costs, raising the total costs to approximately $185 per ton. This still is lower than the $205 per ton for the extended, unheated operation. This impact would be even greater as the ambient temperature and the slurry temperature decrease to 15°C, resulting in even longer batch operating time to achieve treatment criteria. Heating the reactors allowed for continuous operation of the treatment system, reducing the project schedule, increasing equipment utilization, and lowering treatment costs.

CONCLUSIONS

The operating temperature of slurry-phase biological reactors has a significant impact on biodegradation kinetics for the material from the site. At lower temperatures, the biodegradation of the higher-ring PAHs is limited or nonexistent, and the rate of biodegradation is reduced. Heating the full-scale reactors is beneficial for three reasons: (1) biodegradation of the higher-ring (carcinogenic) PAHs, (2) increase in the biodegradation kinetics for PAH removal, and (3) continuous operation of the treatment system. The increase in PAH-biodegradation kinetics offsets the increased cost for heating the reactors, lowering the overall project costs for the system. The effects of operating temperature must be recognized prior to the design and operation of slurry-phase biological reactors.

REFERENCES

Jerger, D. E., D. J. Cady, and J. H. Exner. 1994. "Full-Scale Slurry-Phase Biological Treatment of Wood-Preserving Wastes." In R. E. Hinchee, A. Leeson, L. Sempini, and S.K. Ong (Eds.), *Bioremediation of Chlorinated and Polycyclic Aromatic Compounds*, pp. 480-483. Lewis Publishers, Chelsea, MI.

Jerger, D. E., C. Jespersen, and J. H. Exner. 1993. "The Importance of Treatability Studies in the Development of Commercial Slurry-Phase Biological Treatment Processes." *Proceedings of the Air and Waste Management Association 86th Annual Meeting.*

Verschueren, K. 1983. *Handbook of Environmental Data on Organic Chemicals.* 2nd ed., Van Nostrand Reinhold Company, Inc., New York, NY.

Woodhull, P. M., and D. E. Jerger. 1994. "Bioremediation Using a Commercial Slurry-Phase Biological Treatment System: Site-Specific Applications and Costs." *Remediation* 4(3): 353-362.

Woodhull, P. M., D. E. Jerger, and D. J. Cady. 1993. "Economics of a Full-Scale Slurry-Phase Biological Treatment Process for Wood Preserving Wastes." Presented at Emerging Technologies in Hazardous Waste Management V, American Chemical Society I&EC Special Symposium, Atlanta, GA.

Enhanced Mineral Alteration by Petroleum Biodegradation in a Freshwater Aquifer

Franz K. Hiebert, Philip C. Bennett,
Robert L. Folk, and Sylvia R. Pope

ABSTRACT

Aerobic and anaerobic microbial degradation of petroleum hydrocarbons in shallow aquifers disturbs the geochemical equilibrium between water and mineral phases and accelerates the dissolution and precipitation of minerals. These effects were investigated in an oil-contaminated aquifer near Bemidji, Minnesota, using in situ microcosms with a 14-month reaction period and field column experiments. Minerals recovered from microcosms were studied by scanning electron microscopy (SEM). Groundwater chemistry was characterized before and after the experiment. The in situ microcosms yielded feldspar and quartz grains colonized with a variety of bacteria. On a microscale (1.0 to 10.0 μm), feldspars, quartz, and, in some cases, calcite dissolved in the immediate vicinity of attached bacteria, even though the saturation indices for these minerals in the bulk groundwater indicated that little to no dissolution should occur. On a local scale (the zone of microbial degradation of hydrocarbon within the aquifer), the metabolic production of CO_2 and HCO_3 resulting from acetate methanogenesis and the oxidation of aromatic hydrocarbons coupled with iron reduction provided a supply of reactants to the groundwater. Calcite precipitated in a wide variety of spiky morphologies and clay minerals formed on feldspar minerals. Field column experiments showed that pore fluids from an anaerobic zone within the oil-contaminated aquifer readily precipitated iron hydroxides and calcite upon exposure to aerobic conditions, thus causing a reduction in column permeability.

INTRODUCTION

Biodegradation of petroleum is part of the natural biogeochemical carbon cycle. In situ bioremediation is an applied technology based on elements of natural biodegradation that is rapidly gaining acceptance as an environmental cleanup tool. The study of microbiological and geochemical processes at field sites where natural biodegradation is occurring provides data on fundamental interactions between bacteria, hydrocarbons, and aquifer material that can guide engineered bioremediation projects.

The ecology and biochemistry of indigenous bacteria in hydrocarbon-rich aquifer environments has been studied intensively (e.g., Balkwill and Ghiorse 1985; Ghiorse and Wilson 1988; Chapelle 1993). Many bacteria use hydrocarbons as a carbon source for growth and energy, and both aerobic and anaerobic pathways have been recognized (e.g., Atlas 1984; Gibson 1984; Lovley et al. 1989). Hydrocarbon metabolism results in the production of a variety of metastable intermediate by-products, and chemical gradients of inorganic and organic compounds maintained within and around the cell are not necessarily in equilibrium with the chemistry of the surrounding fluid (e.g., Hiebert and Bennett 1992; Folk 1993). However, little work has focused on the effect of in situ microbial activity on the solid phase of aquifer material.

The purpose of this investigation was to directly observe the effects that indigenous surface-adhering bacteria have on mineral diagenesis in a hydrocarbon-contaminated shallow aquifer.

Site Background

The study site is an oil-contaminated glacial outwash aquifer near Bemidji, Minnesota (Fig. 1), where 400,000 L of light crude oil leaked from a pipeline in

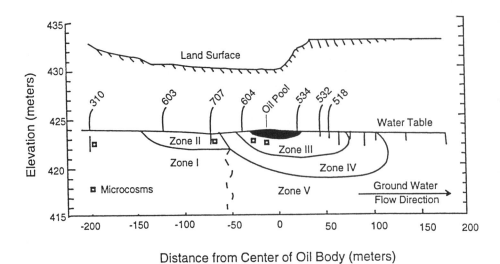

FIGURE 1. Cross section of the flow cell chosen as the location of the experimental microcosms for this study. Microcosm locations are indicated by open squares in Zone 3.

1979 and infiltrated the soil overlying the aquifer. The oil eventually pooled on the water table at 3 to 5 m below the surface. The unconfined aquifer is composed of outwash sediments approximately 57% quartz, 29% feldspars, 2% calcite, 1 to 2% dolomite, 2% hornblende, < 1% clay minerals, and < 0.2% organic carbon (Bennett et al. 1993). The local flow system discharges into a lake 300 m downgradient from the spill site.

Native Microbial Activity. Three categories of biogeochemical reactions are occurring at the Bemidji site and correspond to distinct geochemical zones that have developed since the pulse of crude oil. Aerobic hydrocarbon metabolism occurs in the unsaturated zone and in the oxygenated groundwater at the upgradient edge of the oil-water contact, where molecular oxygen serves as the terminal electron acceptor. In anaerobic waters below and downgradient of the oil body, oxidation continues via the reduction of Fe^{3+} to Fe^{2+} (e.g., Lovley et al. 1989; Chapelle 1993; Bennett et al. 1993). Methane is produced in the most anoxic regions, probably by acetate methanogenesis (Baedecker et al. 1993). The end products of metabolism and intermediates, such as organic acids, react with the mineral solid phase, producing a plume of organic and inorganic solutes that has migrated downgradient with local groundwater flow (Cozzarelli et al. 1990, Baedecker et al. 1993).

Groundwater Geochemistry. The groundwater chemistry at this site has been monitored annually since 1983 as part of a research effort sponsored by the U.S. Geological Survey (e.g. Baedecker et al. 1993; Bennett et al. 1993; Eganhouse et al. 1993), and more than 100 wells have been installed to characterize the chemistry and fate of oil in the aquifer. Baedecker et al. (1993) and Bennett et al. (1993) recognized within the study area five geochemical zones characterized by similarities in physicochemical properties of the groundwater. Zone 1 contains groundwater that is uncontaminated with dissolved hydrocarbons and is aerobic, Zone 2 is lightly contaminated and disaerobic, Zone 3 is highly contaminated and anaerobic, Zone 4 is lightly contaminated and anaerobic to disaerobic, and Zone 5 represents a mixing zone between water that is lightly contaminated disaerobic and uncontaminated aerobic (Fig. 1). A characteristic chemistry of the principal flowpath through the study area is summarized by Bennett et al. (1993, Table 3, p. 535).

Calculations using chemical speciation and equilibrium models show that the contaminated groundwater is everywhere supersaturated with respect to quartz and approaches equilibrium with amorphous silica in the contaminated waters (Bennett et al. 1993). The waters are supersaturated with respect to gibbsite and kaolinite, but probably undersaturated with respect to the primary feldspars, within the uncertainty of the available thermodynamic database. This background database provided a detailed chemical context in which to interpret experimental results and the direct observations of in situ mineral weathering.

METHODS AND MATERIALS

Pure crystals of feldspars, calcite, dolomite, and quartz were separately crushed and cleaned. Aliquots of prepared fragments were placed together in porous polyethylene cylinders (microcosms) and submerged in the aquifer for 14 months. The microcosms were suspended in the screened flowthrough portion of the well approximately 10 cm below the oil layer in Zone 3 (Fig. 1). The wellbore was sealed and left undisturbed for the duration of the experiment. At the time of microcosm emplacement and recovery, groundwater samples were collected from the contaminated and uncontaminated regions. At the end of 14 months, the microcosms were recovered and the minerals were prepared for SEM.

Microcosm work was augmented with flowthrough column experiments carried out in the field. Groundwater was pumped directly from the level at which the microcosms were suspended through columns (4 × 50 cm) of quartz sands for 5 days. The reacted column sands and organomineral precipitates from groundwater samples were analyzed by SEM and x-ray diffractometry (XRD).

RESULTS

Mineral Surface Coatings

Virtually all of the minerals retrieved from the microcosms were coated with organics, Fe, and Si, in a "microcurdled" surface texture (Fig. 2). A red sludge identical in composition and similar in morphology formed on quartz grain surfaces at the inlet and first few centimeters of the field flowthrough columns where anaerobic groundwater mixed with aerobic column water. The red sludge consisted of minor amounts of quartz and calcite silt, authigenic calcite, amorphous iron, kaolinite or chlorite and goethite, and viable bacteria. Large rod-shaped cells (> 1.0 μm in length) were identified as growing bacteria. Nanobodies (0.10 to 0.25 μm in diameter) probably are a mixture of bacterial cells and inorganic precipitates. EDAX analysis resulted in major Fe and minor Si and Ca peaks, consistent with x-ray analysis (Hiebert 1994).

Mineral Colonization by Bacteria

Bacteria colonized the surfaces of minerals recovered from the microcosms. Several colonization morphologies were documented, primarily individual cells and small patches (Fig. 3). The 0.25-μm and smaller spherical nanobodies are abundant and appear similar in overall texture to nanobacteria from acid-etched aragonite needles and oolites described by Folk (1993). Microcolonies (2 to 100 cells clumped together) were common. Unexpectedly, no large-scale biofilms of full-sized bacteria were observed.

FIGURE 2. Electron micrograph of attached bacteria and underlying microcurdled coating on the mineral surface.

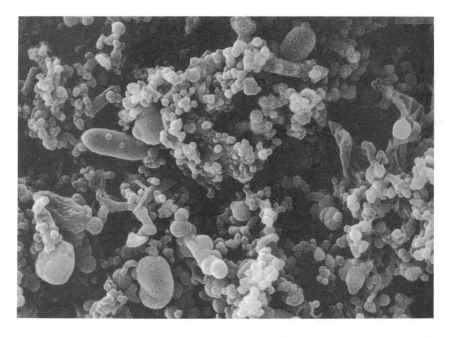

FIGURE 3. Electron micrograph of red sludge with bacteria formed in groundwater samples collected during the recovery of microcosms.

Mineral Alteration

Quartz was etched especially in the vicinity of attached bacterial cells (Fig. 4). Etching of quartz occurred in water many times supersaturated with respect to quartz, and the triangular-shaped etch pits are diagnostic of crystallographic control of quartz dissolution (Bennett and Siegel 1987). Etching effects were observed on cleaned quartz surfaces studied with high-resolution, high-magnification SEM. Open-ended and very regular triangular etch pits ranged from 0.5 m to 1.0 m in diameter. Microrunnels (0.02 μm × 1.0 μm) traversed 1.0-μm steps in the quartz surface between triangular etch pits and appeared to be etching effects as well.

In contrast to the light weathering of quartz, some feldspar surfaces were intensely etched and weathered (Fig. 5). Prismatic etch pits 1 to 5 μm across and more than 1 μm deep are oriented along cleavage and twinning planes, again suggesting crystallographic control of dissolution (Berner and Holdren 1979; Berner et al. 1985). The etch pits observed are similar to those produced in laboratory studies of chemical dissolution of feldspars (e.g., Berner and Holdren 1979).

Calcite crystals showed sparse colonization by vegetative or full-sized bacteria in comparison to the silicate minerals. Short (0.5-μm) rod-shaped bacteria were noted in isolated patches on some flat calcite surfaces. In the immediate

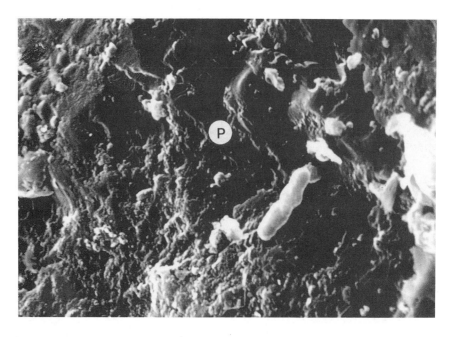

FIGURE 4. Quartz recovered from Zone 3 after 14 months. Distinct triangular etch pits formed on the formerly smooth quartz surface and near various types of attached bacteria.

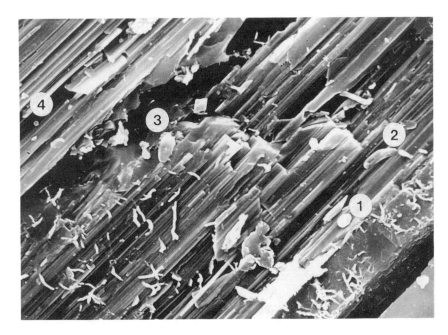

FIGURE 5. Etched microcline recovered from Zone 3 after 14 months. At least four morphologies (1 to 4) of attached bacteria are present in the surface area of this 300-μm^2 micrograph. This indicates the wide diversity of bacteria that colonize feldspar surfaces.

vicinity of these colonizations, deep (1.0 to > 3.0 μm) etching of the calcite surface was observed, with deep irregular prismatic dissolution along cleavage planes.

Dolomite crystals recovered from the Bemidji microcosms showed deep etching, especially along crystal edges. Few bacteria were observed colonizing the dolomite crystals. The microcosms were placed in groundwater that has been shown to be undersaturated with respect to dolomite (Baedecker et al. 1993, Bennett et al. 1993). Evidence of dissolution of dolomite was expected and is consistent with the chemistry of the groundwater.

Honeycomb-shaped secondary mineral precipitate formed on some microcline grains, with individual strings about 0.05 μm $\times$ 1.0 μm (Fig. 6). Based on morphology and analysis of groundwater geochemistry, the material is likely kaolinite, smectite, or possibly gibbsite.

Spikey calcite precipitated on most of the faces of calcite examined. Spikes of precipitate in linear and box patterns grow perpendicularly to the surface and parallel to each other to a uniform height (Fig. 7). The smallest spikes, 0.05 to 0.10 μm in diameter, appear as irregular hemispheres, whereas larger spikes, 0.10 to 0.05 μm in height, occur with both rounded and sharp euhedral points. Oblique EDAX analysis shows only Ca, with no detectable Mg. Some 0.5-μm-high spikes showed flattened tops and evidence that narrow spikes coalesced

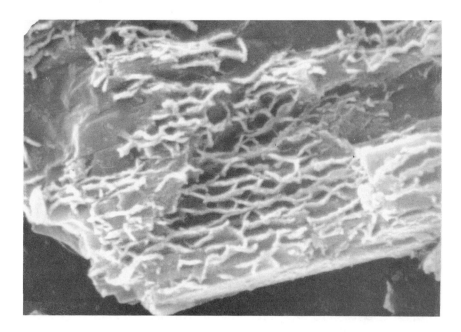

FIGURE 6. Authigenic clay minerals on the surface of a microcline grain.

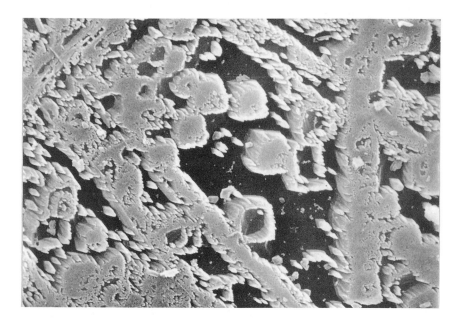

FIGURE 7. Surface of calcite recovered from Zone 3 after 14 months. All calcite fragments showed a carpet of precipitation features with a uniform and
level surface horizon. Morphology of precipitate ranged from narrow
round-tipped pinnacles to positive rhombohedrons and rhombic pyramids.

as they increased in width. A distinct contact between the surface of the mother grain and the beginning of the precipitate is clearly revealed in cross section (Fig. 8). In areas of the calcite where the precipitation has been removed by scraping, the surface of the calcite grain appears smooth and unaltered.

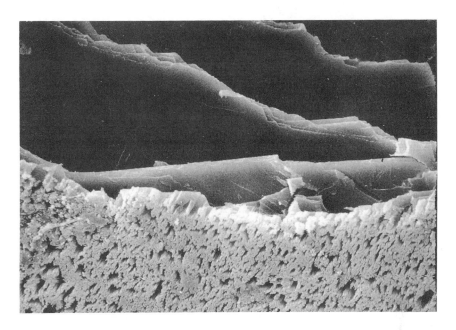

FIGURE 8. Chipped calcite grain showing contact between smooth mineral surface and spiky precipitate coating.

DISCUSSION

Changes in groundwater chemistry and mineral alteration were observed on two distinct scales of reaction: (1) on a microscale in the immediate vicinity of bacteria attached to mineral surfaces, and (2) on a local scale within the area of enhanced microbial activity in the aquifer. The metabolic activity of bacteria is responsible for alteration of chemical conditions on the immediate surface of aquifer matrix materials and in the bulk chemistry of contaminated groundwater. Different reactions are occurring in the aquifer at the microscale and the local scale.

A hypothesis to explain the rapid surface etching of microcline and quartz in spite of the supersaturated condition of the bulk pore water is that surface-adhering bacteria create a chemical reaction zone that is out of equilibrium with the surrounding pore water in the immediate vicinity of the cell/mineral interface (Fig. 9). Intermediate by-products of bacterial metabolism of

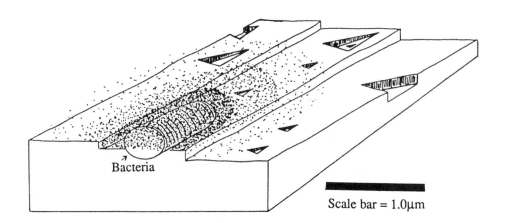

Bacteria

Scale bar = 1.0μm

FIGURE 9. Conceptual model of a bacterially generated zone at the surface of a quartz fragment. Surface-adhering bacteria (represented in cross section here) create a reaction zone in their immediate vicinity by producing and releasing organic acids during metabolism. The organic acids released by the cell create a gradient between the cell surface and the surrounding pore water. In this microenvironment, complex organic acids bind SiO_2 from the mineral surface and cause dissolution despite the quartz-supersaturated bulk pore water.

hydrocarbon, especially organic acids, influence the dissolution of both feldspars and quartz. Organic acids must occur at concentrations of over 1,200 mg/L carbon (i.e., as citrate) in the bulk pore water to have a chelating effect (Bennett 1991). This concentration is far in excess of that measured in bulk samples of groundwater from Zone 3 and from natural waters in general (Cozzarelli et al. 1990). In the near vicinity of carbon-utilizing microbes, however, it is possible that bacteria cause a high concentration of organic acids to occur immediately near their cell wall. Within this microenvironment, defined by the organic acid concentration gradient, complex organic acids bind silica at the mineral surface, thus causing dissolution despite the supersaturated status of the bulk pore water.

Dolomite, with no evidence of direct bacterial involvement, is dissolving as expected from results of geochemical modeling of the bulk pore water and observations of increasing Mg concentrations in the anaerobic zone. Calcite, in contrast, precipitates in the anaerobic Zone 3 waters. Metabolic production of CO_2 and HCO_3, resulting from acetate methanogenesis and the oxidation of aromatic hydrocarbons coupled with iron reduction, provides a supply of these reactants to the local-scale bulk groundwater for calcite precipitation. In addition, nucleation of calcite crystal growth may be catalyzed by the presence of bacteria (e.g., Chafetz and Buczynski 1992; Folk 1993). Subsequent microscale changes in geochemistry at the surface result in secondary mineral precipitation on uncolonized mineral faces.

CONCLUSIONS

Aerobic and anaerobic microbial metabolism of hydrocarbons at this site appear to be causing enhanced silicate dissolution and calcite and clay precipitation. The microbial production of hydrocarbon-related metabolic by-products mobilized sparingly soluble metals and metalloids from aquifer minerals to the point of accelerating natural mineral diagenetic processes.

Microscale geochemical reactions may not be predicted accurately by analysis of the bulk groundwater because nonequilibrium reactions are catalyzed at the cell/mineral contact in microreaction zones.

Although the rates of mineral alteration are significant in a geologic time frame, it is unlikely that enhanced diagenesis will meaningfully alter aquifer flow conditions on a remediation time scale of months to years. However, the bacterially produced mineral-rich anaerobic groundwater developed beneath the oil plume did precipitate enough iron hydroxide and bacterial sludge upon mixing with aerobic conditions that plugging of intergranular pores in the field columns occurred.

Bioremediation engineers are cautioned to evaluate the potential for rapid plugging of intrastratal pores and the effect on remediation efforts before designing the introduction of molecular oxygen to anaerobic zones in contaminated aquifers. Our evidence indicates that oxidation of mineral- and bacteria-rich anaerobic groundwater will cause intrastratal plugging to occur.

ACKNOWLEDGMENTS

We thank Mary Jo Baedecker, Dennis Brown, Henry Chafetz, Isabelle Cozzarelli, Marc Hult, E.F. McBride, Don Siegel, and Dennis Trombatore for their help and friendly advice. This work was supported by the National Science Foundation (EAR 91-05778) and the United States Geological Survey Toxic Substances Hydrology Program.

REFERENCES

Atlas, R. M. (Ed.). 1984. *Petroleum Microbiology,* Macmillan, New York, NY.

Baedecker, M. J., I. M. Cozzarelli, R. P. Eganhouse, D. I. Siegel, and P. C. Bennett. 1993. "Crude oil in a shallow sand and gravel aquifer—III. Biogeochemical reactions and mass balance modeling in anoxic groundwater." *Applied Geochem. 8*: 569-586.

Balkwill, D. L., and W. C. Ghiorse. 1985. "Characterization of subsurface bacteria associated with two shallow aquifers in Oklahoma." *Appl. Environ. Microbiol. 50*: 580-588.

Bennett, P. C. 1991. "Quartz dissolution in organic-rich aqueous systems." *Geochim. Cosmochim. Acta, 49*: 1781-1797.

Bennett, P. C., D. E. Siegel, M. J. Baedecker, and M. F. Hult. 1993. "Crude oil in a shallow sand and gravel aquifer—I. Hydrogeology and inorganic geochemistry." *Applied Geochem. 8*: 529-549.

Bennett, P. C., and D. I. Siegel. 1987. "Enhanced dissolution of quartz by dissolved organic carbon." *Nature. 326*:684-686.

Berner, R. A., and G. R. Holdren. 1979. "Mechanism of feldspar weathering. II. Observation of feldspars from soils." *Geochim. Cosmochim. Acta. 43*: 1173-1186.

Berner, R. A., G. R. Holdren, B. Schott, 1985, "Surface layers on dissolving silicates." *Geochim. Cosmochim. Acta 49*:1657-1658.

Chafetz, H. S. and C. Buczynski. 1992. "Bacterially induced lithification of microbial mats." *Palaios 7*: 277-293.

Chapelle, F. H. 1993. *Ground-Water Microbiology and Geochemistry,* John Wiley and Sons, New York, NY.

Cozzarelli, I. M., R. P. Eganhouse, and M. J. Baedecker. 1990. "Transformations of monoaromatic hydrocarbons to organic acids in anoxic groundwater environment." *Environ. Geol. Water Sci. 16*:135-141.

Eganhouse, R. P., M. J., Baedecker, I. M. Cozzarelli, G. R. Aiken, K. A. Thorn, and T. F. Dorsey. 1993. "Crude oil in a shallow sand and gravel aquifer—II. Organic geochemistry." *Applied Geochem 8*: 551-567.

Folk, R. L., 1993. "SEM imaging of bacteria and nannobacteria in carbonate sediments and rocks." *J. Sed. Pet. 63*: 990-999.

Ghiorse, W. C., and J. T. Wilson. 1988. "Microbial ecology of the terrestrial subsurface." *Advances in Applied Microbiology 33*:107-172.

Gibson, D. T. (Ed.). 1984. *Microbial Degradation of Organic Compounds,* Marcel Dekker, New York, NY.

Hiebert, F. K. 1994. "Microbial diagenesis in terrestrial aquifer conditions: Laboratory and field studies." Ph.D. Dissertation, The University of Texas at Austin, Austin, TX.

Hiebert, F. K., and P. C. Bennett. 1992. "Microbial control of silicate weathering in organic-rich groundwater." *Science 258*: 278-281.

Lovley, D. R., M. J. Baedecker, D. J. Lonergan, I. M. Cozzarelli, E.J.P. Phillips, and D. I. Siegel. 1989. "Oxidation of aromatic contaminants coupled to microbial iron reduction." *Nature, 339*: 297-299.

Microbial Production of 2-Keto-L-Gulonic Acid from L-Sorbose by *Pseudomanas putida*

Stanka S. Stefanova, Zdravka V. Sholeva,
and Ludmila G. Peeva

ABSTRACT

One step in industrial vitamin C production is chemical conversion of L-sorbose to diacetone sorbose and the chemical oxidation of the last to 2-keto-L-gulonic acid (2 KGA). It was found that microorganisms, classified into 12 genera produced 2 KGA. Some of the more productive were *Acetobacter, Alcaligenes, Gluconobacter, Micrococcus, Pseudomonas,* and *Serratia*. In this study the production of 2 KGA by five strains of *Pseudomonas* (*P. aeruginosa*—NBIMCC 647 and NBIMCC 1088; *P. putida*—NBIMCC 1090 and NBIMCC 1133; *P. syringae*—NBIMCC 1310) was determined. *P. putida* 1133 was the most productive strain and was chosen for further investigation. The first trial for strain improvement was done by single colony isolation. Further strain improvement was accomplished by ultraviolet mutagenesis. After slight reversion, the 2 KGA productivity of the U113 mutant of *P. putida* 1133 remains three times more stable than the type strain.

INTRODUCTION

One step in industrial vitamin C production is the chemical conversion of L-sorbose to diacetone sorbose and the chemical oxidation of the latter to 2 KGA. Aggressive reagents such as acetone, H_2SO_4 (oleom), NaOH, and NaClO are used in all chemical reactions of this process. As a result serious environmental protection problems arise. These problems could be avoided by using a biological conversion stage. That is why establishment of an efficient process for microbial production of 2 KGA has long been an important objective in the vitamin C industry (Sugisawa et al. 1990).

It was found that microorganisms classified into 12 genera (some of which were *Acetobacter, Alcaligenes, Gluconobacter, Micrococcus, Pseudomonas,* and *Serratia*) produced 2 KGA (Sugisawa et al. 1990, Sonoyama et al. 1982, and Stroshane and Periman 1977).

Here we describe a strain improvement study of *Pseudomonas putida* strain, a producer of 2 KGA from L-sorbose.

EXPERIMENTAL PROCEDURES AND MATERIALS

Strains and Media

Microorganisms from the collection of the National Bank for Industrial Microorganisms and Cell Cultures (NBIMCC) were used in this study. They belonged to the genus *Pseudomonas*: *P. aeruginosa*—647 and NBIMCC 1088, *P. putida*—1090 and 1133, and *P. syringae*—1310. Mutants obtained under this investigation were tested as well.

Pseudomonas strains were maintained on nutritional broth. Two media were used for studying 2 KGA production. SI medium contained (per liter): 5 g yeast extract, 5 g glycerol, 5 g $MgSO_4 \cdot 7H_2O$, 5 g $CaCO_3$, and 5 g sorbose. Mutants were obtained using SI medium supplemented with 20 g agar-agar /L. The second medium, SII, contained 5 g yeast extract, 0.5 g glycerol, and 1.5 g $CaCO_3$/L.

Cultivation Conditions

The investigated strains were cultivated for 4 to 7 days (up to 10 days in some experiments) on a rotary shaker at 220 rpm, 30°C, in 20-mL test tubes containing 2 mL medium and in 500-mL flasks containing 30 mL medium.

A glass fermentor (Gallenkamp) with 1 L volume and 300 mL of medium SI was used. The inoculum was obtained from rotary shaker flasks. The fermentation was done at 30°C and an aeration of 9 v/v/m.

Analyses

Thin layer chromatography (TLC) was used to detect the 2 KGA production. Cellulose TLC plates and propanol, ethylacetate, water, and acetic acid (20:4:12:4) were used. The plates were sprayed with solution containing 0.82 g phthalic acid and 0.5 mL aniline in 25 mL water-saturated *n*-butanol. After the sprayed plates were heated at 104°C for 4 min, 2 KGA was detected.

The concentration of 2 KGA was determined by the method of Lazarus and Seymour (1986) with high-performance liquid chromatography (HPLC, Perkin Elmer Series 10, supplied with the refractometric detector LC-25). A chromatographic column, type HPX-87H, packed with sulfonated divinyl-benzene-styrene copolymer as a stationary phase, was used. The mobile phase was 0.005 N H_2SO_4 fed at a rate of 0.5 mL/min. The retention time of 2 KGA was 10 min.

RESULTS AND DISCUSSION

2 KGA Production by Various Microorganisms

The production of 2 KGA by five strains (*P. aeruginosa*—NBIMCC 647 and NBIMCC 1088, *P. putida*—NBIMCC 1090 and NBIMCC 1133, and *P. syringae*—

NBIMCC 1310) was determined. The samples were taken at 120, 144, and 168 h and analyzed by HPLC (Table 1). *P. putida* 1133 was the most productive strain and was chosen for further investigation. The chosen strain was cultivated on media SI and SII. Better results were obtained on medium SI (1.06 g/L vs. 0.76 g/L on medium SII for 9 days of cultivation). All further experiments were carried out on medium SI.

TABLE 1. Production of KGA from L-sorbose by various species.

Species	No. in NBIMCC	2 KGA production (g/L)		
		120 h	144 h	168h
P. aeruginosa	647	0.41	0.60	0.62
P. aeruginosa	1088	0.68	0.75	0.75
P. putida	1090	0.24	0.17	0.28
P. putida	1133	0.62	0.74	0.76
P. syringae	1310	0.38	0.16	0.36

Strain Improvement by Single Colony Isolation

The first trial for strain improvement was done by single colony isolation. *P. putida* 1133 was cultivated on sorbose agar, and the cells of one test tube were diluted in saline and spread on the same agar. After 4 days of cultivation, 200 colonies were isolated and cultivated in liquid medium SI. The production of 2 KGA was detected by TLC and, for the three colonies (C143, C144, and C151) that were the most productive, by HPLC (Table 2).

TABLE 2. Production of 2 KGA by various *P. putida* 1133 variants, obtained by the single-colony isolation procedure.

Variants	C143	C144	C158
2 KGA (g/L) in test tubes	1.20	0.98	0.94
2 KGA (g/L) in flasks	1.25	1.23	1.00

Strain Improvement by Mutagenesis

Further strain improvement was done by ultraviolet mutagenesis. The variant C143 was used in this experiment; 200 colonies, formed after UV-radiation on sorbose agar, were isolated and cultivated on liquid medium SI for 2 KGA production. The first stage of cultivation (immediately after irradiation) was carried out in test tubes. The production of 2 KGA by the 200 colonies was detected by TLC, and only 32 were chosen for HPLC analyses. The mutants with the best results were cultivated in flasks. Some of the better results are presented in Table 3. It is seen from the results that the increased productivity of the UV mutants is not too stable. After several cycles of cultivation, reversion of mutants was observed.

TABLE 3. Production of 2 KGA from L-sorbose by various mutants of C143 variant of *P. putida* 1133 in test tube and flasks.

Mutants	2 KGA production (g/L)	
	Test Tube	Flasks
U70	2.9	—
U82	4.5	1.3
U86	2.3	1.4
U88	2.4	—
U113	2.4	1.6
U116	2.4	1.4
U139	2.3	1.2
U150	2.6	—
U158	3.1	1.4
U164	3.2	1.5
U188	2.4	0.2

U113 retained the most stable 2 KGA production and was used in a second UV irradiation procedure. The production of 2 KGA of 60 colonies was determined after a second UV irradiation. The second UV irradiation did not lead to higher 2 KGA production.

The U113 mutant was cultivated in a fermentor. The concentration of 2 KGA was 2.3 g/L. The comparative results of 2 KGA production of *P. putida* 1133 before and after improvement experiments are presented in Table 4.

TABLE 4. Comparative results of 2 KGA production (g/L) by *P. putida* 1133 before and after improvement experiments.

	2 KGA production (g/L)		
	P. putida 1133	C143	C113
In test tube	0.74	1.20	2.4
In flasks	0.82	1.25	1.6
In fermentor	—	—	2.3

The higher 2 KGA production probably was due to better aeration because *P. putida* is a strongly aerophyllic species.

After reversion, the 2 KGA production of the U113 mutant of *P. putida* 1133 remained stable and was three times higher than the initial production.

REFERENCES

Lazarus, R. A., and J. L. Seymour, 1986. "Determination of 2-Keto-L-Gulonic Acid and Other Ketoaldonic and Aldonic Acids Produced by Ketogenic Bacterial Fermentation." *Anal. Biochem.* 157:360-366.

Sonoyama, T., H. Tani, K. Matsuda, B. Kagejama, T. Tanimoto, K. Kobayashi, S. Yagi, H. Kyotani, and K. Mitsushima, 1982. "Production of 2-Keto-L-Gulonic Acid from D-Glucose by Two-Stage Fermentation." *Appl. Environ. Microbiol.* 43:1046-1069.

Stroshane, R. M., and D. Periman, 1977. "Fermentation of Glucose by Acetobacter malanogenus." *Biotechnol. Bioeng.* 19:459-465.

Sugisawa, T., T. Hoshino, S. Masuda, S. Nomura, Y. Setogushi, M. Tazoi, M. Shinjon, S. Sohema, and A. Fujiwara, 1990. "Microbial Production of 2-Keto-L-Gulonic Acid from L-Sorbose by *Gluconobacter melanogenus*." *Agric. Biol. Chem.* 54:1201-1209.

Influence of Sorption on Organic Contaminant Biodegradation

Weixian Zhang, Edward J. Bouwer,
Al B. Cunningham, and Gordon A. Lewandowski

ABSTRACT

Significant progress has been made toward understanding how to stimulate microbial growth in the subsurface during bioremediation by optimizing the chemical conditions. For hydrophobic organic contaminants, mass transfer is likely to control the overall rate of bioremediation. Biodegradation rates in the field are significantly slower than in the laboratory because of reduced bioavailability. The influence of sorption on biodegradation is quantified by defining a bioavailability factor, B_f. The B_f is helpful for determining the extent of mass transfer control during biodegradation of organic contaminants.

INTRODUCTION

Hydrophobic organic contaminants, e.g., polyaromatic hydrocarbons (PAHs), polychlorinated biphenyls (PCBs), and halogenated aliphatic compounds (HACs), tend to distribute among the solid, liquid, and gas phases within the subsurface such that only a small fraction of these compounds may actually be present in the bulk water phase. For example, these compounds have been found sorbed to soils/sediments, associated with colloids and dissolved organic molecules (e.g., natural organic matter), volatilized into the gas phase, and as a separate nonaqueous-phase liquid (NAPL). There is strong evidence suggesting that mass transfer between the aqueous and nonaqueous phases can significantly limit the effectiveness of many widely used remediation technologies (National Research Council 1994).

The conventional pump-and-treat technique has been used at 73% of U.S. Superfund sites for remediating groundwater contamination (National Research Council 1994). Because only a small portion of the contaminants is in the mobile (extractable) water, pump-and-treat approaches have been inefficient and costly for many site cleanups. A recent survey (Bredehoeft 1994) suggests it could take 10s and even 100s of years to reach health-based cleanup objectives by the pump-and-treat method.

Biological treatment is an attractive alternative to conventional pump-and-treat. It offers the prospect of destroying organic contaminants or converting them to less harmful products in situ. However, most evidence indicates that uptake of organic compounds for metabolism by microorganisms occurs via the aqueous phase. Partitioning (sorption) into the nonaqueous phases reduces the availability of organic contaminants to microorganisms (Mihelcic et al. 1993). For hydrophobic organic compounds, the extent and rate of biodegradation in the subsurface can be controlled by their bioavailability (Bouwer and Zehnder 1993).

This paper addresses issues concerning the effect of sorption on biodegradation of hydrophobic organic compounds and the implications for field-scale bioremediation. For soils, sediments, and groundwater aquifers where the solid fraction and surface area are high, sorption is the primary mechanism limiting bioavailability. A method to quantify the effect of sorption on biodegradation is proposed.

EFFECTS OF SORPTION ON BIODEGRADATION

Concentration Effects

A direct impact of sorption on biodegradation is the reduction of organic compounds in the bulk water phase. As sorbed chemicals are effectively protected from direct biodegradation, microbial growth must rely on organic substrate(s) in the bulk water phase. The rate of bulk water uptake and metabolism of the organic compound is generally given by the Monod relationship:

$$\frac{dC}{dt} = -\frac{kXC}{K_s + C} \tag{1}$$

where C is the bulk water concentration of the organic compound, t is time, k is the maximum rate of biodegradation, X is the concentration of microorganisms in water, and K_s is the half-velocity constant. As sorption reduces the bulk water concentration, the consumption rate of the contaminant decreases. Thus, sorption will prolong the time to degrade the same amount of organic pollutant as compared with no soil in the system. At very low contaminant concentrations, often in the range of μg to ng per liter, insufficient energy and carbon may be available for microbial growth and maintenance. Rittmann and McCarty (1980) defined a critical concentration (C_{min}) at which microbial growth is just balanced by decay:

$$C_{min} = \frac{K_s b}{YK - b} \tag{2}$$

where Y is the biomass yield coefficient and b is the microbial decay coefficient. If sorption diminishes the concentration below C_{min}, biodegradation will decrease or even stop with time because there will be net decay of biomass.

On the other hand, at high concentrations, many organic compounds become toxic to microorganisms. One expression for the biodegradation rate is the Haldane equation, which is given below (Andrews 1968):

$$\frac{dC}{dt} = -\frac{kXC}{K_s + C + \dfrac{C^2}{K_i}} \tag{3}$$

where K_i is the inhibition constant. Here sorption can reduce the bulk liquid concentration to lessen the toxic inhibition and increase microbial growth and biodegradation.

Desorption Rate Limitations

Due to geometrical and mass transfer restrictions, most bacteria are present in the external surface of soil particles and in the bulk water. Only limited biological activities exist within the intraparticle pores (Jones et al. 1993). Decontamination of soils and sediments is thus a two-step process: desorption and biodegradation occurring in series. The apparent biodegradation rate in a solid-water system can be controlled by either the desorption rate or the biodegradation rate.

A mass balance for an organic compound in soil/water systems with first-order biodegradation kinetics can be written as:

$$\frac{d}{dt} [VC + mq] = -Vk_bC \tag{4}$$

where V is the volume of bulk water, C is the bulk water concentration of the organic compound, m is the mass of soil, q is the average sorbed concentration, and k_b is the intrinsic first-order biodegradation rate coefficient. The left-hand side of Equation 4 is the change of organic compound in the system, and the right-hand side is the loss due to biodegradation. One approximation for the mass exchange between soil and water (sorption/desorption) is a first-order reaction:

$$\frac{dq}{dt} = -k_m (C_{pore} - C) \tag{5}$$

where k_m is the mass transfer coefficient, C is the bulk water concentration of the organic compound, and C_{pore} is the average organic concentration in the intraparticle pore water:

$$C_{pore} = \frac{q}{K_d} \qquad (6)$$

Here K_d is the soil/water distribution coefficient. Coupling Equations 4, 5, and 6, the overall rate of biodegradation in a soil-water system can be expressed as:

$$\frac{dC}{dt} = -B_f k_b C \qquad (7)$$

where B_f is termed the bioavailability factor

$$B_f = \frac{1}{1 + \dfrac{m}{V} K_d + \dfrac{k_b}{k_m} K_d} \qquad (8)$$

The overall rate of biodegradation in a soil/water slurry is determined by the two factors: B_f and k_b. B_f is evaluated from the extent and rate of sorption, and k_b is determined by the metabolic capability of microorganisms and environmental factors such as temperature. The time to reduce the bulk water concentration by half is given by:

$$T_{50\%} = \frac{1n2}{B_f k_b} \qquad (9)$$

The influence of the B_f on the extent and rate of contaminant biodegradation is illustrated in a plot of the contaminant fraction remaining (C/C_o) versus dimensionless time (product of k_b and time, t) (Figure 1). A B_f value of 1.0 means there is no influence of sorption on the biodegradation, and this situation corresponds to a well-mixed liquid culture system without sorbent. As the B_f decreases below 1.0, the persistence of the chemical increases (Figure 1). Many organic compounds of interest in the subsurface have K_d values much greater than 0.1 kg/L. For example, the K_d of PCE for Borden sand is 0.5 kg/L (Ball and Roberts 1991). Using Equation 8, K_d values > 0.5 kg/L will yield B_f values that are less than 0.3 for biodegradation in the subsurface (m/V ≥ 2 kg/L). Consequently, organic compounds that can be readily biodegraded in liquid culture with half-lives ranging from a few hours to a few days (large k_b) can be very persistent in soils and groundwater aquifers (curves with B_f < 0.3 in Figure 1). Low bioavailability (small B_f) can be the major factor responsible for slow in situ bioremediation.

SIGNIFICANCE OF SORPTION
DURING IN SITU BIOREMEDIATION

A schematic of the processes and contaminant concentrations during in situ bioremediation is given in Figure 2. A typical groundwater plume is

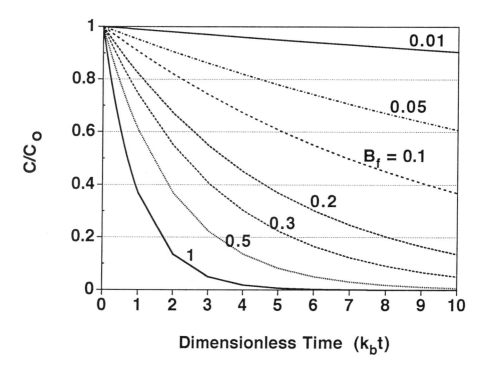

FIGURE 1. Influence of bioavailability factor, B_f, on extent and rate of contaminant biotransformation.

shown in Figure 2a. As chemicals in the source region move with the groundwater, concentrations of the chemicals will undergo physicochemical and biological changes. An engineered bioremediation system usually includes extraction and injection wells and equipment for adding and mixing nutrients.

Two important features of bioremediation (as shown in Figure 2b) are: (1) most bacteria are associated with solid surfaces, and (2) organic compounds are sorbed into the solid phase. Transport of microorganisms from the bulk liquid to solid surfaces can occur by chemotaxis, advection, and diffusion. Once microorganisms contact a solid surface, they may attach permanently to the solid surfaces. As they grow, a surface film of microorganisms will accumulate. Surface growth can be removed by decay and/or detached by fluid shear and sloughing.

Figure 2c presents an idealized model of a soil particle. The solid matrix is porous, and sorption sites are homogeneously distributed. Diffusion is the only major mass transfer mechanism within the solid and occurs only in the radial direction. The pore structure within the particle is assumed to be too small for penetration of bacteria. Thus, bacteria tend to be located on the outer surfaces. Distribution of biomass can be patchy colonies or a continuous surface film dependent upon the net results of growth, decay, and attachment/detachment (Rittmann 1993).

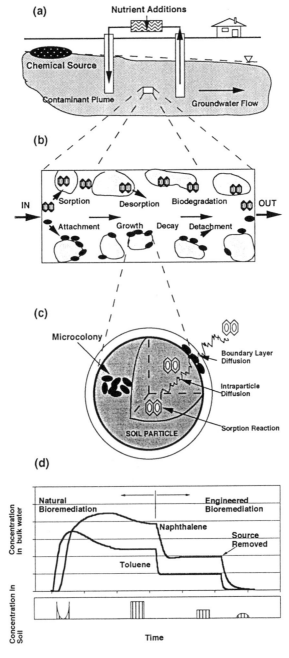

FIGURE 2. The role of sorption during in situ bioremediation: (a) a schematic of in situ bioremediation, (b) major mechanisms involved during in situ bioremediation, (c) an idealized (model) soil particle, and (d) model simulations of sorption and biodegradation of toluene and naphthalene in the subsurface.

Shown in Figure 2d is the theoretical response of organic concentrations in aqueous and solid phases as a function of time during bioremediation in the subsurface. Biotransformation is insignificant initially because the amount of biomass present is very small. Sorption removes a large portion of the organic compounds from the bulk water. As concentration gradients within soil aggregates diminish, organic concentrations in the bulk water increase quickly and finally level off and reach a steady state, with biomass accumulated from use of the organic contaminant.

With engineered bioremediation, microbial activity is enhanced bringing about a rapid decrease in the bulk water concentration (Figure 2d). This lowering of the bulk water concentration will promote desorption of the contaminants by increasing the local concentration gradient for diffusion. The desorbing contaminants can be biodegraded as they pass through the attached biomass, which keeps the bulk water concentrations low. Thus, the rate of desorption is accelerated in the presence of biotransformation. However, if the rate of desorption is much slower than the biodegradation rate (low B_f), stimulating microbial growth will have only a minimal impact on accelerating the rate of soil decontamination. Furthermore, upgradient sources of contamination must be eliminated (e.g., removal and containment of separate phase organic liquids) to achieve effective in situ bioremediation of the contaminant plume.

CONCLUSIONS

Current bioremediation practice originates mainly from knowledge of biostimulation, i.e., stimulating microbial growth in the subsurface by providing adequate nutrients, electron acceptor(s), and/or microorganisms with specific metabolic capabilities. It is clear from both field demonstrations and laboratory studies that bioremediation of highly hydrophobic organic contaminants, such as PCBs and PAHs, is not likely to be controlled by the metabolic rates of microorganisms but by their availability to microorganisms. Sorption tends to reduce the bioavailability of organic compounds. Slow desorption is likely to control the overall performance of bioremediation. The influence of sorption on biodegradation in laboratory or field systems can be quantified with the value of B_f (Equation 8). The practical effect of sorption is to decrease the B_f, which increases the time required to achieve cleanup and the amount of chemicals that must be added to sustain microbial activity (Bouwer and Zehnder 1993).

ACKNOWLEDGMENTS

This work was supported in part by Cooperative Agreement ECD-8907039 between the National Science Foundation and Montana State University, in part by the Hazardous Substance Management Research Center and the New

Jersey Commission on Science and Technology headquartered at the New Jersey Institute of Technology, and in part by the Baltimore Gas and Electric Company.

REFERENCES

Andrews, J. F. 1968. "A mathematical model for the continuous culture of microorganisms utilizing inhibitory substrates." *Biotechnology and Bioengineering, 10*: 707-723.

Ball, W. P. and P. V. Roberts. 1991. "Long-term sorption of halogenated organic chemicals by aquifer materials: I equilibrium studies." *Environ. Sci. Technol. 25*, 1223-1237.

Bouwer, E. J. and A.J.B. Zehnder, 1993. "Bioremediation of organic compounds—putting microbial metabolism to work." *Trends Biotech. 11*:(8), 287-318.

Bredehoeft, J. D. 1994. "Hazardous waste remediation: A 21st century problem." *Groundwater Monit. and Remediation, 14*:(1), 95-100.

Jones, W. L., J. D. Dockery, C. R. Vogel, and P. J. Sturmen. 1993. "Diffusion and reaction within porous packing media: a phenomenological model." *Biotechnology and Bioengineering, 41*, 947-956.

Mihelcic, J. R., D. R. Lueking, R. Mitzell and J. M. Stapleton. 1993. "Bioavailability of sorbed- and separate-phase chemicals." *Biodegradation, 4*:(3), 141-154.

National Research Council, 1994. *Alternatives for Ground Water Cleanup*, National Academy Press, Washington, DC.

Rittmann, B. E. 1993. "The significance of biofilms in porous media." *Wat. Resour. Res. 29*, 2195-2202.

Rittmann, B. E. and P. L. McCarty. 1980. "Model of steady-state biofilm kinetics." *Biotechnology and Bioengineering, 22*, 2343-2357.

Control of Bacterial Exopolysaccharide Production

Thomas F. Hanneman, Donald L. Johnstone,
David R. Yonge, James N. Petersen,
Brent M. Peyton, and Rodney S. Skeen

ABSTRACT

To retain permeability and limit subsurface plugging when bacterial growth is stimulated for groundwater remediation, the ratio of bacterial exopolysaccharides (EPS) to bacterial cells should be minimized. In this study, the effect of the type of carbon source on the EPS/biomass ratio was evaluated. Tricarboxylic acid cycle intermediates and other organic acids were tested to determine total biomass and EPS production by two strains of carbon tetrachloride-degrading, denitrifying bacteria. An inverse correlation was found between EPS/biomass ratios and total biomass production; i.e., organic acid substrates that gave low EPS/biomass ratios produced high total biomass concentrations, whereas organic acids that gave a high ratio of EPS/biomass produced low total biomass concentrations.

INTRODUCTION

Biofouling usually is a problem only when high concentrations of both electron donor and electron acceptor are present; however, these conditions commonly occur near nutrient injection well outlets (Jennings et al. 1995). Bacterial biofilms often include large quantities of EPS (Marshall 1992); excessive EPS production may therefore cause severe biofouling. To limit plugging, the ratio of EPS to bacterial cell mass should be minimized.

In this study, the free and capsular components of EPS in batch cultures were quantified individually. Our objective in this study was to determine how the ratio of EPS to cell mass produced by bacteria grown under anoxic (denitrifying) conditions is affected by the type of organic acid used as an electron donor. Several organic acids that stimulated the growth of bacterial cells while producing relatively low quantities of EPS were identified. Although the presence of sediment may enhance EPS production for the purpose of attachment, in this fundamental study cultures were grown in the absence of sediment to simplify the experimental design and analytical procedures.

In subsequent work, prior to field trials, a porous medium should be used to provide surfaces for attachment. Although the concentration of carbon tetrachloride in the field will be very low relative to that of the carbon source provided as a cometabolite, the effect of carbon tetrachloride on EPS production should be tested in subsequent experiments (prior to field trials).

MATERIALS AND METHODS

To evaluate the effect of carbon source type on EPS and biomass production, organic acid electron donors were tested in batch cultures. Two strains of denitrifying bacteria (DN2 and DN5, identified as *Pseudomonas stutzeri* [Kong et al. 1992]) were isolated from a consortium obtained from carbon tetrachloride- and nitrate-contaminated groundwater at the U.S. Department of Energy's Hanford site (Brouns et al. 1990). DN2 and DN5 have been shown to be capable of cometabolically degrading carbon tetrachloride under anoxic (denitrifying) conditions (Hansen et al. 1994).

Nine organic acid carbon sources were tested (acetate, butyrate, citrate, fumarate, lactate, malate, oxalate, pyruvate, and succinate) in DN2 and DN5 cultures grown in Erlenmeyer flasks under denitrifying conditions. Flasks were filled with a simulated groundwater medium (SGM) (Kong et al. 1992) and inoculated with 1% (by volume) of an unwashed log-phase culture grown on acetate (300 mg/L biomass by dry weight). The SGM pH was adjusted to 7.0 to 7.1 with sterile sodium hydroxide immediately prior to inoculation (during incubation, pH was maintained between 7.0 and 7.4 with a phosphoric acid buffer). Flasks were then sealed and incubated under anoxic conditions at 30°C, and 24 h after the beginning of the early stationary phase, samples were withdrawn for quantification of biomass, total and free EPS, and the remaining carbon source and nitrate.

The SGM was formulated to approximate the major ion concentration of the contaminated Hanford groundwater. However, the nitrate concentration was based on the concentration of nitrate that would be required to completely oxidize the carbon source (calculated from the stoichiometry of the oxidation reaction). A high concentration (~5 g/L as nitrate) was used to ensure that growth was not limited by the electron acceptor (nitrate) or by nitrogen. Immediately prior to inoculation, 100 mM carbon/L (1,200 mg carbon/L) of filter-sterilized organic acid was added aseptically to the nutrient medium.

Biomass concentrations were quantified by suspended solids analysis using 0.45-μm membrane filters (American Public Health Association 1992). Aliquots were dialyzed for 48 h against deionized water prior to EPS determination by the phenol-sulfuric acid (P/S) method (Dubois et al. 1956). Both total and free EPS were quantified directly by P/S; capsular EPS was determined by subtracting the free EPS concentration from the total EPS concentration. To dissolve capsular EPS for determination of total EPS concentrations, aliquots were heated for 60 min in a boiling water bath. Suspensions were filtered through 0.45-μm membranes to remove cells before P/S analysis. For free EPS determination, heat treatment was omitted; capsular EPS was thus left attached to the cells and was removed with the cells by filtration. Calibration curves for EPS quantification were developed by isolating the polysaccharides (Fett et al. 1989) and dissolving known weights in deionized water. Nitrate, nitrite, and carbon source concentrations were determined with a Dionex 4000I ion chromatograph, equipped with a Dionex AS-11 anion-exchange column (Dionex, Sunnyvale, California).

RESULTS

The following basic parameters were determined for each culture: total biomass, total EPS, free EPS, capsular EPS (the difference of total and free EPS), electron donor uptake (as mg C/L), and electron acceptor uptake (as mg NO_3^-/L). For comparison between the carbon sources, ratios of total EPS/total biomass, free EPS/total biomass, and capsular EPS/total biomass were determined (Fig. 1), and final total biomass concentrations were plotted (Fig. 2).

Ratios of total EPS to total biomass were maximal for growth on citrate (DN2, 54.0%; DN5, 39.8%) and minimal for malate (DN2, 11.3%; DN5, 16.0%). On eight of the nine carbon sources tested, ratios of total EPS (capsular EPS plus free EPS) to total biomass were greater for DN2 cultures than for DN5

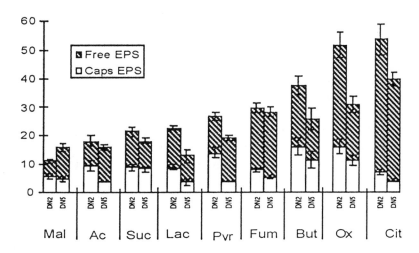

FIGURE 1. DN2 and DN5 total EPS to total biomass ratios showing both capsular and free EPS components (dry w/dry w). Values are based on three analyses of each of three replicate cultures (+ standard deviations [error bars]).

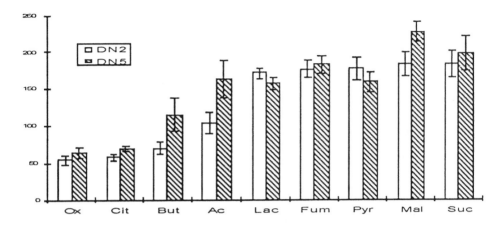

FIGURE 2. Total biomass concentrations 24 h after cultures entered the stationary phase. Initial carbon source concentrations were 100 mM as carbon. Values are based on three analyses of each of three replicate cultures (± standard deviations [error bars]).

cultures (Fig. 1). EPS yields were quantified also as grams (dry weight) of EPS produced per gram of carbon consumed ($Y_{E/C}$). Oxalate gave the highest $Y_{E/C}$ values (approximately 0.3 g/g). Relatively high $Y_{E/C}$ values were also found for fumarate and succinate. $Y_{E/C}$ values for the other carbon sources were below 0.1. Substrates on which Y_{NO_3} (growth efficiency as biomass-produced/mole of nitrate consumed) was low (e.g., oxalate and citrate) produced the highest EPS/biomass ratios, whereas, on EPS/biomass ratios were generally low substrates that gave higher Y_{NO_3} values (e.g., succinate). In addition, DN2 produced higher EPS/biomass ratios than DN5 on eight of the nine carbon sources tested, while exhibiting lower Y_{NO_3}.

Except when grown on lactate (Lac) or pyruvate (Pyr), DN5 consistently produced higher total biomass concentrations than DN2 (Fig. 2). DN5 yielded the greatest mean biomass concentration (approximately 230 mg/L) when grown on malate (Mal); whereas DN2 produced nearly equivalent maximum mean biomass concentrations (approximately 180 mg/L) when grown on either malate or succinate (Suc). Both strains produced minimum biomass (50 to 70 mg/L) when grown on oxalate (Ox) or citrate (Cit).

DISCUSSION

The results of this investigation indicate that the type of cometabolite used to stimulate biomass production under denitrifying conditions will influence EPS production. Both total biomass production and growth efficiency (Y_{NO_3}) were found to be inversely correlated with EPS production. EPS/biomass ratios were low for carbon sources that gave high Y_{NO_3} (e.g., malate and succinate), whereas EPS/biomass ratios were high for carbon sources that gave low Y_{NO_3} (e.g., oxalate, citrate, and butyrate). EPS/biomass ratios and $Y_{E/C}$ values also are related: low ratios of EPS/biomass correspond to low $Y_{E/C}$ values, whereas high EPS/biomass ratios correspond to high $Y_{E/C}$ values.

The results of our work with organic acid carbon sources are consistent with the findings of Gancel and Novel (1994), who demonstrated an inverse relationship between growth efficiency and EPS production for carbohydrate carbon sources. However, our research differs from previous work in that organic acids were used as carbon sources (instead of carbohydrates) and denitrifying conditions were maintained (instead of aerobic conditions). The above results suggest that EPS synthesis may provide a means to dispose of excess carbon and energy. When growth efficiency is low, a higher carbon and energy flux is required to support active cell growth.

The oxidation state of the carbon source can be used also to estimate the relative potential for EPS production of a carbon source. When the organic acid carbon sources were relatively reduced, EPS/biomass ratios were high; relatively oxidized organic acids gave low EPS/biomass ratios. Our results are similar to those of Linton et al. (1987) that suggest the rate of EPS synthesis is determined by adenosine triphosphate (ATP) turnover rates, which in turn depend on the oxidation state of the carbon source provided to the organism. Substrates more reduced than glucose yield more energy than can be taken up in polysaccharide synthesis; ATP therefore is diverted away from EPS production to a metabolic pathway that can cope with the high energy flux (Linton 1990). ATP turnover rates are optimal for EPS production when the oxidation state of the substrate is similar to glucose, but ATP production is limited on substrates more oxidized than glucose. Energy thus is available for cell synthesis but not for production of significant quantities of EPS (Linton et al. 1987).

In this study, we found that DN5 grew more efficiently (higher Y_{NO_3}) and produced lower EPS/biomass ratios than DN2. Other workers have found that strains of aerobic bacteria that grow efficiently (high Y_{O_2}) produce lower EPS/biomass ratios than their less efficient counterparts (Linton and Rye 1989). Because both nitrate and oxygen are terminal electron acceptors used in the tricarboxylic acid cycle, it is reasonable to expect that, under denitrifying conditions, EPS production rates should follow the trends observed under aerobic conditions.

ACKNOWLEDGMENTS

This research was supported by Pacific Northwest Laboratory, Battelle Memorial Institute, which is operated for the U.S. Department of Energy under contract DE-AC06-76RLO.

REFERENCES

American Public Health Association. 1992. *Standard Methods for the Examination of Water and Wastewater*, 18th ed. American Public Health Association, Washington, DC.

Brouns, T.M., S.S. Koegler, W.O. Heath, J.K. Frederickson, H.D. Stensel, D.L Johnstone, and T.L. Donaldson. 1990. *Development of a Biological Treatment System for Hanford Groundwater Bioremediation.* FY 1989 Status Report. PNL 7290. Pacific Northwest Laboratory, Richland, WA.

Dubois, M., K.A. Gilles, J.K. Hamilton, P.A Rebers, and F. Smith. 1956. "Colorimetric Method for Determination of Sugars and Related Substances." *Anal. Chem. 28*:350-356.

Fett, W.F., S.F. Osman, and M.F. Dunn. 1989. "Characterization of Exopolysaccharides Produced by Plant-Associated Fluorescent Pseudomonads." *Appl. Environ. Microbiol. 55*:579-583.

Gancel, F., and G. Novel. 1994. "Exopolysaccharide Production by *Streptococcus salivarius* ssp. *thermophilus* Cultures. 1. Conditions of Production." *J. Dairy Sci. 77*:685-688.

Hansen, E.J., D.L. Johnstone, J.K. Frederickson, and T.M. Brouns. 1994. "Transformation of Tetrachloromethane under Denitrifying Conditions." In R.E. Hinchee, A. Leeson, L. Semprini and S. Ong (Eds.), *Bioremediation of Chlorinated and Polycyclic Aromatic Hydrocarbon Compounds.* Lewis Pub., Chelsea, MI.

Jennings, D.A., J.N. Petersen, R.S. Skeen, B.S. Hooker, B.M. Peyton, D.L. Johnstone, and D.R. Yonge. 1995. "Effects of Slight Variations in Nutrient Loadings on Pore Plugging in Soil Columns." *Appl. Biochem Biotechnol.* (In Press).

Kong, S., D.L. Johnstone, D.R. Yonge, J.N. Petersen, and T.M. Brouns. 1992. "Comparison of Chromium Adsorption to Starved and Fresh Subsurface Bacterial Consortium." *Biotechnol. Tech. 6*:143-148.

Linton, J.D. 1990. "The Relationship between Metabolite Production and the Growth Efficiency of the Producing Organism." *FEMS Microbiol. Rev. 75*:1-18.

Linton, J.D., and A.J Rye. 1989. "The Relationship Between the Energetic Efficiency in Different Microorganisms and the Rate and Type of Metabolite Over-Production." *J. Indust. Microbiol. 4*:85-96.

Linton, J.D., D.S. Jones, and S. Woodward. 1987. "Factors that Control the Rate of Exopolysaccharide Production in *Agrobacterium radiobacter* NCIB11883." *Journal of General Microbiology 133*: 2979-2987.

Marshall, K.C. 1992. "Biofilms: an Overview of Bacterial Adhesion, Activity, and Control at Surfaces." *ASM News 58.*

Temperature Effects on Propylene Glycol-Contaminated Soil Cores

Wendy J. Davis-Hoover and Stephen J. Vesper

ABSTRACT ▬▬▬▬▬▬▬▬▬▬▬▬▬▬▬▬▬▬

We are examining the effect of temperature on the biodegradation of propylene glycol (PPG) in subsurface soil cores. Subsurface soils (glacial till with 27% clay, 5% sand, and 68% silt at 1.5 m in depth at the B_2 horizon) were contaminated in situ with PPG and allowed to diffuse into the soil for 30 days. The treated soil was reexposed, and intact cores were incubated for 30 days at temperatures ranging from 9 to 39°C in a temperature gradient incubator. At 30 days, soil moisture, soil pH, microbial activity [fluorescein diacetate (FDA) test], R2A plate counts, and plate counts of PPG degraders were studied. Although the soil moisture and pH remained relatively unchanged, the parameters of microbial activity varied rather consistently with temperature. Multiple populations or subpopulations of bacteria appear to exist between temperatures of 9 and 39°C in these soils.

INTRODUCTION

Bioremediation of contaminants in subsurface soil is a natural but often slow process (Leahy and Colwell 1990). The natural process often is too slow to allow its effective use in reducing risks. We have been evaluating methods to enhance the rates of in situ subsurface bioremediation (Davis-Hoover et al. 1991, Vesper et al. 1994), because in situ treatment eliminates many risks of human exposure and usually is more cost effective than on-site bioremediation. One possible way is to heat the subsurface, but the optimal temperature for bioremediation of an environment that is normally 12 to 15°C is not known.

EXPERIMENTAL PROCEDURES AND MATERIALS

Test soil was obtained by digging in a glacial till (27% clay, 5% sand, and 68% silt) at the Center Hill Facility in Cincinnati, Ohio, to 1.5-m depth. At this depth, a plastic ring 0.5 m in diameter was driven into the soil about 2 cm deep, and 4 L of a 0.01% solution of PPG in sterile deionized water was added to the surface of this soil and allowed to diffuse. After 30 days, 20 small cores were taken randomly by driving a stainless steel sampling tube lined with a plastic sheath into the soil; 15 of these cores were incubated for 30 days at temperatures from 9 to 39°C in the temperature block gradient (TBG).

Temperature readings were collected every 15 min in the TBG. The other parameters were evaluated after 30 days of incubation. Triplicate grab samples were analyzed by the FDA method (Schnurer and Rosswall 1982) for adenosine triphosphate (ATP) activities and for soil moisture (U.S. EPA 1990), soil pH (U.S. EPA 1990), and PPG-degrading and total organisms as discussed in Vesper et al. (submitted for publication). This was done in June and January.

RESULTS

In all these reported studies, the soil pH remained steady and constantly low (Figure 1), regardless of the temperature. The soil moisture remained relatively constant, although it was much lower in the TBG4 experiment. We would expect to see lower levels of microbial activity, growth, and thus degradation in this lower-moisture experiment.

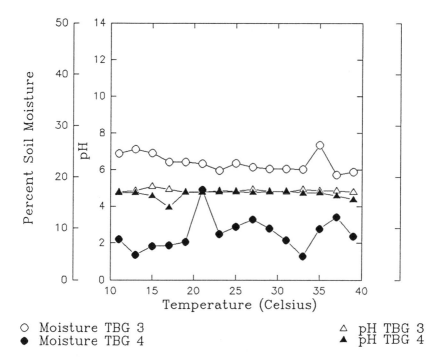

O Moisture TBG 3   △ pH TBG 3
● Moisture TBG 4   ▲ pH TBG 4

FIGURE 1. Percent soil moisture and soil pH varying by temperature.

Interestingly, the number of total heterotrophic and PPG-degrading bacteria was very similar at all temperatures and declined only at the highest temperatures (Figure 2). The microbial activity measured by ATP activity (FDA) was generally highest between 20 to 30°C.

In comparing the first experiment (Figure 2, TBG3) to the second experiment (Figure 3, TBG4), we do see that the lower soil moisture shown in Figure 1 may have led to these lower numbers of bacteria. However, these bacteria are more productive as the microbial activity levels are increased. In each separate experiment the number of total heterotrophic and PPG-degrading bacteria remains relatively constant even as the temperature varies. However, the microbial activity increased mainly from temperatures of about 18 to 25°C.

DISCUSSION

At 30 days, temperature does not seem to affect soil pH or moisture content. The decreased moisture does correlate well with decreased numbers of bacteria. The increased activity levels may reflect this dryness, as the effect of a decreased microbial population of seasonal changes of the subsurface soil

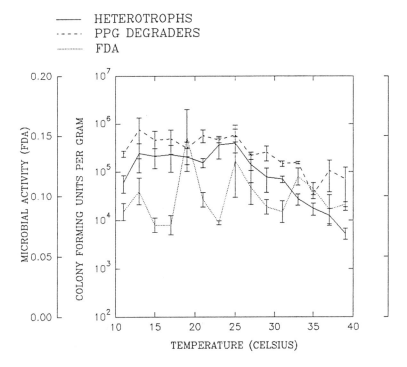

FIGURE 2. Microbial activity and numbers of total heterotrophic and PPG degraders varying by temperature in TBG3.

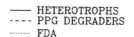

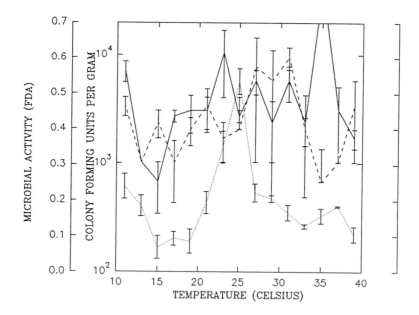

FIGURE 3. Microbial activity and numbers of total heterotrophic and PPG degraders varying by temperature in TBG4.

ecosystem to the microbacteria, because these soil cores were taken in June and January (TBG3 versus TBG4). The effect of dryness or other "winter" seasonal changes may lead to a decrease in the biodiversity but an increase of bioactivity of certain populations, as in TBG4. The moisture or "summerness" of the TBG3 ecosystem may lead to an increase in the biodiversity but a decrease of bioactivity of certain populations.

It is of note that the relative numbers of bacteria may not vary by temperature. The subsurface normally is about 12°C, and one might expect significantly more bacteria at those temperatures, because they would have been naturally selected for this temperature. This was not seen, nor was the highest microbial activity observed at 12°C. Thus, certain populations of bacteria have increased their activity, but not their numbers, in adjusting to the varying temperatures.

These data show that change of temperature activates and inactivates different subpopulations of bacteria. Thus, fluctuations of temperature may temporarily slow down the process, but the microbial subsurface soil ecosystem seems to be able to adjust with another population of organisms more suitable to the new conditions. This could be seen in the field in a Superfund Innovative Technology Evaluation (SITE) demonstration project where biodegradation occurred even under nonoptimized temperatures (J.R. Simplot's Anaerobic Bioremediation of Nitroaromatics) (Kaake et al. 1992) more quickly than anticipated.

ACKNOWLEDGMENT

The authors would like to thank the U.S. EPA's Risk Reduction Engineering Laboratory at the Center Hill Research Facility in Cincinnati, Ohio, for its support. Also we thank R. D. Brand, S. Pfaller, K. P. Paris, D. S. Ebert, and M. C. Davis.

REFERENCES

Davis-Hoover, W. J., L. C. Murdoch, S. J. Vesper, H. R. Pahren, O. L. Sprockel, C. L. Chang, A. Hussain, and W. A. Ritshel. 1991. "Hydraulic Fracturing to Improve Nutrient and Oxygen Delivery for In situ Bioreclamation." In R. E. Hinchee and R. F. Olfenbuttel (Eds.), *In Situ Bioreclamation: Applications and Investigations for Hydrocarbon and Contaminated Site Remediation.* pp. 67-83. Butterworth-Heinemann, Stoneham, MA.

Kaake, R. H., D. J. Roberts, T. O. Stevens, R. L. Crawford, and D. L. Crawford. 1992. "Bioremediation of Soils Contaminated with the Herbicide 2-sec-Butyl-4,6-Dinitrophenol (Dinoseb)." *Appl. Environ. Microbiol. 58*(5) 1683-1689.

Leahy, J. G., and R. R. Colwell. 1990. "Microbial Degradation of Hydrocarbons in the Environment." *Microbiol. Reviews 54*(3):305-315.

Schnurer, J., and T. Rosswall. 1982. "Fluorescein Diacetate Hydrolysis as a Measure of Total Microbial Activity in Soil and Litter." *Appl. Environ. Microbiol. 45*:1256-1261.

U.S. EPA, Office of Solid Waste and Emergency Response. 1990. *Test Methods for Evaluating Solid Waste, Physical/Chemical Methods,* 3rd ed. SW-846. September.

Vesper, S. J., W. J. Davis-Hoover, and L. C. Murdoch. Submitted for publication.

Vesper, S. J., L. C. Murdoch, S. Hayes, and W. J. Davis-Hoover. 1994. "Solid Oxygen Source for Bioremediation in Subsurface Soils." *Journal of Hazardous Materials. 36*: 265-274.

Confirmation Tests Using Column Percolation on Methanogenic Biomass

Luca Di Palma, Carlo Merli, and Roberta Palmieri

ABSTRACT

This study forms part of a research program carried out for the European Community for the purpose of using simple, quick laboratory tests to assess the biodegradability and toxicity of chemical compounds by the microorganisms present in soils affected by landfill leachate. The results of the simple toxicity testing carried out on a batch of enriched microbial cultures by Andreoni and Sorlini (1994) were in need of confirmation. To guarantee the reliability of the laboratory test batches, column tests were set up to confirm the results of the laboratory testing. These columns were used to simulate, in controlled conditions, the phenomena that may take place in anaerobic environments, in such a way as to provide conditions closer to reality than those of laboratory testing. Particular attention was paid to the removal of the toxic compounds investigated while monitoring biogas production in qualitative and quantitative terms. The data obtained by feeding the columns with increasing concentrations of the toxic compounds were compared with the data from the above-mentioned batch laboratory test. The results were found comparable, although the column conditions are more similar to environmental conditions than the laboratory test conditions because of bacterial acclimatization.

INTRODUCTION

One of the most important problems in landfills is contamination of the surrounding environment by toxic compounds in the leachate. Many synthetic organic chemicals are in contact with anaerobic consortia present in the soil. Methanogenic bacteria, in particular, are very sensitive to toxic aromatic compounds. For this reason, environmental studies on the toxicity and biodegradability of xenobiotic compounds are of interest.

In this study, phenol, catechol, 4-*t*-butylphenol, and 2,4,6-trichlorophenol were chosen as model compounds, in accordance with the U.S. Environmental

Protection Agency (EPA) list of priority pollutants. An active methane-producing anaerobic bacteria was fed to a series of completely mixed cells containing a porous inert medium (sand) under saturated conditions. The existence of only two populations (acidogens and acetotrophic methanogens) was assumed because hydrogenotrophic methanogens dominate the early stages of methanogenesis and are succeeded by acetotrophic methanogens (Olie and Taat 1994).

The substrate for the bacteria is an aqueous solution of sodium acetate (used by the methanogenic biomass that produces CO_2 and CH_4) and Todd Hewitt Broth [THB (formula per liter: 500 g infusion from beef heart, 20 g of peptone, 2 g of glucose, 2 g of NaCl, 2 g of $NaHCO_3$, 0.4 g of Na_2HPO_4)]. The latter is used as a source of nitrogen, which can be used by the acidogenic biomass as a second carbon source. The toxic compound is a source of aqueous carbon, which can be used by acidogenic bacteria to form CO_2, acetate carbon, and succinate. The acetate is then used by methanogenic bacteria to produce CO_2 and CH_4 (Figure 1) (El Fadel et al. 1989). There is, however, a competitive inhibition

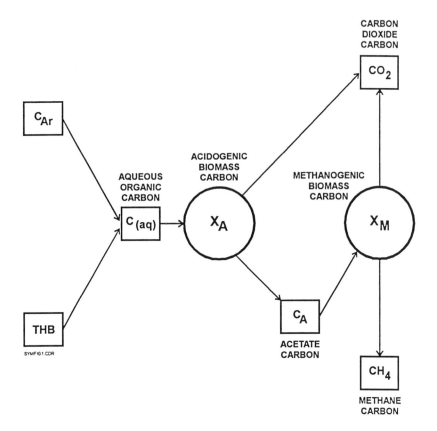

FIGURE 1. Organic carbon reservoirs and pathways for a batch solid-substrate bioreactor.

effect on the maximum growth rates of both bacterial groups. This effect results in an exponential decrease in bacterial growth proportional to the toxic concentration. The methanogenic bacteria are the most sensitive to aromatic compounds (Patel et al. 1991).

During the operation, the environment is favorable with regard to temperature and nutrient concentration. Therefore, these factors are not limiting. Instead, it can be assumed that the biomass is fixed on the sand grains, which have a limited superficial area.

EXPERIMENTAL PROCEDURES AND MATERIALS

Four vertical columns (Figure 2) with an internal diameter (D) of 100 mm were filled (filling height = 445 mm) with a porous inert medium (sand) (porosity = 0.4 m^3/m^3; density = 1,622 [kg/m^3]; organic carbon concentration = 0). These columns were used to evaluate the toxicity of catechol [$C_6H_4(OH)_2$], phenol ($C_6H_5 \cdot OH$), 4-*t*-butylphenol [$(CH_3) \cdot C \cdot C_6H_4 \cdot H$], and 2,4,6-trichlorophenol ($Cl_3C_6H_2 \cdot OH$).

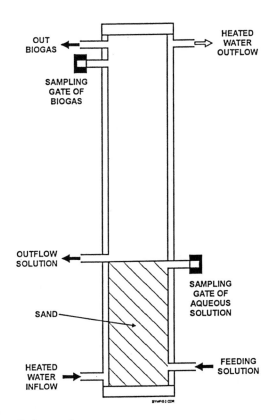

FIGURE 2. Percolation column.

The feeding solution, flowing under hydraulic pressure, was injected into the lower inlet of the column by means a multichannel peristaltic pump, thus producing the necessary upflow in the column (0.4 dm³/day). The columns were fed with a THBA flow [a solution containing 36.4 g of THB + 2 g of CH_3COONa in 1 L of distilled water (pH = 7.8)] to impregnate all the sand, and were inoculated with 200 mL each of a methanogenic consortium (acidogenic bacteria + acetotrophic methanogens) previously prepared in the same medium. A constant temperature of 35°C was maintained.

For 40 days, the concentrations in the feeding solution were 10 g/L of TH and 4 g/L of sodium acetate. When a methane percentage of nearly 50% was reached in each column, toxic compound feeding in the columns had begun. (It was assumed that the methanogenic biomass development was completed and that the biogas flow was proportional to the methanogenic biomass quantity.) The feeding solution consisted of 10 g/L of THB and 4 g/L of sodium acetate together with the toxic compounds at the stated concentration in distilled water.

The toxic compound concentrations in the feeding solution were increased step by step (Table 1), until total inhibition in biogas production was obtained. The concentrations of toxic compound were increased week by week to give the upflowing feeding solution (0.4 dm³/day) sufficient time to impregnate all the sand present in the column, thus reaching the topmost layers to the level of the sampling outlet. The concentrations of toxic compounds taken from the columns were measured three times a week, and the last two figures obtained in the series of three measurements always proved comparable. The period of 1 week was not chosen for the purpose of acclimatizing the bacteria to the toxic compound because it was assumed that this was achieved by increasing the concentrations of toxic compound slowly and gradually over the entire period of the experiment.

Measurements During Column Operation

Daily Volume of Produced Biogas. Biogas was measured by means of a hydraulic gasmeter for low rate (measure cell capacity = 55 mL) connected to the outlet placed at the top of the column.

Concentration of Toxic Compound in the Feeding Solution and in the Aqueous Outflow Solution. Toxic compound concentrations were measured by means of gas chromatography (GC) SE-52 (detector: FID, carrier: helium, injector temperature: 230°C, detector temperature: 240°C). Two volumes of aqueous solution from the toxic compounds and ethyl acetate were mixed and shaken for 10 min. The solution obtained in the ethyl acetate was then analyzed by means of GC.

Concentrations of Methane and Carbon Dioxide in the Biogas Sampled at the Top of the Columns. Methane and CO_2 were measured by using a GC with packed columns (reverse phase) filled with Cromosorb 102, a (thermal conduc-

TABLE 1. Results of concentrations and biogas production.

Step	Concentration in the Influent [mg/L]	Concentration in the Effluent [mg/L]	Removal [mg/L]	Removal [%]	Average Biogas Production [dm³/d]	CH₄ in the Biogas [% v/v]	CH₄ in the Biogas [dm³/d]	CO₂ in the Biogas [% v/v]	CO₂ in the Biogas [dm³/d]
Phenol									
1[a]	128.09	10.28	117.81	91.97	0.560	49.53	0.277	50.47	0.283
2[a]	281.31	4.29	277.02	98.47	0.488	77.41	0.378	22.59	0.110
3[a]	484.75	75.26	409.49	84.47	0.640	57.54	0.368	42.46	0.272
4[a]	1873.13	556.82	1316.31	70.27	0.456	—	—	—	—
5[b]	3399.62	2228.51	1171.11	34.45	0.392	46.72	0.183	53.28	0.209
6[b]	4252.75	2870.89	1381.86	32.49	0.280	54.74	0.153	45.26	0.127
7[c]	6242.25	5566.96	675.29	10.82	0.056	—	—	—	—
Catechol									
1[a]	311.46	216.31	95.15	30.55	0.440	51.53	0.227	48.47	0.123
2[a]	568.32	177.68	390.64	68.74	1.084	64.01	0.694	35.99	0.390
3[a]	1341.01	695.44	645.57	48.14	1.650	61.96	1.022	38.04	0.628
4[a]	2574.86	1896.22	678.64	26.36	1.045	61.27	0.640	38.73	0.405
5[b]	2959.96	2610.03	349.93	11.82	0.935	60.23	0.563	39.77	0.372
6[b]	4800.34	3809.93	990.41	20.63	0.916	60.79	0.557	39.21	0.359
7[c]	6073.53	5586.10	487.42	8.03	0.110	—	—	—	—
2,4,6-Trichlorophenol									
1[a]	16.47	11.26	5.21	31.63	0.526	38.52	0.203	61.48	0.323
2[a]	49.33	40.71	8.62	17.47	0.910	66.51	0.605	33.49	0.305
3[a]	166.61	127.99	38.62	23.18	1.382	61.35	0.848	38.65	0.534
4[a]	259.99	241.99	18.00	6.92	1.052	56.21	0.591	43.79	0.461
5[b]	317.89	231.67	86.22	27.12	0.623	58.92	0.367	41.08	0.256
6[b]	380.43	241.59	138.84	36.50	0.220	57.03	0.125	42.97	0.095
7[c]	454.29	333.43	120.86	26.60	0.176	51.75	0.091	48.25	0.085
4-t-Butylphenol									
1[a]	8.14	2.48	5.66	69.53	0.503	43.41	0.218	56.59	0.285
2[a]	26.35	0.0	26.35	100	0.786	—	—	—	—
3[a]	45.43	6.37	39.06	85.98	1.375	66.19	0.910	33.81	0.465
4[a]	116.23	26.81	89.42	76.93	2.184	65.85	1.438	34.15	0.746
5[b]	180.82	38.61	142.21	78.65	0.069	14.67	0.010	85.33	0.059

(a) 7 days; (b) 6 days; (c) 5 days.

tivity detector, carrier: helium, injector temperature: 150°C, detector temperature: 150°C, oven temperature: 70°C for 8 min, 25°C/min until it reaches 230°C, and then 230°C for 8 min). The final results show the ratio between methane and carbon dioxide concentration related to the total concentration of the two gases. Other gases (nitrogen, vapor, etc.) are naturally present in the chromatographic results.

pH. The pH was measured in the feeding solutions and in the aqueous outflow solutions as a control test.

Temperature. The temperature was measured in the aqueous outflow solution as a control test.

RESULTS

Table 2 and Figure 3 show the following concentration limit values from the column experiments (IC50: 50% inhibition in biogas production, EC50: 50% inhibition in methane production) for the four toxic compounds examined (phenol: IC50 = 3,948.07 mg/L, EC50 = 3,351.59 mg/L; catechol: IC50 = 4,944.08 mg/L, EC50 = 4,456.91 mg/L; 4-*t*-butylphenol: IC50 = 149.58 mg/L, EC50 = 148.75; 2,4,6-trichlorophenol: IC50 = 335.78 mg/L, EC50 = 327.74 mg/L). The toxicity scale for the activity of the entire microbial consortium and for methanogenic activity can be outlined as follows:

4-*t*-butylphenol > 2,4,6-trichlorophenol > phenol > catechol

Table 1 shows the following biodegradability values (peak removal before reaching IC50): phenol: 1,316,.1 mg/L; catechol: 990.41 mg/L; 2,4,6-trichlorophenol: 86.22 mg/L; 4-*t*-butylphenol: 8.42 mg/L. This biodegradability scale can be outlined as:

phenol > catechol > 4-*t*-butylphenol > 2,4,6-trichlorophenol.

The results bear out the choice of the four compounds, which cover the four possible cases for any chemical compound: nontoxic/biodegradable, nontoxic/nonbiodegradable, toxic/biodegradable, and toxic/nonbiodegradable.

DISCUSSION

The toxicity values obtained for the four compounds examined (Table 2) are confirmed by the following EC50 values observed by others (Sierra-Alvarez and Lettinga 1991) without adaptation to the toxic compound during the

TABLE 2. IC50 and EC50 values obtained in the laboratory tests and in the column experiments.

	IC50 [mg/L]		EC50 [mg/L]	
Compounds	Laboratory Tests	Column Experiments	Laboratory Tests	Column Experiments
Phenol	2588.03	3948.07	1223.43	3351.59
Catechol	4624.62	4944.08	1651.65	4456.91
2,4,6-trichlorophenol	1066.23	335.78	<51.34	327.74
4-*t*-butylphenol	1351.98	149.58	<49.57	148.75

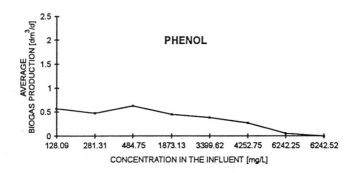

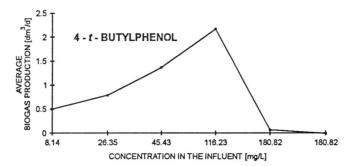

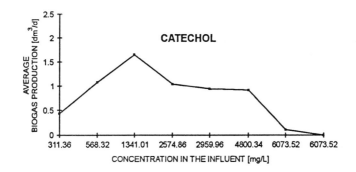

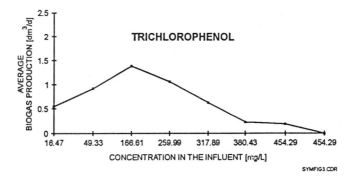

FIGURE 3. Average biogas production.

342 Microbial Processes for Bioremediation

experiments: phenol = 1,100.15 mg/L; catechol 1,813.35 mg/L; 2,4,6-trichlorophenol = 116.58 mg/L.

In fact, the toxicity of the monosubstituted benzenes was found to increase in the following substituent order: $OH < CH_3 < Cl$. In addition to the structural features associated with increasing inhibition, the number and the length of alkyl substitutions increased, as well as the number of chlorine atoms on the aromatic compound (Sierra-Alvarez and Lettinga 1991).

Comparison of the data (Table 2) obtained in the columns with the results obtained in laboratory tests shows that the results are similar, at least for phenol and catechol. For the more toxic compounds, the laboratory tests show a higher resistance of the entire microbial consortium, but methanogenic bacteria appear to be more sensitive. The EC50 values must therefore be taken as the term of reference in assessing the toxicity of chemical compounds by means of laboratory tests.

As regards the toxicity of the four compounds with respect to methanogenic bacteria, the limiting concentration values obtained through the column testing —and hence by subjecting the bacterial consortium to increasing concentrations of toxic compound—proved to be higher than those obtained through laboratory batch testing as the latter method prevented the bacteria from acclimatizing to the toxic compounds. In each batch the bacterial consortium was exposed directly to a constant given concentration of toxic compound, a number of different batches being used to assess the toxicity of the compounds with respect to the different concentrations. Laboratory batch tests can therefore be used to assess the toxicity of chemical compounds but will furnish lower values with respect to reality.

REFERENCES

Andreoni, V., and C. Sorlini. 1994. *Development of Microbiological Test for the Toxicity Assessment of Chemical Compounds in an Anaerobic Environment.* 2nd Report. Contract between the European Economic Community and Castalia-Società Italiana per l'Ambiente.

El Fadel, M., A.N. Findikakis, and J.O. Leckie. 1989. "A Numerical Model for Methane Production in Managed Sanitary Landfills." *Waste Management & Research* 7: 31-42.

Olie, J., and J. Taat. 1994. *Development of Microbiological Test for the Toxicity Assessment of Chemical Compounds in an Anaerobic Environment.* Progress Report. Contract between the European Economic Community and Castalia-Società Italiana per l'Ambiente.

Patel, B.G., B.J. Agnew, and C.J. Dicaire 1991. "Inhibition of Pure Cultures of Methanogens by Benzene Ring Compounds." *Applied Environmental Microbiology* 57 (10): 2969-2974.

Sierra-Alvarez, R., and J. Lettinga 1991. "The Effect of Aromatic Structure on the Inhibition of Acetoclastic Methanogenesis in Granular Sludge." *Applied Microbiology and Biotechnology* 34: 544-550.

AUTHOR LIST

Aitken, Michael D.
University of North Carolina
Dept. of Env. Sci. & Engrg. CB#7400
Chapel Hill, NC 27599-7400 USA

Alvarez, Pedro J. J.
The University of Iowa
Dept. of Civil & Environ. Engineering
1136 Engineering Building
Iowa City, IA 52242-1527 USA

Arvin, Erik
Technical University of Denmark
Inst. of Environ. Science & Engrg.
Building 115
DK-2800 Lyngby
DENMARK

Asaumi, Masayoshi
Marine Biotechnology Institute
3-75-1 Heita
Kamishi City, Iwate 026
JAPAN

Bennett, Philip C.
The University of Texas at Austin
Department of Geological Sciences
Austin, TX 78712-1101 USA

Bertrand, Jean-Louis
Serrener Consultation Inc.
360 Chemin St-Roch Nord
Rock Forest, Québec J1N 2T3
CANADA

Besnaïnou, Bernard
Commissariat à l'Energie Atomique
Centre de Cadarache, DCC, DESS
13108 Saint Paul Lez Durance Cedex
FRANCE

Bicheron, Corinne
Institut National Polytechnique
de Lorraine (INPL)
École Nationale Superieure de
Geologie
Rue du Doyen Roubault BP 40
54501 Vandoeuvre-les-Nancy
FRANCE

Blanchet, Denis
Institut Français du Pétrole
B.P. 311
92506 Rueil-Malmaison Cedex
FRANCE

Bollag, Jean-Marc
The Pennsylvania State University
129 Land and Water Building
University Park, PA 16802-4900 USA

Bombaugh, Karl J.
Radian Corporation
8501 North Mopac Blvd.
P.O. Box 201088
Austin, TX 78720-1088 USA

Bouchez, Murielle
Commissariat à l'Energie Atomique
Centre de Cadarache DCC, DESD
13018 Saint Paul Lez Durance Cedex
FRANCE

Bouwer, Edward J.
The Johns Hopkins University
Dept. of Geog. & Environ. Engrg.
313 Ames Hall
3400 North Charles Street
Baltimore, MD 21218-2686 USA

Breure, Anton M.
RIVM/Natl. Inst. for Public Health
    and the Environment
P.O. Box 1
3720 BA Bilthoven
THE NETHERLANDS

Broholm, Kim
Technical University of Denmark
Inst. of Environ. Science & Engrg.
Building 115
DK-2800 Lyngby
DENMARK

Bruce, Cristin L.
Rice University
Energy and Environ. Systems Inst.
P.O. Box 1892, MS-316
Houston, TX 77251-1892 USA

Buès, Michel A.
Institut National Polytechnique
    de Lorraine (INPL)
École Nationale Superieure de
    Geologie
Rue du Doyen Roubault BP 40
54501 Vandoeuvre-les-Nancy
FRANCE

Carroll, Susan L.
Oak Ridge National Lab
Environmental Sciences Division
Bethel Valley Road
Building 1505 MS 6038
P.O. Box 2008
Oak Ridge, TN 37831-6038 USA

Castaldi, Frank J.
Radian Corporation
8501 North Mopac Blvd.
P.O. Box 201088
Austin, TX 78720-1088 USA

Chavarie, Claude
École Polytechnique de Montréal
Groupe de Recherche Biopro
C.P. Box 6079 Station Centre-ville
Montréal, Québec H3C 3A7
CANADA

Chen, Shu-Hwa
University of North Carolina
Dept. of Environ. Sciences and Engrg.
Chapel Hill, NC 27599-7400 USA

Cook, Peter D.
Exxon Production Research Co.
P.O. Box 2189
Houston, TX 77252-2189 USA

Cox, Chris D.
University of Tennessee
Dept. of Civil and Environ. Engrg.
Knoxville, TN 37996-2010 USA

Crawford, Don L.
University of Idaho
Life Sciences North 132
Moscow, ID 83844-3052 USA

Crawford, Ronald L.
University of Idaho
Dept. of Microbiology, Molecular
    Biology, and Biochemistry
Food Research Center 103
Moscow, ID 83844-1052 USA

Cronkhite, Leslie A.
University of Iowa
Dept. of Civil & Environ. Engrg.
1040 North Governor Street
Iowa City, IA 52242 USA

Cunningham, Al B.
Montana State University
Center for Biofilm Engineering
413 Cobleigh Hall
Bozeman, MT 59717 USA

Davis, Pam S.
Exxon Production Research Co.
P.O. Box 2189
Houston, TX 77252-2189 USA

Davis-Hoover, Wendy J.
U.S. Environ. Protection Agency
Natl. Risk Mgmt. Research Lab
5995 Center Hill Road
Cincinnati, OH 45224 USA

de Bruijne, Jolanda A.
University of Amsterdam
Department of Environmental and
    Toxicological Chemistry
Nieuwe Achtergracht 166
1018 WV Amsterdam
THE NETHERLANDS

Dec, Jerzy
The Pennsylvania State University
129 Land & Water Building
University Park, PA 16802-4900 USA

de Jonge, Hubert
University of Amsterdam
Dept. of Phys. Geography & Soil
    Science
Nieuwe Prinsengracht 130
1018 VZ Amsterdam
THE NETHERLANDS

Deschênes, Louise
INRS – Eau
Université du Québec
2800 rue Einstein, C.P. 7500
Sainte-Foy, Québec G1V 4C7
CANADA

Di Palma, Luca
Università Di Roma La Sapienza
Dipart. Ingegneria
Via Eudossiana 18
Roma 00184
ITALY

Donlon, Brian
Wageningen Agricultural University
Dept. of Environmental Technology
Bomenweg 2
P.O. Box 8129
6700 EV Wageningen
THE NETHERLANDS

Durant, Neal D.
The Johns Hopkins University
313 Ames Hall
3400 North Charles Street
Baltimore, MD 21218 USA

Dyreborg, Søren
Technical University of Denmark
Inst. of Environ. Science & Engrg.
Building 115
DK-2800 Lyngby
DENMARK

Field, Jim A.
Wageningen Agricultural University
Dept. of Environmental Technology
Div. of Industrial Microbiology
P.O. Box 8129
6700 EV Wageningen
THE NETHERLANDS

Findlay, Margaret
Bioremediation Consulting, Inc.
55 Halcyon Road
Newton, MA 02159 USA

Fogel, Samuel
Bioremediation Consulting, Inc.
55 Halcyon Road
Newton, MA 02159 USA

Folk, Robert L.
The University of Texas at Austin
Department of Geological Sciences
Austin, TX 78712-1101 USA

Ghosh, Mriganka M.
University of Tennessee
Dept. of Civil and Environ. Engrg.
219-B Perkins Hall
Knoxville, TN 37996-2010 USA

Ghoshal, Subhasis
Carnegie Mellon University
Dept. of Civil and Environ. Engrg.
Porter Hall 119
Pittsburgh, PA 15213-3890 USA

Goorissen, Heleen
University of Amsterdam
Dept. of Environmental and
    Toxicological Chemistry
Nieuwe Achtergracht 166
1018 WV Amsterdam
THE NETHERLANDS

Gordy, Virginia R.
Rice University
Energy and Environ. Systems Inst.
P.O. Box 1892, MS-316
Houston, TX 77251-1892 USA

Goszczynski, Stefan
University of Idaho
Food Research Center 103
Moscow, ID 83844-1052 USA

Goto, Masafumi
Marine Biotechnology Institute
1900 Sodeshi-cho
Shimizu-Shi, Shizuoka 424
JAPAN

Gray, Nancy R.
American Chemical Society
1155 16th Street, NW
Washington, DC 20036 USA

Grimberg, Stefan J.
University of North Carolina
Dept. of Environ. Science & Engrg.
CB#7400
Chapel Hill, NC 27599-7400 USA

Hanneman, Thomas F.
Washington State University
Dept. of Civil & Environ. Engrg.
Sloan Hall
Pullman, WA 99164-2910 USA

Harayama, Shigeaki
Marine Biotechnology Institute
3-75-1 Heita
Kamaishi City, Iwate 026
JAPAN

Hiebert, Franz K.
RMT/Jones and Neuse, Inc.
912 Capital of Texas Highway South
Suite 300
Austin, TX 78746-5210 USA

Hoshmand, A. Reza
University of Hawaii–West Oahu
96-043 Ala Ike
Pearl City, HI 96782 USA

Hutchins, Stephen R.
U.S. Environ. Protection Agency
R.S. Kerr Environ. Research Lab
P.O. Box 1198
Ada, OK 74820 USA

Hwu, Ching Shyung
Wageningen Agricultural University
Dept. of Environmental Technology
Bomenweg 2
P.O. Box 8129
6700 EV Wageningen
THE NETHERLANDS

Ishihara, Masami
Marine Biotechnology Institute
Kamaishi Laboratory
3-75-1 Heita
Kamaishi City, Iwate 026
JAPAN

Jerger, Douglas E.
OHM Remediation Services Corp.
16406 U.S. Route 224 East
Findlay, OH 45840 USA

Johnstone, Donald L.
Washington State University
Dept. of Civil & Environ. Engrg.
Sloan Hall
Pullman, WA 99164-2910 USA

Kerr, Jill M.
Exxon Production Research Co.
P.O. Box 2189
Houston, TX 77252-2189 USA

Lafrance, Pierre
INRS – Eau
Université du Québec
2800 rue Einstein
C.P. 7500, Sainte-Foy G1V 4C7
CANADA

Lantz, Suzanne E.
SBP Technologies, Inc.
1 Sabine Island Drive
Gulf Breeze Environ. Research Lab
Gulf Breeze, FL 32561-3999 USA

Lee, Kenneth
Fisheries & Oceans Canada
Maurice Lamontagne Institute
P.O. Box 1000
Mont-Joli, Québec G5H 3Z4
CANADA

Leonard, Alfred C.
ChemCycle Corporation
129 South Street
Boston, MA 02111-2820 USA

Lettinga, Gatze
Agricultural University Wageningen
Dept. of Environmental Technology
Bomenweg 2
P.O. Box 8129
6703 EV Wageningen
THE NETHERLANDS

Lewandowski, Gordon A.
New Jersey Institute of Technology
Dept. of Chemical Engineering,
    Chemistry, and Environ. Science
University Heights
Newark, NJ 07102 USA

Lin, Jian-Er
Sybron Chemicals, Inc.
Biochemical and Environ. Services
111 Kesler Mill Road
P.O. Box 808
Salem, VA 24153 USA

Löfvall, Michael
Technical University of Denmark
Inst. of Environ. Science & Engrg.
Building 115
DK-2800 Lyngby
DENMARK

Luthy, Richard G.
Carnegie Mellon University
Dept. of Civil and Environ. Engrg.
Pittsburgh, PA 15213-3890 USA

Marcandella, Elise
Institut National Polytechnique
    de Lorraine
École Nationale Superieure
    de Geologie (ENSG)
Rue du Doyen Roubault BP 40
54501 Vandoeuvre-les-Nancy
FRANCE

Mayer, Raymond
École Polytechnique de Montréal
Groupe de Recherche Biopro
C.P. Box 6079 Station Centre-ville
Montréal, Québec H3C 3A7
CANADA

McFarland, Beverly L.
Chevron Research & Technology Co.
100 Chevron Way
P.O. Box 1627
Richmond, CA 94802-0627 USA

McMillen, Sara J.
Exxon Production Research Co.
P.O. Box 2189
Houston, TX 77252-2189 USA

Mergeay, Max
VITO/Flemish Institute for
    Technological Research
Lab of Genetics and Biotechnology
Boeretang 200
B-2400 Mol
BELGIUM

Merli, Carlo
Universita Di Roma La Sapienza
Dipart. Ingegneria
Via Eudossiana 18
Roma 00184
ITALY

Miller, Dennis E.
U.S. Environ. Protection Agency
Robert S. Kerr Environ. Research
    Laboratory
P.O. Box 1198
Ada, OK 74820 USA

Mitchell, Jr., William H.
Clean Soils Environmental, Ltd.
P.O. Box 591
Ipswich, MA 01938 USA

Mueller, James G.
SBP Technologies Inc.
1 Sabine Island Drive
Gulf Breeze, FL 32561-3999 USA

Palmieri, Roberta
Universita Di Roma La Sapienza
Dipart. Ingegneria
Via Eudossiana 18
Roma 00184
ITALY

Palumbo, Anthony V.
Oak Ridge National Laboratory
Environmental Sciences Division
P.O. Box 2008
Oak Ridge, TN 37831-6038 USA

Panneton, Carol
École Polytechnique de Montréal
Groupe de Recherche Biopro
C.P. Box 6079 Station Centre-ville
Montréal, Québec H3C 3A7
CANADA

Parsons, John R.
University of Amsterdam
Dept. of Environ. and Toxicological
    Chemistry
Nieuwe Achtergracht 166
1018 WV Amsterdam
THE NETHERLANDS

Paszczynski, Andrzej
University of Idaho
Food Research Center Room 103
Moscow, ID 83844-1052 USA

Peeva, Ludmila G.
Institute of Chemical Engineering
Bulgarian Academy of Science
1113 Acad. G. Bonchev str., bl. 103
Sofia
BULGARIA

Petersen, James N.
Washington State University
Chemical Engineering Dept.
118 Dana Hall
Pullman, WA 99164-2710 USA

Peyton, Brent M.
Battelle Pacific Northwest
P.O. Box 999, MS P7-41
Richland, WA 99352 USA

Pope, Sylvia R.
The University of Texas at Austin
Department of Geological Sciences
Austin, TX 78712-1101 USA

Pritchard, P. H. (Hap)
U.S. Environ. Protection Agency
Microbial Ecology & Biotechnol. Br.
Gulf Ecology Division
1 Sabine Island Drive
Gulf Breeze, FL 32561-5299 USA

Ramaswami, Anuradha
Colorado School of Mines
Environ. Science & Engineering Dept.
Boulder, CO 80401 USA

Ramsay, Juliana
École Polytechnique de Montréal
Groupe de Recherche Biopro
C.P. Box 6079 Station Centre-ville
Montréal, Québec H3C 3A7
CANADA

Razo-Flores, Elias
Wageningen Agricultural University
Dept. of Environmental Technology
Bomenweg 2
P.O. Box 8129
6700 EV Wageningen
THE NETHERLANDS

Renoux, Agnès Y.
Université du Québec
Institut National de Recherche
    Scientifique sur l'Eau
c/o Biotechnology Research Institute
6100 Royalmount Avenue
Montréal, Québec H4P 2R2
CANADA

Requejo, Adolpho G.
Texas A&M University
Geochemical & Environ. Research
    Group
833 Graham Road
College Station, TX 77845 USA

Rittmann, Bruce E.
Northwestern University
Dept. of Civil and Environ. Engrg.
2145 Sheridan Road
Evanston, IL 60208-3109 USA

Robinson, Kevin G.
University of Tennessee
Dept. of Civil and Environ. Engrg.
219-A Perkins Hall
Knoxville, TN 37996-2010 USA

Rotert, Kenneth H.
University of Iowa
Dept. of Civil & Environ. Engrg.
1136 Engineering Building
Iowa City, IA 52242 USA

Roy, Yves
Analex Inc.
915 rue Cunard
Chomedey, Québec H7S 2H6
CANADA

Rulkens, Wim H.
Agricultural University Wageningen
Department Milieutechnologie
P.O. Box 8129
6700 EV Wageningen
THE NETHERLANDS

Rutgers, Michiel
Natl. Institute of Public Health &
    Env. Protection
P.O. Box 1
3720 BA Bilthoven
THE NETHERLANDS

Samson, Réjean
École Polytechnique de Montréal
Chemical Engineering Dept.
C.P. Box 6079 Station Centre-Ville
Montréal, Québec H3C 3A7
CANADA

Sasaki, Etsuro
Marine Biotechnology Institute
1900 Sodeshi-cho
Shimizu-Shi, Shizuoka 424
JAPAN

Shi, Zhou
University of Tennessee
Dept. of Civil and Environ. Engrg.
219-B Perkins Hall
Knoxville, TN 37996-2010 USA

Sholeva, Zdravka V.
National Bank of Industrial
    Microorganisms & Cell Cultures
Blvd Tsarigradsko Shose 125 bl. 2
Sofia 1113
BULGARIA

Sinclair, James L.
U.S. Environ. Protection Agency
RS Kerr Environ. Research Lab
P.O. Box 1198
Ada, OK 74820 USA

Siron, Robert
INRS – Oceanologie
310 Allée des Ursulines
Rimovski, Québec G5L 3A1
CANADA

Skeen, Rodney S.
Battelle Pacific Northwest
P.O. Box 999, MS P7- 41
Richland, WA 99352 USA

Springael, Dirk
VITO/Flemish Institute for
    Technological Research
Lab of Genetics and Biotechnology
Boeretang 200
B-2400 Mol
BELGIUM

Stefanova, Stanka S.
National Bank of Industrial
    Microorganisms & Cell Cultures
Blvd Tsarigradsko Shose 125/ bl2
Sofia 1113
BULGARIA

Stringfellow, William T.
University of California, Berkeley
Dept. of Civil Engineering
631 Davis Hall
Berkeley, CA 94720 USA

Strong-Gunderson, Janet M.
Oak Ridge National Laboratory
Environmental Sciences Division
Bethel Valley Road
Building 1505 MS 6038
P.O. Box 2008
Oak Ridge, TN 37831-6038 USA

Sugiura, Keiji
Marine Biotechnology Institute
3-75-1, Heita
Kamaishi City, Iwate 026
JAPAN

Thomas, J. Michele
Rice University
7307 Churchill Road
McLean, VA 22101 USA

Tremblay, Gilles H.
Fisheries & Oceans Canada
Maurice Lamontagne Institute
P.O. Box 1000
Mont-Joli, Québec G5H 3Z4
CANADA

Tyagi, Rajeswar D.
Université du Québec
INRS – Eau
2700 Rue Einstein
C.P. 7500
Sainte Foy, Québec G1V 4C7
CANADA

van Andel, Johan G.
RIVM/National Institute for Public
    Health and the Environment
P.O. Box 1
3720 BA Bilthoven
THE NETHERLANDS

Vandecasteele, Jean-Paul
Institut Français du Pétrole
BP 311
92506 Rueil Malmaison, Cedex
FRANCE

van der Lelie, Daniel
VITO/Flemish Institute for
    Technological Research
Lab of Genetics and Biotechnology
Boeretang 200
B-2400 Mol
BELGIUM

van de Wiel, Rike
Agricultural University Wageningen
P.O. Box 8129
6700 EV Wageningen
THE NETHERLANDS

Verstraten, J.M.
University of Amsterdam
Dept. of Physical Geography &
    Soil Science
Nieuwe Prinsengracht 130
1018 VZ Amsterdam
THE NETHERLANDS

Vesper, Steve J.
University of Cincinnati
Dept. of Civil and Environ. Engrg.
1275 Section Road
Cincinnati, OH 45237-2615 USA

Villeneuve, Jean-Pierre
INRS-Eau
Université du Québec
2800 rue Einstein C.P. 7500
Sainte-Foy Québec
G1V 4C7
CANADA

Volkering, Frank
RIVM/LAE
P.O. Box 1
3720 BA Bilthoven
THE NETHERLANDS

Waltz, Michael D.
REMTECH, Inc.
98 Vanadium Road
Bridgeville, PA 15017 USA

Ward, C. Herb
Rice University
Energy & Environ. Systems Institute
P.O. Box 1892 MS-316
Houston, TX 77251-1892 USA

Weiland, Arjan R.
University of Amsterdam
Department of Environmental and
    Toxicological Chemistry
Nieuwe Achtergracht 166
1018 WV Amsterdam
THE NETHERLANDS

West, Candida C.
U.S. Environ. Protection Agency
RS Kerr Environ. Research Lab
P.O. Box 1198
Ada, OK 74820 USA

Wheelis, Susan
Wesleyan College
Department of Biology
Macon, GA 31297 USA

Wilson, Barbara H.
U.S. Environ. Protection Agency
RS Kerr Laboratory
P.O. Box 1198
Ada, OK 74820 USA

Wilson, Liza P.
The Johns Hopkins University
Dept. of Geog. & Environ. Engrg.
313 Ames Hall
3400 North Charles Street
Baltimore, MD 21218 USA

Woodhull, Patrick M.
OHM Remediation Services Corp.
16406 U.S. Route 224 East
Findlay, OH 45840 USA

Yeom, Ick Tae
University of Tennessee
Dept. of Civil and Environ. Engrg.
219-B Perkins Hall
Knoxville, TN 37996-2010 USA

Yonge, David R.
Washington State University
Dept. of Civil & Environ. Engrg.
Sloan Hall
Pullman, WA 99164-2910 USA

Young, Gary N.
Exxon Production Research Co.
P.O. Box 2189
Houston, TX 77252-2189 USA

Zhang, Weixian
Johns Hopkins University
Dept. of Geography & Environmental
   Engineering
3400 North Charles Street
Baltimore, MD 21218-2686 USA

INDEX